T0274187

COMPACT STAR PHYSICS

This self-contained introduction to compact star physics explains key concepts from areas such as general relativity, thermodynamics, statistical mechanics, and nuclear physics. Containing many tested exercises, and written by an international expert in the research field, the book provides important insights on the basic concepts of compact stars and discusses white dwarfs, neutron stars, quark stars, and exotic compact stars. The topics covered also include a discussion of astrophysical observations of compact stars, and present and future terrestrial experiments with strong relations to the study of compact stars, as experiments on exotic nuclei and relativistic heavy-ion collisions probing the equation of state of dense matter. Major developments in the field such as the discovery of massive neutron stars and a discussion of the recent gravitational-wave measurement of a neutron star merger are also presented. This book is ideal for graduate students and researchers working on the physics of compact stars, general relativity, and nuclear physics.

JÜRGEN SCHAFFNER-BIELICH is a professor in theoretical astrophysics at Goethe University, Frankfurt. Since completing his PhD, he has worked at the Niels Bohr Institute; the Lawrence Berkeley National Laboratory as Feodor-Lynen fellow of the Humboldt Foundation; the RIKEN BNL Research Center at the Brookhaven National Laboratory, Columbia University; and as a professor at the University of Heidelberg.

COMPACT STAR PHYSICS

JÜRGEN SCHAFFNER-BIELICH
Goethe University Frankfurt

CAMBRIDGE
UNIVERSITY PRESS

University Printing House, Cambridge CB2 8BS, United Kingdom

One Liberty Plaza, 20th Floor, New York, NY 10006, USA

477 Williamstown Road, Port Melbourne, VIC 3207, Australia

314–321, 3rd Floor, Plot 3, Splendor Forum, Jasola District Centre, New Delhi – 110025, India

79 Anson Road, #06–04/06, Singapore 079906

Cambridge University Press is part of the University of Cambridge.

It furthers the University's mission by disseminating knowledge in the pursuit of education, learning, and research at the highest international levels of excellence.

www.cambridge.org
Information on this title: www.cambridge.org/9781107180895
DOI: 10.1017/9781316848357

First published 2020

Printed in the United Kingdom by TJ International Ltd, Padstow Cornwall

A catalogue record for this publication is available from the British Library.

ISBN 978-1-107-18089-5 Hardback

To

Joran, Laurin, and Annkatrin

Contents

Preface

Compact stars are stars in which effects from general relativity become important. Commonly, they are associated with white dwarfs and neutron stars. Compact stars in general are more than that and encompass quark stars and hybrid stars, as well as hypothetical boson and fermion stars made of exotic particles. This textbook is about those types of compact stars.

The research area of compact stars experiences exciting new developments. Massive neutron stars with a mass of more than two solar masses have been detected. There is the first direct detection of gravitational waves produced by the merger of two neutron stars by the gravitational wave detectors LIGO and VIRGO. Numerical simulation codes have reached a new status of maturity to compute the birth of neutron stars in core-collapse supernova and to compute the merger of neutron stars. There has been tremendous progress in the microphysical modeling of compact stars in recent years, bringing to light new insights for the behavior of matter under extreme conditions. New space-based missions are planned to explore compact stars in much more detail, as well as new ground-based facilities in the near future, such as the X-ray satellite eROSITA, the James Webb Space Telescope (JWST), the Square Kilometer Array (SKA), and the Extremely Large Telescope (ELT).

There is a rapidly growing scientific community consisting of astrophysicists, numerical relativists, and nuclear physicists who perform research related to compact stars worldwide. There are the excellent classic textbooks by Shapiro and Teukolsky (1983) with an emphasis on relativistic astrophysics and by Glendenning (2000) with a focus on the field theoretical description of compact stars. In view of the rapid development in the research field, this textbook is intended to give an updated introduction to the physics of compact stars, covering many new facets and developments in the field since Shapiro and Teukolsky, and Glendenning. It is based on courses given at Goethe University Frankfurt, at the Ruprecht Karl University of Heidelberg, and several lectures given at summer and winter schools.

The intended readership is advanced undergraduate students and graduate students with a basic knowledge in mechanics, electrodynamics, quantum mechanics, and statistical mechanics. The key concepts of general relativity necessary for compact stars and the basis of dense matter and quantum statistics will be worked out in the textbook. Unfortunately, in view of the large amount of material and the limits of space and time, many topics had to be omitted, such as rotation, cooling of neutron stars, proto-neutron stars, and core-collapse supernovae.

I thank my colleagues, coauthors, students, and friends Mark Alford, Almudena Arcones, Krešo Baotic, Andreas Bauswein, Gordon Baym, David Blaschke, Ignazio Bombaci, Michael Buballa, Fiorella Burgio, Debarati Chatterjee, Jan-Erik Christian, Alessandro Drago, Tobias Fischer, Avraham Gal, Norman Glendenning, Carsten Greiner, Pawel Haensel, Matthias Hanauske, Kai Hebeler, Matthias Hempel, Oliver Lawrence Hoffmann, Thomas Janka, Burkhardt Kämpfer, Aleksi Kurkela, Jim Lattimer, Stefan Leupold, Matthias Liebendörfer, Gabriel Martínez-Pinedo, Bruno Mintz, Igor Mishustin, Guiseppe Pagliara, Chris Pethick, Rob Pisarski, Jose Pons, Madappa Prakash, Sanjay Reddy, René Reifarth, Luciano Rezzolla, Dirk Rischke, Stephan Rosswog, Stefan Rüster, Irina Sagert, Klaus Schertler, the late Stefan Schramm, Achim Schwenk, Hans-Josef Schulze, Armen Sedrakian, Igor Shovkovy, Andrew Steiner, Rainer Stiele, Horst Stöcker, Christian Sturm, Friedel Thielemann, Markus Thoma, Laura Tolos, Stefan Typel, Isaac Vidaña, Aleksi Vuorinen, Fridolin Weber, Simon Weissenborn, and Andreas Zacchi for the numerous discussions on compact star physics and related topics. I apologize to those whom I forgot to mention.

I am indebted to Eduardo Fraga for bringing up my name for a textbook on compact stars to Simon Capelin from Cambridge University Press, and to Simon for following up on it. I thank him, Sarah Lambert, Henry Cockburn, Roisin Munnelly, and the staff at Cambridge University Press for their help and assistance in the production process. I am grateful to Joe Kapusta for helpful advice on how to write a textbook. I thank Jan-Erik Christian, Eduardo Fraga, Matthias Hanauske, Irina Sagert, Laura Tolos, Fridolin Weber, and Andreas Zacchi for a critical reading of selected chapters. Finally, I thank my family, my wife Annkatrin and my two sons Laurin and Joran, for their love. This book is dedicated to them.

1

Introduction

Compact objects are the end points of stellar evolution and comprise white dwarfs, neutron stars, and black holes. There is a crucial difference from ordinary stars: compact stars do not burn nuclear fuel, so they are not supported by thermal pressure against the pull of gravity.

White dwarfs are supported by the degeneracy pressure of electrons. Neutron stars are supported by the repulsive interactions between nucleons. Black holes are completely collapsed objects, they contain a singularity. They are fully described by the Einstein equations in vacuum, so no matter is needed for the black hole solution.

All compact objects are essentially static over the lifetime of the universe and small in size compared to other astronomical objects such as ordinary stars or galaxies. Exceptions are mini black holes, which are subject to Hawking evaporation and evaporate in a finite time. And supermassive black holes have sizes comparable to ordinary stars.

Compact stars involve a huge range of densities. Also they involve all four fundamental forces: the strong, the weak, the electromagnetic force, and gravity, in many cases under extreme conditions. Table 1.1 lists the properties of compact objects together with the ones of our Sun and Earth. The mean mass density is defined as

$$\bar{\rho} = \frac{3M}{4\pi R^3}, \tag{1.1}$$

with the mass M and the radius R, and the surface gravity as

$$g = \frac{GM}{R^2}, \tag{1.2}$$

with the gravitational constant $G = 6.67408(31) \times 10^{-11}\,\mathrm{m^3\,kg^{-1}\,s^{-2}}$. For white dwarfs, the mass and radius is known in a few cases. For neutron stars, masses can be determined quite accurately. The range in neutron star radius comes from theoretical constraints, as its precise determination from observations at the time

Table 1.1 *Comparison of compact objects.*

	Mass	Radius (km)	Compactness	Surface gravity ($m\,s^{-2}$)	Mean density ($kg\,m^{-3}$)
Earth	$5.9724(3) \times 10^{24}$ kg	6 378.1[a]	6.9×10^{-10}	9.80665[b]	5,495
Sun	$1.98848(9) \times 10^{30}$ kg	695 700[c]	2.1×10^{-6}	274.2	1,410
White dwarf	$(0.5 - 1)\, M_\odot$	5,000–10,000	$(0.7\text{–}3) \times 10^{-4}$	$\sim 10^6$	10^8–10^9
Neutron star	$(1\text{–}2)\, M_\odot$	10–15	0.1–0.3	$\sim 10^{12}$	10^{17}–10^{18}
Stellar mass black hole	$\sim 10\, M_\odot$	~ 30	0.5	$\sim 10^{12}$	
Supermassive black hole	$(10^6\text{–}10^{10})\, M_\odot$	3×10^6–3×10^{10}	0.5	10^3–10^7	

[a] Nominal Earth equatorial radius
[b] Defined as standard acceleration due to gravity
[c] Nominal Solar equatorial radius

of writing is still rather model dependent. The satellite mission NICER measured the radius of the neutron star PSR J0030 + 0451 to be $12.71^{1.19}_{-1.14}$ km (Riley, T. E. et al. 2019. A NICER View of PSR J0030+451: Millisecond Pulsar Parameter Estimation. Astrophys. J. **887**, L21) and $13.02^{+1.24}_{-1.06}$ km (Miller, M. C. et al. 2019. PSR J0030+0451 Mass and Radius from NICER Data and Implications for the Properties of Neutron Star Matter. Astrophys. J. **887**, L24). For black holes, two classes are listed in Table 1.1 that are known from astrophysical observations. There exist several candidates for stellar mass black holes in our galaxy with rough mass estimates.

The radius listed for black holes is the Schwarzschild radius, named after Karl Schwarzschild:

$$R_s = \frac{2GM}{c^2}. \tag{1.3}$$

We will define the ratio of the gravitational mass to the radius as the compactness of a star

$$C = \frac{GM}{Rc^2}, \tag{1.4}$$

which is a measure how close the radius of the star is to the size of a black hole of the same mass. For nonrotating black holes, the compactness is one half. For a maximally rotating Kerr black hole, named after Roy Kerr, the event horizon is located at $R = GM/c^2$, so that its compactness is equal to 1. The mean density for black holes is undefined as no matter is present. The surface gravity for black holes is taken from the expression for Schwarzschild black holes at the horizon defined as $\kappa = 1/(4GM)$, which amusingly coincides with the Newtonian definition when using the Schwarzschild radius in Eq. (1.2). The surface gravity of a black hole is the local acceleration experienced by an observer hoovering at the black hole horizon. One should keep in mind, however, that the acceleration at the black hole horizon seen by a distant observer diverges.

White dwarfs, neutron stars, and stellar mass black holes have masses similar to that of our Sun. The sizes of white dwarfs are similar to that of Earth, while those of neutron stars are much smaller and are comparable to the size of a city on Earth. Supermassive black holes are as massive as globular clusters, with typical masses of $10^6 M_\odot$, and can nearly reach the masses of small galaxies. For comparison, the mass of our galaxy is estimated to be $10^{11} M_\odot$. From the first detection of gravitational waves in 2015, we know that stellar mass black holes exist with masses of $36^{+5}_{-4} M_\odot$, $29^{+4}_{-4} M_\odot$, and $62^{+4}_{-4} M_\odot$ (Abbott et al., 2016a). Several black hole mergers have been measured since then, observing black holes in the mass range of about $M = 10$–$80 M_\odot$ (Abbott et al., 2018a). Masses of many supermassive black holes sitting at the center of galaxies are known, in particular the one in our own galaxy with a mass of $(4.3 \pm 0.5) \times 10^6 M_\odot$ (Gillessen et al., 2009). In 2019, a picture of the supermassive black hole at the center of galaxy M87 was taken by the Event Horizon Telescope, showing the accretion disk around the supermassive black hole and its shadow. The mass of black hole M87* was determined to be $(6.5 \pm 0.7) \times 10^9 M_\odot$ (Akiyama et al., 2019).

The radii of black holes span from a size slightly larger than that of our Sun to the size of our solar system – about 100 AU (by definition 1AU = 149 597 870 700 m, which is about the average distance of the Earth to our Sun). The compactness shows how close the star is in size to half its own Schwarzschild radius. The closer the compactness gets to 1, the more important will the effects be from general relativity. White dwarfs have a rather small compactness of about 10^{-4}. Neutron stars have a compactness that is not too far from 1, so it is essential to describe neutron stars within general relativity. By looking at the mean density, one realizes that compact stars, white dwarfs, and neutron stars involve huge mass densities that are several orders of magnitude larger than our Sun and Earth. In fact, one teaspoon of neutron star material is equal to the mass of all humans together.[1] Also, the surface gravity is several orders of magnitude larger than our Earth and Sun and is similar to those of astrophysical black holes.

It is interesting to look at the origin of the names, the etymology, of compact objects.

White dwarfs were first characterized by their appearance. As they have high effective temperatures at their surfaces, they appear much 'whiter' than ordinary stars. The first white dwarf that was discovered was Sirius B, the companion to the brightest star in the sky, Sirius A, or simply Sirius. The deduced radius of Sirius B was much smaller than that of ordinary stars (see Table 1.1). Therefore, the name 'white dwarf' was given and is still used today.

Neutron stars were first characterized by their properties postulated before their actual discovery. The term 'neutron stars' was introduced for the first time by Walter Baade and Fritz Zwicky in the proceedings of the American Physical Society meeting at Stanford, December 15–16, 1933:

> With all reserve we advance the view that supernovae represent the transitions from ordinary stars into *neutron stars*, which in their final stages consist of extremely closely packed neutrons. (Baade and Zwicky, 1934a)

However, the concept of neutron stars was envisioned earlier by Lev Landau (Landau, 1932), probably as early as 1931 in a discussion with Niels Bohr and Leon Rosenfeld in Copenhagen. We refer the interested reader to the excellent historical treatise (Yakovlev et al., 2013). Neutron stars consist mainly of neutrons in the interior, due to the inverse β-decay of $e^- + p \to n + \nu_e$. Hence, neutron stars can be considered as 'giant nuclei' with mass densities comparable to and exceeding

[1] Just imagine having all humans packed onto a single teaspoon!

that of the nucleus of an atom of $\rho_{\text{nuclei}} = 2.5 \times 10^{17}\,\text{kg}\,\text{m}^{-3}$. However, while ordinary stable nuclei have mass numbers of up to $A = 208$ for ^{208}Pb, neutron stars accommodate mass numbers of the order of $A = 10^{57}$. Neutron stars do not shine as ordinary stars. However, neutron stars have been observed in the optical, in X-rays, and in the radio band. Young neutron stars have been observed directly and they have the highest surface temperatures known for astrophysical objects. They radiate mainly in X-rays with temperatures in the range of $1,000,000°$C. Pulsars are rotation-powered neutron stars and are regularly observed by radio telescopes.

Black holes were characterized by their theoretically derived appearance and properties. Neither matter nor light can escape from black holes, so they will appear 'black' to an observer. Anything falling into a black hole will disappear. The naming seems to be obvious and is attributed to John Wheeler, who first introduced it into scientific literature. However, as he recalls in his autobiography, someone in the audience suggested the name to him during a talk on the discovery of pulsars in the fall of 1967 instead of 'gravitationally completely collapsed object' (Wheeler and Ford, 1998, p. 296).

Nowadays, black holes are indirectly observed as astrophysical phenomena in the form of, for example, active galactic nuclei or X-ray bursters. The former imply supermassive black holes at the center of galaxies, the latter, stellar mass black holes with a companion star. In both cases, the black holes accrete gas from the surroundings, which is then heated up. The heated matter radiates, so black holes can be observed through their impact on ordinary matter. The black hole in the center of our galaxy in the constellation Sagittarius is observed by the Keplerian motion of nearby stars, revealing a huge gravitational pull of a supermassive object within a small volume. The accretion disk around the supermassive black hole at the center of galaxy M87 was made directly visible by the very long baseline interferometry of the Event Horizon Telescope collaboration.

White dwarfs and neutron stars are strongly connected with stellar evolution. Both are the endpoints of stellar evolution. Whether a white dwarf or a neutron star is formed at the end of the life of a star depends on the initial mass of the star, to be more precise, the mass of the star when it reaches stable hydrogen burning. The luminosity of the stars can be plotted against the observed surface temperature, which is equivalent to the spectral classification scheme OBAFGKM, in the Hertzsprung–Russell diagram. One observes that most of the stars are located along a line called the main sequence of the Hertzsprung–Russell diagram. The stars burning hydrogen spent most of their time along the main sequence. The initial mass of a star is then called the zero age main sequence (ZAMS) mass.

Black holes could be produced in a variety of ways: as the endpoint of stellar evolution of sufficiently massive stars (including the first stars formed after the big bang), in the early universe forming primordial black holes ('mini black holes' with $M \sim 10^{15}$ g), or by the collision of compact stars in binary systems.

The formation of compact objects as endpoints of stellar evolution can be summarized as follows. As pointed out earlier, the final endpoint of stellar evolution depends on the ZAMS mass.

$M < 0.01 M_\odot$: This is the realm of planets, which are not sufficiently massive to generate any nuclear burning at all. For comparison, the mass of Jupiter is about $M_J \approx 0.001 M_\odot$.

$0.01 M_\odot \leq M < 0.08 M_\odot$: The star is not sufficiently massive for stable nuclear burning in the core. There is some burning of deuterons, which gives the failed star a reddish-brownish color. This is the realm of brown dwarfs.

$0.08 M_\odot \leq M < 0.4 M_\odot$: Stable hydrogen burning is present in the core of the star. The star will eventually exhaust its hydrogen supply and will end up in a hydrogen-helium white dwarf.

$0.4 M_\odot \leq M < 8 M_\odot$: The star is sufficiently massive to start helium burning during its evolution. Heavier elements are produced so that the final white dwarfs contain carbon and oxygen. Shells of matter will be blown into interstellar space and will be illuminated by the ultraviolet glow of the white dwarf, generating a so-called planetary nebula. Our Sun will eventually end as such a carbon-oxygen white dwarf.

$8 M_\odot \leq M < 10 M_\odot$: Carbon burning will be possible during the final stages of stellar evolution. A degenerate neon-oxygen-magnesium core forms, which will eventually collapse to a neutron star. The initially hot proto-neutron star emits neutrinos. The outgoing shock wave releases energy visible in a so-called core-collapse supernova.

$10 M_\odot \leq M < 25 M_\odot$: Above $10 M_\odot$ ZAMS mass silicon burning sets in as the final burning stage of the star, forming a degenerate iron core. The iron core collapses to a neutron star, which is visible as a core-collapse supernova.

$M > 25 M_\odot$: This mass limit is not well known and depends on the unknown maximum mass of the neutron star formed in the collapse of the degenerate iron core. The still hot neutron star is too massive to withstand the pull of gravity and collapses to a black hole. This could lead to a failed supernova or in the other extreme to a particularly bright supernova, a hypernova.

$M \sim 100 M_\odot$: This is about the upper mass limit for ordinary stars. They are so hot in the core that they produce electron-positron pairs that destabilize the star. The first stars formed in the early universe can be much more massive due to a lack of elements that are heavier than helium (metals in the jargon of astrophysicists).

Table 1.2 *The endpoints of stellar evolution based on the zero age main sequence (ZAMS) mass, the initial mass of the star.*

ZAMS mass	0.04–$8M_\odot$	8–$25M_\odot$	$> 25M_\odot$
	White dwarf	Neutron star	Black hole

This picture has been simplified, as it neglects effects from magnetic fields and rotation, for example. As a short summary, Table 1.2 shows the compact objects that are the endpoints of stellar evolution based on the ZAMS mass of the star.

It should be clear from this that the different endpoints in stellar evolution are controlled by the ZAMS mass not the mass of the compact object itself. In standard stellar evolution, there is not a path from one compact object to the other. However, there are possibilities of a compact object to transform to another one for accreting systems, that is, for a white dwarf or a neutron star to accrete matter from a companion star. An accreting neutron star will eventually collapse to a black hole when reaching the maximum possible mass. For white dwarfs the issue is more subtle.

A white dwarf reaches a critical density for igniting thermonuclear burning when accreting matter. The standard scenario is that this burning is a runaway reaction, leading to an explosion of the whole white dwarf with no remnants. This explosion is well known as a type Ia supernova and used as a standard candle in cosmology. There is the possibility that the accreting white dwarf experiences a collapse instead of an explosion, a so-called accretion-induced collapse. The collapsing white dwarf would then form a neutron star during the collapse. For iron white dwarfs this would be an option. However, as mentioned earlier, the stellar burning stops before carbon fusion for stars ending in a white dwarf, so that the standard white dwarf contains carbon and oxygen but not iron in its core.

Exercises

(1.1) Calculate the Schwarzschild radius of the Sun, the Earth, and your own one.

(1.2) Calculate the gravitational acceleration for the Sun, white dwarfs, and neutron stars.

(1.3) Assume that every 100 years a supernova produces a neutron star in our galaxy. Estimate the present number of neutron stars in our galaxy.

(1.4) How many nucleons are in the Sun?

2
General Relativity

This chapter is a brief introduction to some of the basic concepts of general relativity. It is not meant to provide a thorough introduction to general relativity and we refer the reader to the many excellent textbooks written on that subject. The main purpose of this chapter is to derive the results of general relativity necessary for our discussion of the properties of compact stars and its astrophysical applications in the later chapters. A separate chapter is devoted to the subject of gravitational waves (see Chapter 10).

2.1 Gravity and the Equivalence Principle

Let us introduce the concept of general relativity by looking at Newton's law and Newton's expression for the gravitational force. According to Newton's law a massive particle feels a kinetic force

$$F_{\text{kin}} = m_i \cdot a \tag{2.1}$$

that is proportional to the acceleration a with the coefficient m_i. We will denote the coefficient m_i as the inertial mass of the particle. According to Newton, the classical gravitational force

$$F_g = -m_g \cdot \nabla\phi \tag{2.2}$$

is proportional to the gradient of the gravitational potential ϕ with the coefficient m_g. We will denote the coefficient m_g as the gravitational mass of the particle. In experiments first performed by Galileo Galilei, it was found out that every object falls with the same acceleration in a gravitational field so that

$$a = -\nabla\phi \tag{2.3}$$

independent of its inertial and gravitational mass. Hence, the inertial and the gravitational mass have to be the same

$$m_g = m_i. \tag{2.4}$$

Eq. (2.4) constitutes the weak equivalence principle (or WEP).

Weak equivalence principle: All particles experience the same acceleration in a gravitational field irrespective of their masses.

WEP has been tested to a high accuracy in many different experiments. Modern torsion balance experiments have tested the WEP down to a level of 10^{-13} (Wagner et al., 2012). The satellite experiment MICROSCOPE has tested the weak equivalence principle at a level of 10^{-15} (Touboul et al., 2017). There are equivalent formulations of the WEP. For example, WEP implies that there is a preferred trajectory of particles through spacetime on which particles move that is determined just by gravity.

An extension of the WEP is the Einstein equivalence principle (EEP), which includes any form of matter, not only particles. Consider an observer in a sealed box performing experiments. From the equivalence of the inertial and the gravitational mass, an observer cannot distinguish whether an object is accelerated by the acceleration of the box or whether it is accelerated by the presence of a gravitational field. It is impossible to detect the existence of a gravitational field by local experiments. It is important to add the condition of locality to the experiments as gradients of the gravitational field will lead to tidal forces on larger scales that would be measurable. The observer will measure locally only the laws of physics as they would be in a spacetime without gravity, so that the kinematics are governed by special relativity. The EEP can be stated as follows (see e.g., Schutz, 2009):

Einstein equivalence principle: In small enough regions of spacetime, any physical experiment not involving gravity will have the same result if performed in a freely falling inertial frame as if performed in the flat spacetime of gravity.

We will heavily use EEP in the following sections. It allows the transfer of equations valid in special relativity to its general form in general relativity as both equations have to be of the same form in a small enough region of spacetime.

EEP has important implications for the way gravity couples to matter. As all physical laws, except gravity, are included in defining the EEP, it implies that gravity couples to the other forces, of nature in a special way. Let us be more specific here. An atom consists of electrons and a nucleus bound by electromagnetic forces, which is stabilized by quantum mechanics. The mass of an atom is not the sum of the mass of electrons and the nucleus but slightly less. The difference in mass is dubbed the mass defect or simply the binding energy of the atom. Seemingly, gravity couples not only to the masses of the constituents of the atom but it is also sensitive to the mass defect generated by the quantum electromagnetic forces. Hence, gravity couples to electromagnetic energy. The same story goes for the nucleus of an atom. The nucleus consists of protons and neutrons. The mass of the nucleus is not the sum of the masses of the protons and neutrons in the nucleus but somewhat less, which is just the mass defect or the binding energy of the nucleus. In the nucleus, the nuclear forces or the strong interactions are at work. So gravity couples to energy generated by strong interactions. The same reasoning should also apply for the other fundamental force of the standard model left, weak interactions. The masses of the elementary particle, quarks and leptons, are generated by the Higgs mechanism. So gravity couples also to energy (mass) generated by the weak interactions. The reader is invited to fancy at a system consisting of leptons only that is bound by weak interactions only, as, for example, a hypothetical neutrino ball, and perform the same reasoning for nuclei and strong interactions.

The equivalence of performing experiments in an accelerating system and in one at rest being exposed to a gravitational field can be exploited in Einstein's famous way via thought experiments. Hereby one imagines certain experimental situations to arrive at statements about physical laws. Let us imagine a rocket that is constantly accelerating in free space that shall be equal to the gravitational acceleration on Earth. According to EEP, experiments performed inside the rocket are equivalent to the ones performed in a rocket standing on Earth as the systems are exposed to the same acceleration. From this setting one can immediately derive two important consequences of EEP that form the basis of general relativity: the gravitational redshift of light and the gravitational bending of light in a gravitational field.

Imagine an electromagnetic wave with a fixed frequency (e.g., from a laser) being emitted from the bottom of the rocket and being measured at the top of the rocket. If the rocket is continuously accelerating, the speed of the rocket at the time of the detection will be correspondingly larger compared to the one at the time of emission for each crest (or trough) of the wave. The time of arrival of the wave crests of the light at the detector will be correspondingly larger than the one at constant speed. This effect leads to a decrease of the frequency of the wave being measured at the top of the rocket compared to the original frequency, that is, a redshift of the electromagnetic wave. Switching the emitter with the detector, so that the wave travels from the top of the rocket to its bottom, the electromagnetic wave will be blueshifted when measured at the bottom of the rocket. EEP now states that the experimental situation is equivalent to the one in a gravitational field with the same (gravitational) acceleration. Hence, an electromagnetic wave being emitted within a gravitational potential to the outside will be measured to have a lower frequency compared to original frequency by an outside observer. This is the famous gravitational redshift effect predicted by general relativity. In fact, as we have seen, the gravitational redshift is more fundamental as it is a consequence of EEP.

We can derive an approximate expression for the gravitational redshift by using the Doppler shift formula for nonrelativistic velocities. The Doppler shift depends on the velocity difference of the emitting source and the receiver Δv. The relative change in the wavelength $\Delta\lambda$ measured by the receiver in comparison to the original wavelength of the emitter λ is given by

$$\frac{\Delta\lambda}{\lambda} = \frac{\Delta v}{c}, \tag{2.5}$$

where we put in the velocity of light c as the velocity of the wave. For the thought experiment of the rocket, the difference in the velocity comes from the constant acceleration a of the rocket within the time span the light is traveling from the bottom of the rocket to the top $\Delta t = \Delta x/c$, where Δx is the distance between the emitter and the detector. The difference in the arrival of wave crests corresponds now to

$$\frac{\Delta\lambda}{\lambda} = \frac{a \cdot \Delta t}{c} = a \cdot \frac{\Delta x}{c^2}. \tag{2.6}$$

The equivalence principle now demands that the shift in wavelength is exactly the same as for a gravitational field with the same (gravitational) acceleration due to the gradient of the gravitational potential $\nabla\phi$:

$$\frac{\Delta\lambda}{\lambda} = \nabla\phi \frac{\Delta x}{c^2} = \frac{\Delta\phi \cdot \Delta x}{\Delta x \cdot c^2} = \frac{\Delta\phi}{c^2}. \tag{2.7}$$

There is a change in the sign from Eq. (2.3) as the gravitational acceleration is pointing inward and the wave is traveling outward. In the last step, we assumed a gravitational potential with a constant gradient. However, the expression holds also for the general case. Note that $\Delta\phi$ for the case studied here is a positive quantity so that the observed wavelength will be longer than it should be. Specifying the wavelength for the emitting source by λ_e and the one observed at the detector by λ_d one finds

$$\frac{\lambda_d}{\lambda_e} = 1 + \frac{\Delta\phi}{c^2}. \tag{2.8}$$

As the dispersion relation for light reads

$$c = \lambda \cdot \nu, \tag{2.9}$$

with the frequency ν, the corresponding expression for the ratio of the frequencies is

$$\frac{\nu_d}{\nu_e} = \left(1 + \frac{\Delta\phi}{c^2}\right)^{-1} \approx 1 - \frac{\Delta\phi}{c^2} \tag{2.10}$$

to first order in $\Delta\phi/c^2$. The redshift factor z is defined as

$$1 + z = \frac{\lambda_d}{\lambda_e} \tag{2.11}$$

so that the gravitational redshift corresponds to a redshift factor of

$$z = \frac{\Delta\phi}{c^2}. \tag{2.12}$$

For light emitted from the surface of a massive body with mass M and a radius R, the gravitational potential is

$$\phi = \frac{GM}{R}. \tag{2.13}$$

For an outside observer at infinity, the observed redshift factor is then

$$z = \frac{GM}{Rc^2}, \tag{2.14}$$

where R is the radius of the massive body. We notice that the redshift factor is equal to the compactness of an object, see Eq. (1.4). Be aware that the expression equation (2.14) only holds for weak gravitational fields.

The second prediction of EEP is the bending of light in a gravitational potential. Imagine a laser beam being emitted from one side of the rocket to a detector at the other side of the rocket. If the rocket is moving with a constant velocity, the path of the laser beam will be a straight line. If the rocket is accelerating, there will be a difference in the velocity between the time of the emission and the time of detection. The detector needs to be moved down toward the bottom of the rocket to measure the laser beam compared to the nonaccelerating case. The path of the laser beam will be no longer a straight one, instead the laser beam is bent due to the acceleration of the rocket. According to EEP, the same experimental situation will be present for a gravitational field with the same (gravitational) acceleration. Hence, a gravitational field will bend the path of light.

The strong equivalence principle (SEP) is an extension of EEP. While the EEP explicitly dismisses experiments involving gravity, SEP explicitly includes all experiments, including those involving gravity.

Strong equivalence principle: In small enough regions of spacetime, any physical experiment *including gravity* will have the same result if performed in a freely falling inertial frame as if performed in the flat spacetime of gravity.

SEP not only includes self-gravitating systems, where the gravitational binding energy is important, but also gravitational systems as a gravitational two-body system moving in the presence of a gravitational field from a third body. Extensions of general relativity can be compatible with EEP but can violate SEP, for example, by the presence of a fifth force.

Tests of SEP have been performed within the solar system by lunar laser ranging experiments. However, the gravitational fields involved are rather weak. A more stringent test of SEP would be provided by the study of three compact stars orbiting each other. In fact, such a system has been observed. The pulsar PSR J0337+1715 is part of a triple system together with two white dwarfs that are gravitationally bound to each other (Ransom et al., 2014). The universality of free fall has been tested for the white dwarf and the pulsar in the gravitational potential of the second white dwarf, providing a test of SEP at the level of 2.6×10^{-6} (Archibald et al., 2018).

2.2 Special Relativity and the Metric

The basic building principle of special relativity is the constancy of the speed of light c, independent of the coordinate system. For a light ray, the relative difference in time $\mathrm{d}t$ and the relative differences in three-dimensional space $\mathrm{d}x$, $\mathrm{d}y$, and $\mathrm{d}z$ are related such that the difference of the squared quantities

$$\mathrm{d}s^2 = -c^2 \cdot \mathrm{d}t^2 + \mathrm{d}x^2 + \mathrm{d}y^2 + \mathrm{d}z^2 \tag{2.15}$$

is zero. This condition has to be fulfilled in any coordinate system to ensure the constancy of the speed of light. Hence, the quantity $\mathrm{d}s^2$ is an invariant quantity. Moreover, it connects time with space coordinates so that one has to consider coordinates in the combined spacetime that is now four-dimensional. Any measurement, where $\mathrm{d}s^2 < 0$, is allowed as the corresponding velocity would be lower than the speed of light. Those distances in spacetime are called timelike and are causal in character. The path of particles through spacetime will be timelike to ensure causality. Measurements with $\mathrm{d}s^2 >$ are not allowed as it would involve velocities larger than the speed of light. Such distances are called spacelike and are acausal in character. In summary, distances in the four-dimensional spacetime can be classified as

$$\mathrm{d}s^2 \begin{cases} = 0 & \text{lightlike or null} \\ < 0 & \text{timelike} \\ > 0 & \text{spacelike} \end{cases} . \tag{2.16}$$

We note that the sign of Eq. (2.15) can be chosen arbitrarily. While textbooks in general relativity prefer the same sign as chosen here, textbooks in particle physics usually adopt the other sign. While this would change the conditions for timelike and spacelike distances, it would not change the classification itself. Eq. (2.15) can be written in matrix form as

$$ds^2 = \sum_{\mu,\nu} \eta_{\mu\nu} dx^\mu dx^\nu = \eta_{\mu\nu} dx^\mu dx^\nu, \tag{2.17}$$

where $\mu = 1, \dots 4$ and $\nu = 1, \dots 4$. Here we introduced the Einstein's summation convention where one has to sum over repeated upper and lower indices. The matrix $\eta_{\mu\nu}$ stands for the metric of the spacetime of special relativity. The quantity x^μ is a four-vector in spacetime, where the 0-component corresponds to the time coordinate and the other three components to the space coordinates. For Cartesian coordinates

$$\eta_{\mu\nu} = \begin{pmatrix} -1 & 0 & 0 & 0 \\ 0 & 1 & 0 & 0 \\ 0 & 0 & 1 & 0 \\ 0 & 0 & 0 & 1 \end{pmatrix}. \tag{2.18}$$

For general coordinates, the metric $\eta_{\mu\nu}$ will be different. For spherical coordinates, for example, one finds

$$ds^2 = -c^2 dt^2 + dr^2 + r^2 \cdot \left(d\theta^2 + \sin^2\theta d\phi^2\right) \tag{2.19}$$

$$= -c^2 dt^2 + dr^2 + r^2 \cdot d\Omega^2, \tag{2.20}$$

where $d\Omega$ stands for the differential solid angle. In a general curved spacetime of general relativity, the four-dimensional distance is defined as

$$ds^2 = g_{\mu\nu} dx^\mu dx^\nu, \tag{2.21}$$

where $g_{\mu\nu}$ is the metric defining curved spacetime distances and thereby spacetime itself. The metric depends on the spacetime coordinates and is a tensor of rank two. As the metric is symmetric in the indices $g_{\mu\nu} = g_{\nu\mu}$, it has ten independent components in four dimensions ($D(D+1)/2$ components for D dimensions).

Let us look at a four-vector in spacetime A^μ and define the following quantity:

$$A_\mu = g_{\mu\nu} A^\nu \tag{2.22}$$

with a lower index. The components of A_μ are the ones of a so-called one-form. A one-form maps a vector to a real number as

$$A_\mu A^\mu = g_{\mu\nu} A^\nu A^\mu. \tag{2.23}$$

If we choose $A^\mu = x^\mu$, we know from the definition of the metric that the expression in Eq. (2.23) is a scalar quantity independent of the choice of the coordinate system. This should be true then for any choice of the vector A^μ. Let us denote g as the determinant of the metric $g_{\mu\nu}$. The determinant g must be nonvanishing to ensure a well-defined coordinate system, where the coordinates are providing independent directions in spacetime. Then one can define the inverse relation

$$A^{\mu} = g^{\mu\nu} A_{\nu}. \tag{2.24}$$

The metric can then be used to lower or raise indices. Combining Eqs. (2.22) and (2.24) one finds

$$A^{\mu} = g^{\mu\nu} g_{\nu\rho} A^{\rho}. \tag{2.25}$$

As this holds for any vector, the metric has to fulfill the condition that

$$g^{\mu\nu} g_{\nu\rho} = \delta^{\mu}_{\rho}, \tag{2.26}$$

where δ^{μ}_{ρ} is the Kronecker symbol

$$\delta^{\mu}_{\nu} = \begin{cases} 1 & \text{for } \mu = \nu \\ 0 & \text{for } \mu \neq \nu. \end{cases} \tag{2.27}$$

The metric is its own inverse and can be used to lower or raise its own indices.

For an observer at rest, the time measured must be a coordinate invariant quantity. The metric for an observer at rest reads

$$\mathrm{d}s^2 = g_{00} \cdot c^2 \cdot \mathrm{d}t^2. \tag{2.28}$$

One can define a coordinate independent scalar quantity, the proper time τ of an observer at rest by

$$c^2 \cdot \mathrm{d}\tau^2 = -\mathrm{d}s^2 = -g_{00} \cdot c^2 \cdot \mathrm{d}t^2. \tag{2.29}$$

Eigentime differences are then related to the 00-component of the metric tensor

$$\Delta\tau = \sqrt{-g_{00}} \cdot \Delta t. \tag{2.30}$$

Note, that t is a time coordinate but not the physical time being measured. Only the proper time has a physical meaning.

For weak gravitational fields, the metric has to recover the Newtonian limit. For this we take a look at the gravitational redshift discussed in connection with the equivalence principle. Recall that there is a light beam emitted from the bottom of a rocket to the top of it within a gravitational field characterized by the gravitational potential ϕ. The observed frequency ν_d will be shifted compared to the original frequency of the emitter ν_e by the difference in the gravitational potential, see Eq. (2.10):

$$\frac{\nu_d}{\nu_e} \approx 1 - \frac{\Delta\phi}{c^2}. \tag{2.31}$$

The frequency of the wave is the inverse of the time difference between two consecutive crests of the wave. Hence, the ratio of the frequencies is given by the inverse proper time difference at the location of the emitting source and the detector

$$\frac{\nu_d}{\nu_e} = \frac{\Delta\tau_e}{\Delta\tau_d} = \frac{\left(-g_{00}^{(e)}\right)^{1/2}}{\left(-g_{00}^{(d)}\right)^{1/2}}, \tag{2.32}$$

where we have chosen to have the same time coordinates within the rocket. Hence, the relative change in the frequencies is solely determined by the square root of the 00-component of the metric tensor. Comparing Eq. (2.31) with Eq. (2.32), it is apparent that for weak gravitational fields

$$g_{00} = -\left(1+\phi\right)^2 \approx -\left(1+2\phi\right), \tag{2.33}$$

which gives correctly

$$\frac{\nu_d}{\nu_e} = \frac{1+\phi_e/c^2}{1+\phi_d/c^2} \approx 1 + \frac{\phi_e - \phi_d}{c^2} = 1 - \frac{\Delta\phi}{c^2}. \tag{2.34}$$

As general relativity has to recover the correct limit for weak gravitational fields, Eq. (2.33) can be used to fix undetermined constants of the full metric tensor.

2.3 Einstein's Equation

Here we give heuristic arguments for the form of the Einstein equations. From here on we will use natural units by setting formally $c = 1$. Newton's gravitational law can be written in a form to describe a distribution of mass characterized by a mass density distribution $\rho(r)$. The gravitational potential is given by the Laplace equation:

$$\nabla^2\phi(r) = 4\pi G \cdot \rho(r), \tag{2.35}$$

where G is the gravitational constant. From the equivalence principle it is known that space tells matter how to move. The characteristic quantity to describe spacetime is the metric of the spacetime $g_{\mu\nu}$, which has to replace the gravitational potential ϕ. The differential equation of general relativity has to involve then the metric tensor and its derivatives up to second order on the left-hand side.

In a general curved spacetime, the partial derivative has to be replaced by the covariant derivative, which takes into account the curvature of spacetime. The covariant derivative of a vector is defined as

$$\nabla_\mu V^\nu = \partial_\mu V^\nu + \Gamma^\nu_{\mu\lambda} V^\lambda, \tag{2.36}$$

where $\Gamma^{\lambda}_{\mu\nu}$ stands for the Christoffel symbols that are a measure of the curvature of spacetime. Here ∂_μ stands for the shorthand notation

$$\partial_\mu = \frac{\partial}{\partial x^\mu}. \tag{2.37}$$

In the following, we will also denote the partial derivative with respect to some index with a comma followed by the index:

$$\partial_\mu V^\nu = V^\nu{}_{,\mu} \tag{2.38}$$

and for the covariant derivative we use a semicolon instead:

$$\nabla_\mu V^\nu = V^\nu{}_{;\mu} = V^\nu{}_{,\mu} + \Gamma^{\nu}_{\mu\lambda} V^\lambda. \tag{2.39}$$

For a product of vectors, we demand the Leibniz rule to hold:

$$\nabla_\kappa \left(V^\mu \cdot W^\nu\right) = \left(\nabla_\kappa V^\mu\right) \cdot W^\nu + V^\mu \cdot \left(\nabla_\kappa W^\nu\right). \tag{2.40}$$

The Christoffel symbols should be given in terms of the metric and its first derivatives. They can be fixed by demanding that the covariant derivative of the metric vanishes

$$\nabla_\kappa g_{\mu\nu} = 0, \tag{2.41}$$

which is equivalent to saying that the covariant derivative is metric compatible. It ensures that one can lower and raise indices of the argument of a covariant derivative. Hence,

$$\nabla_\kappa g^{\mu\nu} = 0. \tag{2.42}$$

The metric can be considered as a map of a product of two vectors to the real numbers from its definition (Eq. (2.21)). Considering the metric tensor as a product of two vectors, the Leibniz rule, Eq. (2.40), inquires that the covariant derivative has to act on both indices, so that it involves two Christoffel symbols:

$$\nabla_\kappa g^{\mu\nu} = \partial_\kappa g^{\mu\nu} + \Gamma^{\mu}_{\kappa\lambda} g^{\lambda\nu} + \Gamma^{\nu}_{\kappa\lambda} g^{\mu\lambda}. \tag{2.43}$$

Using the condition of metric compatibility, Eq. (2.41), and the definition of the covariant derivative, Eq. (2.36), one arrives at an expression for the Christoffel symbols:

$$\Gamma^{\lambda}_{\mu\nu} = \frac{1}{2} g^{\lambda\kappa} \left(g_{\kappa\nu,\mu} + g_{\kappa\mu,\nu} - g_{\mu\nu,\kappa}\right), \tag{2.44}$$

where

$$g_{\mu\nu,\kappa} = \frac{\partial g_{\mu\nu}}{\partial x^\kappa} \tag{2.45}$$

are the partial derivative of the metric tensor in the comma notation. We note that the Christoffel symbols are symmetric in the lower two indices (which is attributed as being torsion-free).

The quantity to describe a general curved spacetime should be a tensor that describes the curvature of spacetime. Hence, the tensor shall depend on the derivatives of the metric tensor up to second order. This can be achieved by either using the derivative of the Christoffel symbol or by the Christoffel symbol squared. Let us have a look at the second covariant derivative of a vector. If we look at the antisymmetrized version of the second covariant derivative, the partial derivatives will cancel out. The result shall then be proportional to the vector again

$$\nabla_\mu \nabla_\lambda V^\kappa - \nabla_\lambda \nabla_\mu V^\kappa = R^\kappa{}_{\mu\lambda\nu} V^\nu, \tag{2.46}$$

where we introduced a new tensor of rank four $R^\kappa{}_{\mu\nu\lambda}$, which is called the Riemann curvature tensor. It will depend only on the derivative of the Christoffel symbols and the Christoffel symbols squared. However, it has more independent components than we need to fix the ten components of the metric tensor $g_{\mu\nu}$. In fact, the Riemann curvature tensor has 20 independent components. We can transform the Riemann curvature tensor from a rank four tensor to a tensor of rank two by summing over one upper and one lower index. This is called a contraction of a tensor. The contraction of the Riemann curvature tensor gives

$$R_{\mu\nu} = R^\lambda{}_{\mu\lambda\nu} \tag{2.47}$$

and the resulting tensor is called the Ricci tensor. The antisymmetrized second covariant derivative of a vector involving the Ricci tensor is given by contracting the corresponding lower and upper index of Eq. (2.46)

$$\nabla_\mu \nabla_\lambda V^\lambda - \nabla_\lambda \nabla_\mu V^\lambda = R_{\mu\nu} V^\nu. \tag{2.48}$$

The Ricci tensor is symmetric in its indices and given by the derivative of the Christoffel symbols and the Christoffel symbol squared. Indeed, it has the right properties describing the curvature of spacetime. Using the definition of the covariant derivative, Eq. (2.36), one finds the expression of the Ricci tensor in terms of the Christoffel symbols and its derivative as

$$R_{\mu\nu} = \Gamma^\alpha_{\mu\alpha,\nu} - \Gamma^\alpha_{\mu\nu,\alpha} - \Gamma^\alpha_{\mu\nu}\Gamma^\beta_{\alpha\beta} + \Gamma^\alpha_{\mu\beta}\Gamma^\beta_{\nu\alpha}. \tag{2.49}$$

The comma in the subscript stands again for the partial derivative with respect to the index following the comma

$$\Gamma^\alpha_{\mu\nu,\kappa} = \frac{\partial}{\partial x^\kappa}\Gamma^\alpha_{\mu\nu}. \tag{2.50}$$

In vacuum, the Einstein equations read then simply

$$R_{\mu\nu} = 0. \tag{2.51}$$

These are 10 coupled nonlinear differential equations as the Ricci tensor is symmetric in its lower indices. Due to the nonlinear character of the differential equation (2.51), there are nontrivial solutions to the Einstein equations even in vacuum. The most famous one is the Schwarzschild metric, which will be discussed in the following section.

2.4 The Schwarzschild Metric

Shortly after Einstein has written down the differential equations of general relativity (Einstein, 1916a), Schwarzschild found an analytic solution for the Einstein equations in vacuum in 1916, the famous Schwarzschild metric (Schwarzschild, 1916b). As already discussed, the Einstein equations in vacuum read

$$R_{\mu\nu} = 0. \tag{2.52}$$

The starting point to solve the Einstein equation in vacuum, Eq. (2.52), is the ansatz for the metric, which is chosen to be static and spherical symmetric. Extending the metric for spherical coordinates to the general case, there are in principle three different metric functions possible:

$$\mathrm{d}s^2 = -A(r) \cdot \mathrm{d}t^2 + B(r) \cdot \mathrm{d}r^2 + C(r) \cdot r^2 \mathrm{d}\Omega^2. \tag{2.53}$$

We have the freedom to choose the coordinates, the physics must be invariant under coordinate transformations. The angular coordinates are fixed by assuming spherical symmetry. One can define a new radial coordinate by absorbing the function $C(r)$ via

$$\bar{r} = C(r) \cdot r \tag{2.54}$$

so that

$$\mathrm{d}s^2 = -A(\bar{r}) \cdot \mathrm{d}t^2 + \bar{B}(\bar{r}) \cdot \mathrm{d}\bar{r}^2 + \bar{r}^2 \mathrm{d}\Omega^2. \tag{2.55}$$

The radial coordinate can be relabeled back to $\bar{r} \to r$. Hence, there are only two metric functions left that need to be determined from the Einstein equations. For later convenience, we write those metric functions in exponential form:

$$\mathrm{d}s^2 = -e^{2\alpha(r)}\mathrm{d}t^2 + e^{2\beta(r)}\mathrm{d}r^2 + r^2 \cdot \left(\mathrm{d}\theta^2 + \sin^2\theta \mathrm{d}\phi^2\right). \tag{2.56}$$

The components of the metric are then given by

$$g_{00} = -e^{2\alpha(r)} \tag{2.57}$$
$$g_{11} = e^{2\beta(r)} \tag{2.58}$$
$$g_{22} = r^2 \tag{2.59}$$
$$g_{33} = r^2 \sin^2\theta \tag{2.60}$$

and all other components are zero. The components of the inverse metric are just given by inverse of the components, as there are only diagonal components:

$$g^{00} = -e^{-2\alpha(r)} \tag{2.61}$$
$$g^{11} = e^{-2\beta(r)} \tag{2.62}$$
$$g^{22} = r^{-2} \tag{2.63}$$
$$g^{33} = \left(r^2 \sin^2\theta\right)^{-1} \tag{2.64}$$

and all other components are zero. Here, we need to demand that $r \neq 0$. The procedure is now to calculate the components of the Ricci tensor from the Christoffel symbols. The Christoffel symbols as defined in Eq. (2.44) have the following nonzero components

$$\begin{array}{ll} \Gamma^1_{00} = \alpha' \cdot e^{2(\alpha-\beta)} & \Gamma^0_{10} = \alpha' \\ \Gamma^1_{11} = \beta' & \Gamma^2_{12} = \Gamma^3_{13} = r^{-1} \\ \Gamma^1_{22} = -r \cdot e^{-2\beta} & \Gamma^3_{23} = \cot\theta \\ \Gamma^1_{33} = -r \cdot \sin^2\theta \cdot e^{-2\beta} & \Gamma^2_{33} = -\sin\theta\cos\theta. \end{array} \tag{2.65}$$

The prime indicates the partial derivative with respect to r:

$$\alpha' = \frac{\partial\alpha}{\partial r} \qquad \beta' = \frac{\partial\beta}{\partial r}. \tag{2.66}$$

Note that the Christoffel symbols are symmetric in the lower indices, giving other nonzero components that we have not written down explicitly here. Now one can compute the components of the Ricci tensor. Only the diagonal components are nonzero:

$$R_{00} = e^{2(\alpha-\beta)} \cdot \left(-\alpha'' + \alpha' \cdot \beta' - \alpha'^2 - \frac{2}{r} \cdot \alpha'\right) \tag{2.67}$$
$$R_{11} = \alpha'' - \alpha' \cdot \beta' + \alpha'^2 - \frac{2}{r} \cdot \beta' \tag{2.68}$$
$$R_{22} = e^{-2\beta} \cdot \left(1 + r \cdot \alpha' - r \cdot \beta'\right) - 1 \tag{2.69}$$
$$R_{33} = \sin^2\theta \cdot R_{22}, \tag{2.70}$$

where double primes indicate the second derivative with respect to r:

$$\alpha'' = \frac{\partial^2 \alpha}{\partial r^2}. \tag{2.71}$$

In vacuum, all components are vanishing according to the Einstein equations (Eq. (2.51)). Inspection of Eqs. (2.67) and (2.68) shows that similar terms are present in those equations. One can take a look at the combination

$$e^{2(\alpha+\beta)} \cdot R_{00} + R_{11} = \frac{2}{r}\left(\alpha' + \beta'\right) = 0, \tag{2.72}$$

which leads to

$$\alpha' + \beta' = 0. \tag{2.73}$$

So up to a constant and a sign the functions $\alpha(r)$ and $\beta(r)$ are equal. The constant can be absorbed either by redefining the time coordinate or by setting it to zero by demanding that for $r \to \infty$ one has to recover flat spacetime. In any case, we can set

$$\alpha(r) = -\beta(r). \tag{2.74}$$

Let us look now at the Einstein equation for the 22-component of the Ricci tensor (Eq. (2.69)):

$$\begin{aligned} R_{22} &= e^{-2\beta} \cdot \left(1 + r \cdot \alpha' - r \cdot \beta'\right) - 1 \\ &= e^{2\alpha} \cdot \left(1 + 2r \cdot \alpha'\right) - 1 \\ &= \frac{\mathrm{d}}{\mathrm{d}r}\left(r \cdot e^{2\alpha}\right) - 1 = 0, \end{aligned} \tag{2.75}$$

where we used Eq. (2.73). The differential equation can be readily solved to lead to the solution

$$e^{2\alpha} = 1 - \frac{R_s}{r} \tag{2.76}$$

with an integration constant R_s, which is called the Schwarzschild radius. Finally, the metric for a static, spherical symmetric spacetime reads

$$\mathrm{d}s^2 = -\left(1 - \frac{R_s}{r}\right)\mathrm{d}t^2 + \left(1 - \frac{R_s}{r}\right)^{-1}\mathrm{d}r^2 + r^2 \cdot \mathrm{d}\Omega^2 \tag{2.77}$$

and is called the Schwarzschild metric. One can convince oneself that the Schwarzschild metric is also a solution for the other components of the Einstein equation in vacuum individually ($R_{00} = 0$ and $R_{11} = 0$) not only of the combination used in Eq. (2.72).

The integration constant R_s can be fixed by looking at the weak field limit.

For weak gravitational fields, the 00-component of the metric is given by the Newtonian gravitational potential ϕ:

$$g_{00} = -(1 + 2\phi) = -\left(1 - \frac{2GM}{r}\right). \tag{2.78}$$

One sees that one can interpret the Schwarzschild radius as

$$R_s = 2GM. \tag{2.79}$$

Note that Eq. (2.79) is the definition of the quantity M, which can be interpreted as a conventional Newtonian mass for an observer at asymptotically flat spacetime. We recall from the equivalence principle that the concept of a mass is absent in general relativity. In the limit of vanishing mass M, one recovers flat spacetime. In the limit $r \to \infty$, one recovers asymptotically flat spacetime. The Schwarzschild metric has two singularities. The one for $r = 0$ is a true singularity. The one for $r = R_s$, however, is a coordinate dependent singularity. One can find different coordinates, where this singularity is absent. However, $r = R_s$ marks a special place in spacetime. As one can read off from the Schwarzschild metric, for both radii below and above $r = R_s$, the timelike and the spacelike components of the metric change signs. So the Schwarzschild time coordinate will be the new radial coordinate inside the event horizon and vice versa. This condition marks the event horizon as the point of no return for anything going to $r < R_s$. A timelike trajectory will turn beyond the event horizon to one which inadvertently has to go to the singularity at $r \to 0$. The coordinate r switches to the timelike coordinate within the event horizon and, as time just goes on and on, the radial coordinate r will go to zero, inadvertedly reaching the singularity at $r = 0$.

2.5 Energy-Momentum Tensor

In the following, we discuss the properties of matter in a four-dimensional spacetime. From special relativity, one knows that a massive particle moves along a timelike path through spacetime, as every particle can not move faster than the speed of light. A massless particle, as light, moves along lightlike (null) curves. This path is called a worldline. A worldline can be considered as a map of real numbers λ onto the curved spacetime: the worldline is a parameterized curve given by $x^\mu(\lambda)$. For massive particles, one can use the proper time as parameter. We define the four-velocity as a tangent vector of the wordline

$$u^\mu = \frac{\mathrm{d}x^\mu}{\mathrm{d}\tau}. \tag{2.80}$$

The proper time τ is defined by

$$d\tau^2 = -g_{\mu\nu} \cdot dx^\mu dx^\nu. \tag{2.81}$$

Division by the proper time difference squared results in

$$g_{\mu\nu} \cdot u^\mu u^\nu = -1. \tag{2.82}$$

This is the definition of u^μ being a timelike vector. For a null vector, the right-hand side would be zero, for a spacelike vector, the right-hand side would be $+1$. In the rest frame of the particle, only the time component is nonvanishing. Hence, the components of the four-velocity in the rest frame are given by

$$u^\mu = \left((-g_{00})^{-1/2}, 0, 0, 0\right), \tag{2.83}$$

where we used Eq. (2.82). The corresponding one-form for the four-velocity is then

$$u_\mu = \left(-(-g_{00})^{1/2}, 0, 0, 0\right). \tag{2.84}$$

For flat spacetime, the expressions are particular simple. Replacing $g_{\mu\nu}$ with $\eta_{\mu\nu}$, Eq. (2.82) reads

$$\eta_{\mu\nu} \cdot u^\mu u^\nu = -1 \tag{2.85}$$

and the four-velocity is given by

$$u^\mu = (1, 0, 0, 0) \tag{2.86}$$

as $\eta_{00} = -1$ in flat space. The corresponding components of the one-form are

$$u_\mu = (-1, 0, 0, 0). \tag{2.87}$$

A general four-velocity in flat spacetime can be written as

$$u^\mu = \gamma\,(1, \boldsymbol{v}), \tag{2.88}$$

where $\boldsymbol{v}$ is the three-dimensional velocity. The components of the one-form are

$$u_\mu = \gamma\,(-1, \boldsymbol{v}). \tag{2.89}$$

The prefactor γ can be determined from the requirement that the four-velocity is a timelike vector (Eq. (2.85)):

$$\gamma = \frac{1}{\sqrt{1 - v^2}}, \tag{2.90}$$

where $v = |\boldsymbol{v}|$. One recovers the standard gamma-factor of special relativity for a relativistic boost.

We define now the four-momentum by

$$p^\mu = m \cdot u^\mu = (E, \boldsymbol{p}), \tag{2.91}$$

where m is the rest mass of the particle. The energy of the particle E appears as the 0-component of the four-momentum, the three-momentum $\boldsymbol{p}$ as the spatial components of the four-momentum. Using Eq. (2.82), the norm of the four-momentum squared is

$$g_{\mu\nu} \cdot p^\mu p^\nu = p_\mu \cdot p^\mu = -m^2, \tag{2.92}$$

which is a scalar quantity and thereby coordinate independent. For a flat spacetime, one recovers the relativistic energy-momentum equation from Eq. (2.92) when plugging in the components of the four-momentum (Eq. (2.91))

$$E = \sqrt{p^2 + m^2}, \tag{2.93}$$

where $p = |\boldsymbol{p}|$. One can make also a connection to the components of a boosted four-velocity of Eq. (2.88). The boost factor γ and the three-velocity v in terms of the components of the four-momentum are given by

$$\gamma = \frac{E}{m} \qquad \boldsymbol{v} = \frac{\boldsymbol{p}}{E}, \tag{2.94}$$

which can be derived from the definition of the four-momentum equation (2.91) and the properties of the four-velocity.

For many particles we want to describe the system as a fluid with macroscopic quantities, which are the energy density ε, the pressure P, and the flow of the bulk matter u^μ. We stress that thermodynamic quantities, such as the energy density and the pressure, are only well defined in the rest frame of the bulk matter. However, to couple matter to gravity, we need to extend the description of the properties of matter to spacetime quantities, that is, to vectors and tensors. In addition to the bulk properties, we need to define the flux of energy and momentum in a certain direction. In four dimensions, this corresponds to define the flux of the four-momentum p^μ along a four-dimensional direction. The spacetime quantity to describe matter must therefore be a product of the four-momentum and the four-velocity, that is, a tensor. Indeed, this is the energy-momentum tensor $T^{\mu\nu}$ that in symbolic notation we can write as

$$T^{\mu\nu} \propto p^\mu \cdot p^\nu, \tag{2.95}$$

where we replaced the four-velocity with the four-momentum to arrive at the correct dimensionality of the energy-momentum tensor. For a distribution of particles, we need to sum up over all four-momenta. The integral needs to be weighted by a function $f(p)$, which describes the distribution of the energy and momentum of the particles. The distribution function $f(p)$ shall be a scalar quantity. Energy and momentum of the particles are connected by the energy-momentum relation (Eq. (2.93)) for locally flat spacetimes, which can be ensured by a δ-distribution. The energy-momentum tensor can be defined then by the following integral over the flat four-dimensional space of four-momentum:

$$T^{\mu\nu} = \frac{g}{(2\pi)^4} \int d^4p \, p^\mu p^\nu \cdot f(p) \cdot \delta\left(p^0 - \sqrt{p^2 + m^2}\right) \cdot \theta(p^0), \tag{2.96}$$

where g stands for the degeneracy factor of the particles. The θ-function ensures that the energy of the particles has to be positive. Integration over p^0 with $p^0 = E(p)$ results in

$$T^{\mu\nu} = \frac{g}{(2\pi)^3} \int d^3p \, \frac{p^\mu p^\nu}{E(p)} \cdot f(p), \tag{2.97}$$

which is still a well-defined tensorial quantity. Let us look at the diagonal components of the energy-momentum tensor, assuming an isotropic distribution of the three-momentum p. The 00-component is

$$T^{00} = \frac{g}{(2\pi)^3} \int d^3p \, \frac{p^0 p^0}{E} \cdot f(p) \tag{2.98}$$

$$= \frac{g}{(2\pi)^3} \int d^3p \, E(p) \cdot f(p). \tag{2.99}$$

In fact, it corresponds to the integral of all single particle energies $E(p)$. We can interpret the 00-component of the energy-momentum tensor as the energy density ε. The diagonal ii-components are equal for an isotropic momentum distribution and read

$$T^{11} = T^{22} = T^{33} \tag{2.100}$$

$$= \frac{g}{(2\pi)^3} \int d^3p \, \frac{p^i p^i}{E(p)} \cdot f(p) \tag{2.101}$$

$$= \frac{g}{(2\pi)^3} \int d^3p \, \frac{p^2}{E(p)} \cdot f(p), \tag{2.102}$$

as $p^1 = p^2 = p^3 = p$. Note that the repeated spatial indices are not meant to be summed up. The ii-components of the energy-momentum tensor can be interpreted as a pressure P, as we will show now. In fact, the integral as defined in Eq. (2.102) is known as the pressure integral. The first part of the integrand can be interpreted as the product of the three-momentum times the relativistic three-velocity of the particle v:

$$\frac{p^2}{E(p)} = p \cdot \frac{p}{E(p)} = p \cdot v, \tag{2.103}$$

see Eq. (2.94). In Newtonian physics, the pressure is the force per unit area, that is, the change of momentum per unit area. Imagine a unit area that is hit by particles with a momentum p per unit time resulting in a force on that unit area. The number of particles hitting the unit area per unit time are within a volume given by the unit area times the path length the particle can travel within a unit time. The path length per unit time is nothing else than the velocity of the particles. Hence, the pressure is proportional to the momentum times the velocity of the particle summed over all particles. This is exactly what Eq. (2.102) is expressing.

We construct now the energy-momentum tensor by looking at bulk matter at rest. Then the 00-component is the energy density ε ('flux of energy along time'). The mixed $0i$-components T^{0i} are describing the three-dimensional flow of energy, which we will set to zero in the rest frame of the bulk matter. The diagonal spatial components must be the pressure along the three different spatial directions of the flow ('flux of three-momentum along space'). The remaining off-diagonal components of T^{ij} with $i \neq j$ describe a flux of momentum with a three-dimensional flow in a perpendicular direction, that is, they stand for shear forces of matter. In summary, we have the following components of the energy-momentum tensor

$$\begin{aligned} T^{00} &= \varepsilon && \text{energy density} \\ T^{0i} &= T^{i0} && \text{flux of energy density} \\ T^{ii} &= P^i && \text{pressure in direction } i \text{ (no summation)} \\ T^{ij} &= T^{ji} && \text{off-diagonal } (i \neq j) \text{ momentum flux (shear).} \end{aligned} \tag{2.104}$$

We consider in the following a so-called ideal fluid, which has no shear forces and no heat flux, so we set all off-diagonal elements to zero. Also, matter is considered to be isotropic so that the spatial diagonal elements in the rest frame are equal. This is the equivalent of having equal pressure in all three-dimensional directions. The energy-momentum tensor for an ideal fluid in the rest frame as constructed in this way is then given by

$$T^{\mu\nu}_{\text{rest}} = \begin{pmatrix} \varepsilon & 0 & 0 & 0 \\ 0 & P & 0 & 0 \\ 0 & 0 & P & 0 \\ 0 & 0 & 0 & P \end{pmatrix}. \tag{2.105}$$

We need now to cast the expression of the energy-momentum at rest in a covariant form. Let us start with flat space. The covariant tools at hand are, besides the four-velocity of bulk matter u^μ, just the metric tensor $\eta_{\mu\nu}$. A combination of these covariant quantities is

$$T^{\mu\nu} = (\varepsilon + P) \cdot u^\mu u^\nu + P \cdot \eta^{\mu\nu}, \tag{2.106}$$

which gives indeed in the rest frame of the bulk matter the correct answer as $u^\mu = \delta_0^\mu$ and

$$(\varepsilon + P)\delta_0^\mu \delta_0^\nu + P\eta^{\mu\nu} = \begin{pmatrix} \varepsilon + P & 0 & 0 & 0 \\ 0 & 0 & 0 & 0 \\ 0 & 0 & 0 & 0 \\ 0 & 0 & 0 & 0 \end{pmatrix} + \begin{pmatrix} -P & 0 & 0 & 0 \\ 0 & P & 0 & 0 \\ 0 & 0 & P & 0 \\ 0 & 0 & 0 & P \end{pmatrix}.$$

The expression of the energy-momentum tensor in Eq. (2.106) is in a covariant form. Hence, it is valid in any reference frame and constitutes the correct covariant expression for the energy-momentum tensor of an ideal fluid in special relativity. Energy-momentum conservation is ensured by the vanishing derivative of the energy-momentum tensor:

$$\partial_\mu T^{\mu\nu} = 0 \tag{2.107}$$

For a boosted four-velocity, the 0-component of Eq. (2.107) gives

$$\partial_t \varepsilon + \nabla (\varepsilon \cdot \boldsymbol{v}) = 0, \tag{2.108}$$

which is the continuity equation. The spatial components of Eq. (2.107) give

$$\varepsilon \cdot [\partial_t \boldsymbol{v} + (\boldsymbol{v} \cdot \nabla)\boldsymbol{v}] = -\nabla P, \tag{2.109}$$

which is the Euler equation of hydrodynamics. In fact, Eq. (2.107) is the basic equation defining relativistic hydrodynamics.

The extension of relativistic hydrodynamics to hydrodynamics in curved spacetime is now straight forward. The general recipe is

(1) write down a covariant expression using special relativity
(2) replace the metric of flat spacetime $\eta_{\mu\nu}$ with the general one $g_{\mu\nu}$
(3) replace all partial derivatives ∂_μ with the covariant derivative ∇_μ.

Let us apply this recipe to the energy-momentum tensor. The first step is done with Eq. (2.106). The second step results in

$$T^{\mu\nu} = (\varepsilon + P) \cdot u^\mu u^\nu + P \cdot g^{\mu\nu} \tag{2.110}$$

and the third step in the energy-momentum conservation

$$\nabla_\mu T^{\mu\nu} = 0. \tag{2.111}$$

Indeed, Eqs. (2.110) and (2.111) are the correct expressions for the energy-momentum tensor and for energy-momentum conservation in a general curved spacetime.

2.6 The Full Einstein Equation with Matter

For including matter to the Einstein equations, we know from the equivalence principle that gravity couples to all forms of energy. The characteristic quantity to describe the energy of matter is the symmetric energy-momentum tensor $T^{\mu\nu}$, which replaces the mass density distribution ρ on the right-hand side of the full equations of general relativity that we are looking for. In addition, energy-momentum conservation is ensured in flat space by the condition

$$\partial_\mu T^{\mu\nu} = T^{\mu\nu}{}_{,\mu} = 0. \tag{2.112}$$

For curved spacetime, we have to use the covariant derivative. In Eq. (2.36), the covariant derivative of vector was written as

$$\nabla_\mu V^\nu = \partial_\mu V^\nu + \Gamma^\nu_{\mu\lambda} V^\lambda. \tag{2.113}$$

A tensor can be considered as a product of two vectors. Actually, this is how we defined the energy-momentum tensor in Eq. (2.96), as a product of two four-momenta. Hence, we need to take the covariant derivative with respect to both vectors. This implies that for a tensor one has to repeat the formula for the covariant derivative for each index of the tensor. Then, the energy-momentum conservation in a curved spacetime reads

$$\nabla_\mu T^{\mu\nu} = \partial_\mu T^{\mu\nu} + \Gamma^\mu_{\mu\lambda} T^{\lambda\nu} + \Gamma^\nu_{\mu\lambda} T^{\mu\lambda}. \tag{2.114}$$

We can denote the covariant derivative with a colon instead of a comma

$$T^{\mu\nu}{}_{;\mu} = \nabla_\mu T^{\mu\nu}. \tag{2.115}$$

However, the covariant derivative of the Ricci tensor does not vanish as

$$\nabla_\mu R^{\mu\nu} = \frac{1}{2}\nabla_\mu \left(g^{\mu\nu} R\right) = \frac{1}{2} g^{\mu\nu} \nabla_\mu R \tag{2.116}$$

with the Ricci scalar or curvature scalar R defined as

$$R = g_{\mu\nu} R^{\mu\nu} = R^\mu{}_\mu. \tag{2.117}$$

The second equality of Eq. (2.116) follows from the property of the covariant derivative that the covariant derivative of the metric vanishes.

The relation equation (2.116) is also called the twice-contracted Bianchi identity. It is clear from Eq. (2.116) that a combination of the Ricci tensor and the curvature scalar R with the metric tensor will do the job. The resulting tensor describing the curvature of spacetime with a vanishing derivative is the Einstein tensor

$$G^{\mu\nu} = R^{\mu\nu} - \frac{1}{2} g^{\mu\nu} R \tag{2.118}$$

for which indeed

$$\nabla_\mu G^{\mu\nu} = \nabla_\mu \left(R^{\mu\nu} - \frac{1}{2} g^{\mu\nu} R \right) = 0. \tag{2.119}$$

The measure of spacetime curvature on the left-hand side of the differential equation we are looking for should be the Einstein tensor. The measure of the energy content of matter is the energy-momentum tensor on the right-hand side of the differential equation. The constant of proportionality has to be the gravitational constant. The Einstein equations are then given by

$$G^{\mu\nu} = 8\pi G \cdot T^{\mu\nu}. \tag{2.120}$$

where $G_{\mu\nu}$ is the Einstein tensor, which depends on the metric and its derivatives up to second order, and $T^{\mu\nu}$ is the energy-momentum tensor that describes the coupling to matter. The factor 8π instead of the factor 4π comes from the condition to arrive at the correct Newtonian limit, that is, the Newtonian law for weak gravitational fields.

We close by showing that the full Einstein equations in vacuum

$$G_{\mu\nu} = R_{\mu\nu} - \frac{1}{2} g_{\mu\nu} R = 0 \tag{2.121}$$

can be cast in the form

$$R_{\mu\nu} = 0 \tag{2.122}$$

used to derive the Schwarzschild metric. One just needs to show that the curvature scalar R vanishes in vacuum. This can be seen immediately by contracting Eq. (2.121) with the metric

$$\begin{aligned} g^{\mu\nu} G_{\mu\nu} &= g^{\mu\nu} R_{\mu\nu} - \frac{1}{2} g^{\mu\nu} g_{\mu\nu} R \\ &= R - \frac{1}{2} \delta^\mu_\mu R = R - 2R = -R = 0, \end{aligned} \tag{2.123}$$

where we used Eq. (2.26) for the inverse metric. Note that the indices of the Kronecker delta needs to be summed according to the summation convention.

2.7 Tolman–Oppenheimer–Volkoff Equation

We consider now static spherically symmetric objects (a star) consisting of matter. The full Einstein equations need to be solved now:

$$G_{\mu\nu} = R_{\mu\nu} - \frac{1}{2} g_{\mu\nu} R = 8\pi G \cdot T_{\mu\nu}. \tag{2.124}$$

The star shall be a sphere with radius R, which has to be larger than the Schwarzschild radius $R_s = 2GM$. The solution outside the star for $r > R$ has no matter and has to be the solution in vacuum. The Birkhoff theorem states that the outside solution has to be the Schwarzschild metric.

Birkhoff's theorem: Any spherically symmetric solution of the Einstein equations in vacuum must be static and asymptotically flat and is given by the Schwarzschild metric.

So the interior solution of the full Einstein equation has to be matched to the Schwarzschild metric as the exterior solution at the radius R of the star. For the interior solution, we are looking for static, spherically symmetric solutions of the full Einstein equation. For the metric, we choose the ansatz

$$\mathrm{d}s^2 = -e^{2\alpha(r)} \cdot \mathrm{d}t^2 + e^{2\beta(r)} \cdot \mathrm{d}r^2 + r^2 \cdot \mathrm{d}\Omega^2 \tag{2.125}$$

according to our discussion for the static, spherical symmetric solution of the vacuum Einstein equation. For the full Einstein equations, we need to determine now the components of the Einstein tensor $G_{\mu\nu}$. The Ricci scalar R can be calculated from the components of the Ricci tensor, Eqs. (2.67)–(2.70):

$$R = 2e^{-2\beta}\left[-\alpha'' - (\alpha')^2 + \alpha'\beta' + \frac{2}{r}\left(\beta' - \alpha'\right) - \frac{1}{r^2}\right] + \frac{2}{r^2}, \tag{2.126}$$

where primes indicate the derivative with respect to r. The components of the Einstein tensor are then given by

$$G_{00} = \frac{1}{r^2}e^{2(\alpha-\beta)} \cdot \left(2r\beta' - 1 + e^{2\beta}\right) \tag{2.127}$$

$$G_{11} = \frac{1}{r^2} \cdot \left(2r\alpha' + 1 - e^{2\beta}\right) \tag{2.128}$$

$$G_{22} = r^2 e^{-2\beta} \cdot \left[\alpha'' + (\alpha')^2 - \alpha'\beta' + \frac{1}{r}\left(\alpha' - \beta'\right)\right] \tag{2.129}$$

$$G_{33} = \sin^2\theta \cdot G_{22}. \tag{2.130}$$

For the right-hand side of the Einstein equation, we need to specify now the energy-momentum tensor $T_{\mu\nu}$. We model the matter inside the star as a perfect fluid:

$$T_{\mu\nu} = (\varepsilon + P)\, u_\mu u_\nu + P \cdot g_{\mu\nu} \tag{2.131}$$

and fix the four-vector to be $u^\mu = (1, 0, 0, 0)$, that is, at rest with respect to the matter. Note that the energy-momentum tensor is written down with lower indices in Eq. (2.131), so that there appears the metric coefficients squared for the explicit expression of its components:

$$T_{\mu\nu} = \begin{pmatrix} e^{2\alpha} \cdot \varepsilon & 0 & 0 & 0 \\ 0 & e^{2\beta} \cdot P & 0 & 0 \\ 0 & 0 & r^2 \cdot P & 0 \\ 0 & 0 & 0 & r^2 \sin^2\theta \cdot P \end{pmatrix}. \tag{2.132}$$

Putting in the expressions for the Einstein tensor and the energy-momentum tensor into the full Einstein equations, three independent differential equations appear:

$$\frac{1}{r^2} e^{-2\beta} \cdot \left(2r\beta' - 1 + e^{2\beta}\right) = 8\pi G \cdot \varepsilon \tag{2.133}$$

$$\frac{1}{r^2} e^{-2\beta} \cdot \left(2r\alpha' + 1 - e^{2\beta}\right) = 8\pi G \cdot P \tag{2.134}$$

$$e^{-2\beta} \cdot \left[\alpha'' + (\alpha')^2 - \alpha'\beta' + \frac{1}{r}\left(\alpha' - \beta'\right)\right] = 8\pi G \cdot P. \tag{2.135}$$

We can rewrite the metric function $\beta(r)$ in such a form that we can easily recover the exterior Schwarzschild metric by introducing a mass function $m_r(r)$ via

$$e^{2\beta(r)} = \left[1 - \frac{2Gm_r(r)}{r}\right]^{-1}. \tag{2.136}$$

We will denote the expression on the right-hand side also as the Schwarzschild factor. At the radius of the star $r = R$, the metric component equals the one of the Schwarzschild metric with the total mass of the star $m_r(r = R) = M$. Replacing $\beta(r)$ in Eq. (2.133) results in the differential equation

$$\frac{\mathrm{d}m_r(r)}{\mathrm{d}r} = 4\pi r^2 \varepsilon(r) \tag{2.137}$$

or in integral form

$$m_r(r) = 4\pi \int_0^r \mathrm{d}r'\, r'^2 \cdot \varepsilon(r'), \tag{2.138}$$

which can be interpreted as mass conservation. The total mass of the star is then given by

$$M = m_r(r = R) = 4\pi \int_0^R \mathrm{d}r'\, r'^2 \cdot \varepsilon(r'). \tag{2.139}$$

Eq. (2.138) looks like an integral over the energy density within the radius r, that is, the mass within a sphere of radius r. However, the expression of the integral on the right-hand side does not involve the proper spatial integration measure. In fact, the integration measure is not an invariant quantity.

For finding the proper volume element for an integral in curved spacetime, let us have a look at the static, spherical symmetric expression for the metric ansatz, Eq. (2.125), and set $\mathrm{d}t = 0$:

$$\mathrm{d}s^2 = g_{\mu\nu}\mathrm{d}x^\mu \mathrm{d}x^\nu = e^{2\beta}\mathrm{d}r^2 + r^2\mathrm{d}\theta^2 + r^2 \sin^2\theta \mathrm{d}\phi^2. \tag{2.140}$$

Let us take the determinant of this expression that should be an invariant quantity

$$\det(g_{\mu\nu}\mathrm{d}x^\mu \mathrm{d}x^\nu) = \det(g_{\mu\nu}) \cdot \det(\mathrm{d}x^\mu \mathrm{d}x^\nu) \tag{2.141}$$

$$= \left(\sqrt{-g}\,\mathrm{d}r\,\mathrm{d}\theta\,\mathrm{d}\phi\right)^2 \tag{2.142}$$

$$= \left(e^\beta\, r^2 \sin\theta\,\mathrm{d}r\,\mathrm{d}\theta\,\mathrm{d}\phi\right)^2, \tag{2.143}$$

where we introduced the determinant of the metric g:

$$g = \det(g_{\mu\nu}). \tag{2.144}$$

The final expression looks like the familiar integral measure for spherical coordinates squared. However, there appears an additional factor from the metric function $\beta(r)$, which makes the expression a coordinate invariant quantity. Hence, the corresponding integral of the energy density of a star in curved spacetime reads

$$\overline{M} = 4\pi \int_0^R \mathrm{d}r'\, r'^2\, e^{\beta(r)} \cdot \varepsilon(r') = 4\pi \int_0^R \mathrm{d}r'\, r'^2 \left(1 - \frac{2Gm_r(r')}{r'}\right)^{-1/2} \cdot \varepsilon(r'). \tag{2.145}$$

The difference of the mass of the star M and the integral of the energy density $\overline{M}$ is only given by the square-root of the Schwarzschild factor, which gives a different weight of the integral. Note that always $\overline{M} \geq M$ as the radius is larger than the Schwarzschild radius of matter contained within the sphere of radius r: $r > 2Gm_r(r)$. The difference between M and $\overline{M}$ is due to the gravitational binding energy. Therefore, the mass of the star M is also called the gravitational mass of the star. It is the mass of the star measured by an exterior observer at asymptotically flat spacetime distance.

We continue now to solve the Einstein equation and take a look at the 11-component of the Einstein equation, see Eq. (2.134). We can rewrite that equation by using the expression for the mass function $m_r(r)$:

$$\frac{\mathrm{d}\alpha(r)}{\mathrm{d}r} = \frac{Gm_r(r)}{r^2}\left(1 + \frac{4\pi r^3 P(r)}{m_r(r)}\right)\left(1 - \frac{2Gm_r(r)}{r}\right)^{-1}, \tag{2.146}$$

which fixes the metric function $\alpha(r)$ now in terms of the mass function $m_r(r)$ and the pressure $P(r)$. One could now tackle the last equation, Eq. (2.135), but it turns out to be easier to use the equation of energy-momentum conservation instead

$$\nabla_\mu T^{\mu\nu} = 0. \tag{2.147}$$

For our purpose, only the $\mu = \nu = 1$ component is relevant, which leads to the differential equation for the pressure

$$\frac{\mathrm{d}P(r)}{\mathrm{d}r} = -\left(\varepsilon(r) + P(r)\right)\frac{\mathrm{d}\alpha(r)}{\mathrm{d}r}. \tag{2.148}$$

The two equations, Eqs. (2.146) and (2.148), can be combined to give

$$\frac{\mathrm{d}P(r)}{\mathrm{d}r} = -\frac{Gm_r(r)\varepsilon(r)}{r^2}\left(1 + \frac{P(r)}{\varepsilon(r)}\right)\left(1 + \frac{4\pi r^3 P(r)}{m_r(r)}\right)\left(1 - \frac{2Gm_r(r)}{r}\right)^{-1}, \tag{2.149}$$

which is called the Tolman–Oppenheimer–Volkoff equation or TOV equation in short. The first term on the right-hand side is the Newtonian term from hydrostatic equilibrium:

$$\frac{dP(r)}{\mathrm{d}r} = -\frac{Gm_r(r)\rho(r)}{r^2}, \tag{2.150}$$

where $\rho(r)$ is the mass density. The three additional terms of the TOV equation are corrections from general relativity. The first correction term modifies the mass density $\rho(r)$ and takes into account that gravity couples to the energy density $\varepsilon(r)$ and the pressure $P(r)$ of matter. The second correction term modifies the mass function $m_r(r)$ and adds another correction term from the pressure of matter. The third correction term modifies the radius and takes into account the warpage of spacetime that is described by the Schwarzschild factor.

The Einstein equations provide only three independent equations, Eqs. (2.137), (2.146), and (2.148). It can be shown that the 22-component of the Einstein equation do not provide an additional constraint once the condition of energy-momentum conservations is used instead. As there are four independent quantities in the equations, the two metric functions $\alpha(r)$ and $\beta(r)$ (or $m_r(r)$), the energy density $\varepsilon(r)$ and the pressure $P(r)$, an additional equation is needed to close the system of equations. The missing equation is the relation between the pressure and energy density of matter

$$P = P(\varepsilon), \tag{2.151}$$

which fixes the properties of matter of which the star is made. This relation is called the equation of state (or EOS). The EOS describes the physics beyond gravity, the quantum physics of matter. We note that the TOV equation is not sensitive to any other detail from the properties of matter than the bulk quantities energy density and pressure. The number density or the composition of matter is not relevant for the properties of static spherical symmetric stars in general relativity. The EOS, Eq. (2.151), together with the TOV equation, Eq. (2.149), and the equation of mass conservation, Eq. (2.137), fixes all quantities $m_r(r)$, $\varepsilon(r)$, and $P(r)$ to determine the bulk properties of compact stars.

Let us close with a brief historical outline of the TOV equation. Tolman wrote down the equations for static spheres of matter in his textbook in 1934 (Tolman, 1934). Motivated by the work of Landau on neutron stars (Landau, 1932), Oppenheimer started to work out the maximum mass of neutron stars with his graduate student Volkoff. Oppenheimer was in close contact with his colleague Tolman at Caltech, so they published back-to-back two papers in 1939 referring to each other. In the first paper, Tolman worked out analytic solutions for static spheres of fluid for various forms of the metric functions that relate to certain forms of EOS (Tolman, 1939). In the second paper, Oppenheimer and Volkoff set up the equation in the form of Eq. (2.149) and solved it numerically for an EOS of a free gas of neutrons (Oppenheimer and Volkoff, 1939). However, neither Tolman nor Oppenheimer and Volkoff were the first to find a solution to the static Einstein equation for a sphere of fluid; that was Schwarzschild in (1916a).

2.8 The Schwarzschild Solution for a Sphere of Fluid

For solving the TOV equation, one needs to define the EOS. Schwarzschild considered a particular simple ansatz for the EOS. He used as an EOS one of an incompressible fluid, that is, one where the energy density is constant throughout the sphere with radius R:

$$\varepsilon(r) = \begin{cases} \varepsilon_*, & r \leq R \\ 0, & r > R. \end{cases} \tag{2.152}$$

The integration of the mass function, Eq. (2.138), is easily solved to give

$$m_r(r) = \begin{cases} \frac{4\pi}{3} r^3 \varepsilon_*, & r \leq R \\ M, & r > R. \end{cases} \tag{2.153}$$

where M is the total mass of the sphere

$$M = \frac{4\pi}{3} R^3 \varepsilon_* . \tag{2.154}$$

Also, the TOV equation can now be solved analytically to give the pressure as a function of r:

$$P(r) = \frac{R\sqrt{R - 2GM} - \sqrt{R^3 - 2GMr^2}}{\sqrt{R^3 - 2GMr^2} - 3R\sqrt{R - 2GM}} \cdot \varepsilon_* . \tag{2.155}$$

The pressure increases continuously toward the center of the sphere $r \to 0$ and the maximum pressure is given by

$$P(0) = \frac{R\sqrt{R - 2GM} - \sqrt{R^3}}{\sqrt{R^3} - 3R\sqrt{R - 2GM}} \cdot \varepsilon_* . \tag{2.156}$$

The pressure will diverge at the center of the star when the denominator is becoming zero which relates to the following condition for the critical mass and radius of the star:

$$M_c = \frac{4}{9G} R_c \tag{2.157}$$

or in terms of the Schwarzschild radius R_s:

$$R_c = \frac{9}{8} R_s . \tag{2.158}$$

The critical radius is just slightly larger than the radius of a nonrotating black hole. In order to have well-behaved pressure throughout the star, the radius of the star has to be above the critical radius

$$R > \frac{9}{8} R_s , \tag{2.159}$$

which has been noted by Schwarzschild in his original paper (Schwarzschild, 1916a). Buchdahl showed several years later that this limit on the radius of a star holds for any EOS where the energy density is not increasing with the radius (Buchdahl, 1959). Hence, Eq. (2.159) holds for any reasonable EOS and is known as the Buchdahl limit:

Buchdahl limit: All static fluid spheres where the energy density is not increasing outward cannot be more compact than

$$C_{\text{max}} = \frac{4}{9} .$$

The limit can be slightly strengthened by arguing that the pressure cannot exceed the energy density, so that the maximum pressure allowed at the center corresponds to $p_{\text{max}} = \varepsilon_*$. As we will see in the chapter on compact stars, Chapter 4, this represents the causal limit above which the speed of sound in the medium is larger than the speed of light. In this case, the maximum allowed compactness relates to

$$C_{\text{max, causal}} = \frac{3}{8} \tag{2.160}$$

as you are invited to show in the exercises (see also Buchdahl [1959] for a general discussion).

Exercises

(2.1) Show that the equation for the gravitational shift of the wavelength, Eq. (2.7), holds also for a general gravitational potential with an arbitrary dependence on the distance (we assumed a constant gradient in the gravitational potential in the derivation).

(2.2) Show that the Schwarzschild metric, Eq. (2.77), is a solution of $R_{00} = 0$ and $R_{11} = 0$.

(2.3) Derive the relations for the relativistic boost factors, Eq. (2.94), from the definition of the four-momentum.

(2.4) Derive the continuity equation, Eq. (2.108), and the Euler equation, Eq. (2.109), from the conservation of the energy-momentum tensor, Eq. (2.147), for the nonrelativistic limit. Use the nonrelativistic expression for the four-vector $u^\mu = (1, \boldsymbol{v})$ for velocities much smaller than the velocity of light c. Which terms have to be neglected to result in the continuity and the Euler equations, and can you find arguments for neglecting them?

(2.5) Show that the condition of metric compatibility of the metric tensor, Eq. (2.41), results in the expression of the Christoffel symbols in terms of the partial derivatives of the metric tensor, Eq. (2.44).

(2.6) Derive the differential equation for the pressure in terms of the metric function $\alpha(r)$, Eq. (2.148), from the equation of energy-momentum conservation, Eq. (2.147).

(2.7) For a general metric in spherical symmetry, the metric functions could also depend on the time coordinate. Birkhoff's theorem states that the time dependence drops out in the equation so that the general solution for a spherical symmetry is the Schwarzschild solution. Prove Birkhoff's theorem by starting with the general ansatz for the metric in the form

$$\mathrm{d}s^2 = -A(r,t)\mathrm{d}t^2 + B(r,t)\mathrm{d}r^2 + C(r,t)r^2\left(\mathrm{d}\theta^2 + \sin^2\theta\mathrm{d}\phi^2\right) \quad (2.161)$$

and choose proper coordinates to end up in the form

$$\mathrm{d}s^2 = -e^{2\alpha(r,t)}\mathrm{d}t^2 + e^{2\beta(r,t)}\mathrm{d}r^2 + r^2\cdot\left(\mathrm{d}\theta^2 + \sin^2\theta\mathrm{d}\phi^2\right), \quad (2.162)$$

where the metric functions are assumed to depend on time also. Derive the time-dependent Christoffel symbols and the Ricci tensor (note: there are off-diagonal terms). Show that Einstein's equations result in a metric that is time independent.

(2.8) Solve the TOV equation for an incompressible fluid of the form

$$\varepsilon(r) = \begin{cases} \varepsilon_*, & r \le R \\ 0, & r > R, \end{cases} \tag{2.163}$$

where R is the radius of the star. Show that the solution is given by

$$P(r) = \frac{R\sqrt{R - 2GM} - \sqrt{R^3 - 2GMr^2}}{\sqrt{R^3 - 2GMr^2} - 3R\sqrt{R - 2GM}} \cdot \varepsilon_*, \tag{2.164}$$

which is called the Schwarzschild solution.

(2.9) Derive the Buchdahl limit from Eq. (2.156), which states that the maximum compactness of the star is given by

$$C_{\max} = \frac{4}{9}$$

and corresponds to the case of infinite pressure of the Schwarzschild solution.

(2.10) Derive the condition for the maximum compactness of a sphere $C_{\max} = 3/8$ from the Schwarzschild solution, Eq. (2.156), for an incompressible causal fluid where the pressure is not allowed to violate the causal limit $P = \varepsilon$.

3

Dense Matter

3.1 Thermodynamic Potentials

In the following we summarize the thermodynamic description for the microcanonical, the canonical, and the grand canonical ensembles. We consider a system of several species with the particle number N_i and the corresponding chemical potential μ_i.

In the microcanonical ensemble, the internal energy U is given by the fundamental equation of thermodynamics

$$dU = -P \cdot dV + T \cdot dS + \sum_i \mu_i \cdot dN_i, \tag{3.1}$$

with the pressure P, the volume V, the temperature T, and the entropy S. The internal energy is a function of the volume, the entropy, and the particle number

$$U = U(V, S, N_i). \tag{3.2}$$

The other thermodynamic quantities are given in terms of partial derivates as

$$-P = \left.\frac{\partial U}{\partial V}\right|_{S,N_i} \tag{3.3}$$

$$T = \left.\frac{\partial U}{\partial S}\right|_{V,N_i} \tag{3.4}$$

$$\mu_i = \left.\frac{\partial U}{\partial N_i}\right|_{V,S,N_{i\neq j}}, \tag{3.5}$$

keeping the other quantities fixed. The internal energy is an extensive quantity, it is additive and scales linear with the size of the system. Also the volume, the entropy, and the particle number are extensive quantities. By contrast, the pressure, the temperature, and the chemical potential are intensive quantities and do not depend on the size of the system. Hence, the internal energy is a homogeneous function and can be integrated using the Euler theorem to give the Euler equation

$$U = -P \cdot V + T \cdot S + \sum_i \mu_i \cdot N_i. \tag{3.6}$$

The thermodynamic potential in the canonical ensemble is the free energy F, which is given by the Legendre transformation

$$F = F(V, T, N_i) \equiv U - T \cdot S. \tag{3.7}$$

The free energy is a function of the volume, the temperature, and the particle numbers N_i and its total differential reads

$$\mathrm{d}F = -P \cdot \mathrm{d}V - S \cdot \mathrm{d}T + \sum_i \mu_i \cdot \mathrm{d}N_i. \tag{3.8}$$

The remaining thermodynamic quantities are fixed by the partial derivatives

$$-P = \left.\frac{\partial F}{\partial V}\right|_{T, N_i} \tag{3.9}$$

$$-S = \left.\frac{\partial F}{\partial T}\right|_{V, N_i} \tag{3.10}$$

$$\mu_i = \left.\frac{\partial F}{\partial N_i}\right|_{V, T, N_{i \neq j}} . \tag{3.11}$$

Finally, in the grand canonical ensemble, the thermodynamical potential follows from the Legendre transformation

$$\Omega = \Omega(V, T, \mu_i) \equiv F - \sum_i \mu_i \cdot N_i = U - T \cdot S - \sum_i \mu_i \cdot N_i, \tag{3.12}$$

which is equivalent to

$$\Omega = \Omega(V, T, \mu_i) = -P \cdot V \tag{3.13}$$

by using the Euler equation, Eq. (3.6). The grand canonical potential is a function of the volume, the temperature, and the chemical potentials μ_i. The total derivative reads

$$\mathrm{d}\Omega = -P \cdot \mathrm{d}V - S \cdot \mathrm{d}T - \sum_i N_i \cdot \mathrm{d}\mu_i \tag{3.14}$$

and the thermodynamic quantities are given by the following partial derivatives:

$$-P = \left.\frac{\partial \Omega}{\partial V}\right|_{T,\mu_i} \tag{3.15}$$

$$-S = \left.\frac{\partial \Omega}{\partial T}\right|_{V,\mu_i} \tag{3.16}$$

$$-N_i = \left.\frac{\partial \Omega}{\partial \mu_i}\right|_{V,T,\mu_{i\neq j}} . \tag{3.17}$$

Note that the first relation is a triviality by looking at the definition of the grand canonical potential equation (3.13) by virtue of the Euler equation.

Let us consider now the thermodynamic limit, which is given by letting the volume grow to infinity $V \to \infty$. There are no effects from the volume left and all thermodynamic quantities should be independent on the volume. Then matter is referred to as bulk matter. Let us have a look at the thermodynamic relations for bulk matter. We define the energy density ε, the entropy density s, and the number density n_i as

$$\varepsilon \equiv \frac{U}{V} \quad , \qquad s \equiv \frac{S}{V} \quad , \qquad n_i \equiv \frac{N_i}{V}, \tag{3.18}$$

so that they constitute intensive variables as the pressure, the temperature, and the chemical potentials.

First of all, the Euler equation can be recast to

$$\varepsilon = -P + T \cdot s + \sum_i \mu_i \cdot n_i . \tag{3.19}$$

It constitutes an important check of thermodynamic consistency for all thermodynamics quantities derived from the chosen thermodynamic potential.

As the energy density is an intensive quantity, it should only depend on the entropy density and the number densities:

$$\varepsilon = \varepsilon(s, n_i). \tag{3.20}$$

The temperature and the chemical potentials follow from the thermodynamic relations equations (3.4) and (3.5):

$$T = \left.\frac{\partial \varepsilon}{\partial s}\right|_{n_i} \tag{3.21}$$

$$\mu_i = \left.\frac{\partial \varepsilon}{\partial n_i}\right|_{s,n_{j\neq i}} . \tag{3.22}$$

Note that the expression of the pressure, Eq. (3.3), involves the derivative with respect to the volume, which is not well defined in the thermodynamic limit. Alternatively, one can use the Euler equation for the pressure in bulk matter:

$$P = -\varepsilon + T \cdot s + \sum_i \mu_i \cdot n_i, \tag{3.23}$$

which fully defines the pressure as a function of n_i and s for a given energy density in the form of Eq. (3.20) together with the relations equations (3.21) and (3.22). One can also resort to a trick by looking at the derivative of the energy per particle with respect to the total particle number $N = \sum_i N_i$ and the total particle number density $n = \sum_i n_i$ in the following form:

$$n \cdot n_i \cdot \frac{\partial}{\partial n_i}\left(\frac{\varepsilon}{n}\right) = n_i \cdot \frac{\partial \varepsilon}{\partial n_i} - \varepsilon \frac{n_i}{n} = n_i \cdot \mu_i - \varepsilon \frac{n_i}{n}, \tag{3.24}$$

where we used the expression for the chemical potential equation (3.22) in the last step. Summing over the particle species i, it follows that

$$\sum_i n_i \cdot \mu_i = \sum_i n \cdot n_i \cdot \frac{\partial}{\partial n_i}\left(\frac{\varepsilon}{n}\right) + \varepsilon, \tag{3.25}$$

and using the Euler equation (3.19), one finds for the pressure

$$P = T \cdot s + \sum_i n \cdot n_i \cdot \frac{\partial}{\partial n_i}\left(\frac{\varepsilon}{n}\right). \tag{3.26}$$

In practice, one chooses N to be an overall conserved number as, for example, the baryon number.

The expressions for the canonical ensemble in the thermodynamic limit follow accordingly and are left as an exercise for the reader.

The case of the grand canonical ensemble is of special importance for describing hot and dense matter in bulk. The grand canonical potential in the thermodynamic limit is just the negative value of the pressure as

$$P = P(T, \mu_i) = -\frac{\Omega}{V}, \tag{3.27}$$

which is a function of the temperature and the chemical potentials only. The thermodynamic relations in bulk matter are

$$s = \left.\frac{\partial P}{\partial T}\right|_{\mu_i} \tag{3.28}$$

$$n_i = \left.\frac{\partial P}{\partial \mu_i}\right|_{T, \mu_{i \neq j}}. \tag{3.29}$$

The energy density can be calculated by using the Euler equation, Eq. (3.19).

3.2 Chemical Equilibrium

The properties of matter, as present in white dwarfs, are described by the relation between pressure and energy density, the equation of state (EOS). It determines the global properties of bulk matter as the reaction of matter to gravity, rotation, and electromagnetic fields. It also has an impact on local properties of matter as the internal stress (shear and bulk viscosity), dissipation, and emissivities.

Let us consider general matter that consists of different species with a corresponding particle number N_i and the total particle number N. The first law of thermodynamics for the internal energy $U = U(V, S, N_i)$ reads

$$\mathrm{d}U = -P \cdot \mathrm{d}V + T \cdot \mathrm{d}S + \sum_i \mu_i \cdot \mathrm{d}N_i, \tag{3.30}$$

with the pressure P, the volume V, the temperature T, the entropy S and the chemical potentials μ_i for each species i. Note that the internal energy is a function of the volume, the entropy, and the number of particles of the species i.

In chemical equilibrium, the reactions between particles are in detailed balance, that is, each reaction is balanced by its backreaction and N_i is constant. We will take a close look now at the implications of chemical equilibrium for the chemical potentials μ_i. Consider a system that is not doing any work ($\mathrm{d}W = P\mathrm{d}V = 0$) and which is not exchanging any heat ($\delta Q = T\mathrm{d}S = 0$). The first law of thermodynamics, Eq. (3.30), reduces to the condition

$$\mathrm{d}U = \sum_i \mu_i \cdot \mathrm{d}N_i, \tag{3.31}$$

which relates the change of the particle number N_i for a species i to the chemical potentials μ_i. It is apparent that the chemical potential μ_i is just the energy needed to create or destroy a particle species i in the system. In equilibrium, the internal energy does not change, so $\mathrm{d}U = 0$. Hence, chemical equilibrium corresponds to the condition

$$\sum_i \mu_i \cdot \mathrm{d}N_i = 0. \tag{3.32}$$

As an explicit example, take a look at the following weak reaction where an electron and a proton react to a neutron and an electron-neutrino

$$e^- + p \longleftrightarrow n + \nu_e. \tag{3.33}$$

This is the reaction that neutronizes matter, forming neutron star matter. For this reaction, the changes of the particle number of the different species are related by

$$\mathrm{d}N_{e^-} = \mathrm{d}N_p = -\mathrm{d}N_n = -\mathrm{d}N_{\nu_e}. \tag{3.34}$$

The condition of chemical equilibrium, Eq. (3.32), then reads

$$\mu_{e^-} + \mu_p = \mu_n + \mu_{\nu_e}, \tag{3.35}$$

which gives a relation between the chemical potentials of the species involved in the reaction equation (3.33). Hence, the chemical potentials cannot be chosen arbitrarily but are related to each other in chemical equilibrium.

In principle, one has to write down all possible reactions that can happen in matter that consists of certain particle species. For a system of nucleons, electrons, and neutrinos there are additional reactions that involve the antiparticles of the electron and the electron-neutrino, the positron, and the anti-electron-neutrino. For example, the neutron decay

$$n \longrightarrow e^- + p + \bar{\nu}_e \tag{3.36}$$

gives the following condition on particle number changes

$$\mathrm{d}N_n = -\mathrm{d}N_{e^-} = -\mathrm{d}N_p = -\mathrm{d}N_{\bar{\nu}_e} \tag{3.37}$$

and the condition of chemical equilibrium equation (3.32) reads

$$\mu_n = \mu_{e^-} + \mu_p + \mu_{\bar{\nu}_e}. \tag{3.38}$$

The two conditions on the chemical potential, Eqs. (3.35) and (3.38), combine to a condition on the chemical potentials for neutrinos and anti-neutrinos:

$$\mu_{\bar{\nu}_e} = -\mu_{\nu_e}. \tag{3.39}$$

Similar considerations involving reactions with positrons lead to a condition on the chemical potential for electrons and positrons:

$$\mu_{e^+} = -\mu_{e^-}. \tag{3.40}$$

This is not accidental, it is a generic feature: the chemical potential of the antiparticle is given by the negative value of the chemical potential of the particle. It gives also an interesting additional feature. As the particles and antiparticles considered here annihilate to pure energy in the form of photons, the chemical potential of photons must be 0.

Now imagine that there are many different particles in the system all with their chemical potentials. How can one be sure that one has found all possible chemical reactions that are in equilibrium and correspondingly all possible relations between the chemical potentials involved?

Let us be more specific and ask for the number of independent quantities required to describe a mixture of baryons (neutrons, protons...) and leptons ($e^-, \mu^-, \nu_e, \nu_\mu \ldots$). The possible reactions between baryons and leptons are described by the fundamental forces of electrodynamics, weak and strong interactions. Electrodynamics conserves the charge, which is a conserved quantum number. Weak interactions conserve the baryon B and lepton number L. In principle, there are three different lepton numbers associated with the electron, the muon, and the tau and the corresponding neutrinos. Flavor mixing between the three different neutrino flavors will break the individual lepton flavor, but not the total lepton number.[1] Strong interactions conserve the baryon number B and also the flavor number of quarks (up-, down-, strange-, charm-, bottom-, and top-quark number).

So all possible reactions between leptons and baryons conserve the baryon number (B), lepton number (L), and charge (Q). By choosing the corresponding numbers N_B, N_L, and N_Q, the quantum system is completely specified. All particle numbers and their chemical potentials have to be fixed in terms of the corresponding chemical potentials μ_B, μ_L, and μ_Q. One can decompose the chemical potential of a particle i in terms of the quantum numbers of the particle

$$\mu_i = B_i \cdot \mu_B + L_i \cdot \mu_L + Q_i \cdot \mu_Q, \tag{3.41}$$

where B_i, L_i, and Q_i are the baryon number, lepton number, and charge of the particle i, respectively. Hence, indeed, all we need are the independent numbers of conserved charges or their corresponding chemical potential to fully define the particle composition or all chemical potentials.

To make the connection between the chemical potential and the particle composition clear, it is better to switch to the grand canonical potential $\Omega = \Omega(V, T, \mu_i)$, which is a function of volume, temperature, and the chemical potential. Equivalently, one can consider the pressure as

$$P = -\frac{\Omega}{V} = P(T, \mu_i) = P(T, \mu_B, \mu_L, \mu_Q). \tag{3.42}$$

As the pressure is an intrinsic quantity, it does not depend on the volume. Eq. (3.41) implies that the pressure and hence all thermodynamic quantities are a function of T, μ_B, μ_L, and μ_Q alone. There are additional constraints for the astrophysical systems considered here. In astrophysics, matter is globally charge neutral so that the total charge density vanishes $n_Q = N_Q/V = 0$. This condition fixes the charge chemical potential μ_Q, ensuring overall charge neutrality by a vanishing overall charge density n_Q:

$$n_Q = \frac{\partial P}{\partial \mu_Q} = \sum_i \frac{\partial P}{\partial \mu_i} \frac{\partial \mu_i}{\partial \mu_Q} = \sum_i n_i \cdot Q_i = 0, \tag{3.43}$$

[1] Nonperturbative instanton interactions in electroweak theory break baryon and lepton number conservation while conserving the difference $B - L$. Those interactions appear at energy scales that are much higher than the ones relevant for compact stars. They can occur in the early universe at the time of electroweak symmetry breaking.

with

$$n_i = \frac{\partial P}{\partial \mu_i} \tag{3.44}$$

being the number density of particle i. Note, that the connection between the chemical potential and the overall charge density is a nontrivial one and depends not only on the particles considered but also on the interactions between the particles, as all terms in the thermodynamic potential have to be considered that depend on μ_Q.

In the same way, the lepton number and baryon number fix the corresponding lepton and baryon chemical potential. For cold matter, neutrinos escape freely as their interaction rate is small. Neutrinos are not trapped and their numbers are not conserved in matter. Their chemical potentials have to be set to 0, then: $\mu_{\nu_e} = \mu_{\nu_\mu} = \mu_{\nu_\tau} = 0$. This implies that the lepton number is not conserved and the lepton chemical potential can be set to 0, $\mu_L = 0$. For cold degenerate matter, the temperature is much smaller than the Fermi energy $T \ll E_F$ and one can set $T = 0$. Finally, the whole system can be described just in terms of the baryon chemical potential μ_B or in terms of the baryon density n_B alone.

3.3 Matter in β-Equilibrium

We consider in the following an ideal cold gas of electrons, protons, and neutrons in chemical equilibrium, where only the baryon number and the charge are conserved quantities. The lepton number is not conserved and the corresponding chemical potential is set to zero. Hence, we do not need to consider a chemical potential for neutrinos. This type of chemical equilibrium is called β-equilibrium as it involves the chemical reactions of nuclear β-decay.

Let us treat neutrons and protons as free particles first (actually in white dwarfs they are bound in nuclei). In the transformation of protons to neutrons, the weak reaction,

$$e^- + p \longleftrightarrow n + \nu_e, \tag{3.45}$$

proceeds, when the electron has sufficient kinetic energy to compensate for the mass difference of neutrons and protons, which is $m_n - m_p = (939.57 - 928.28)$ MeV$= 1.29$ MeV. On the other hand, the inverse reaction, the transformation of neutrons to protons,

$$n \longrightarrow e^- + p + \bar{\nu}_e, \tag{3.46}$$

is blocked, if the density of electrons is high enough, so that all available energy levels are occupied. If we assume chemical equilibrium (setting $T = 0$, $\mu_\nu = 0$), then the condition on the chemical potentials of electrons, protons, and neutrons reads

$$\mu_e + \mu_p = \mu_n. \tag{3.47}$$

For fermions, all energy levels are occupied until the highest energy level, the Fermi energy. For zero temperature, the energy needed to take away a fermion is given by the Fermi energy $E_{F,i}$, which then has to be equal to the chemical potential

$$\mu_i = E_{F,i} = \sqrt{k_{F,i}^2 + m_i^2}, \tag{3.48}$$

with the Fermi momenta $k_{F,i}$ and the mass m_i of the particle i. The equilibrium condition (3.47) implies for the Fermi momenta of electrons, protons, and neutrons that:

$$\sqrt{k_{F,e}^2 + m_e^2} + \sqrt{k_{F,p}^2 + m_p^2} = \sqrt{k_{F,n}^2 + m_n^2}. \tag{3.49}$$

The number densities are given by the integral over all occupied states, which for a degenerate Fermi gas is given by the Fermi sphere

$$n_i = g_i \int_0^{k_{F,i}} \frac{\mathrm{d}^3 k}{(2\pi)^3} = \frac{g_i}{6\pi^2} k_{F,i}^3, \tag{3.50}$$

with the degeneracy factor $g_i = 2s_i + 1$ of particle i with the spin s_i.

In addition to the condition of β-equilibrium, matter is charge neutral, which fixes the charge chemical potential. The number densities of electrons and protons must be equal to ensure the charge neutrality condition:

$$n_e = n_p \longrightarrow g_e \cdot k_{F,e}^3 = g_p \cdot k_{F,p}^3. \tag{3.51}$$

As the degeneracy factors are $g_e = g_p = 2$, the Fermi momenta of electrons and protons must be equal:

$$k_{F,e} = k_{F,p}. \tag{3.52}$$

This determines all thermodynamic quantities in terms of the baryon chemical potential μ_B or equivalently the baryon number density n_B. The thermodynamic quantities of interest are the total pressure P, the total energy density ε, the total baryon number density n_B, and the charge density n_Q, which are given by the sum of the contributions from the individual particles:

$$P = P_e + P_p + P_n \tag{3.53}$$

$$\varepsilon = \varepsilon_e + \varepsilon_p + \varepsilon_n \tag{3.54}$$

$$n_B = n_p + n_n \tag{3.55}$$

$$n_Q = n_p - n_e = 0. \tag{3.56}$$

As the sum of the proton and the electron mass is smaller than that of the neutron ($m_p + m_e = 938.79\,\mathrm{MeV} < m_n = 939.57\,\mathrm{MeV}$), a gas of protons and electrons is energetically favored at low densities. At some critical electron density, corresponding to a critical proton or baryon number density, the Fermi energies of electrons and protons sum up to the rest mass of the neutron. Then neutrons start to appear in the dense medium. The onset of neutrons is characterized by the condition that their Fermi momentum is $k_{F,n} = 0$. Then Eq. (3.49) reads

$$\sqrt{k_{F,e}^2 + m_e^2} + \sqrt{k_{F,p}^2 + m_p^2} = m_n, \tag{3.57}$$

or for the nonrelativistic limit for protons $k_{F,p} \ll m_p$:

$$\sqrt{k_{F,e}^2 + m_e^2} \approx m_n - m_p. \tag{3.58}$$

The critical electron density (which is equal to the proton density) is then mainly given by the mass difference of neutrons and protons

$$n_c = \frac{1}{3\pi^2}\left((m_n - m_p)^2 - m_e^2\right)^{3/2} \approx 7.4 \times 10^{30}\,\text{cm}^{-3}, \tag{3.59}$$

which amounts to a critical mass density of

$$\rho_c = m_p \cdot n_c \approx 1.2 \times 10^7\,\text{g cm}^{-3}. \tag{3.60}$$

For higher densities, the neutron fraction increases drastically so that eventually $n_n \gg n_p$ and neutron matter forms.

In the ultra-relativistic limit ($k_F \gg m_N$), it follows from the equilibrium condition (3.49) and charge neutrality

$$k_{F,n} \approx k_{F,e} + k_{F,p} = 2k_{F,p}. \tag{3.61}$$

With $n \propto g \cdot k_F^3$ (and $g_n = g_p = 2$), the neutron number density approaches in the high density limit

$$n_n = 8n_p \quad \text{or} \quad \frac{n_n}{n_B} = \frac{8}{9}. \tag{3.62}$$

Despite the fact that the neutron is heavier than the proton, dense matter in β-equilibrium consists mainly of neutrons above a critical density. The reason is that it is energetically favored to add a charge neutral particle to the system instead of two charged particles with opposite charges to ensure charge neutral matter.

In reality, the nucleons in their lowest energy state are not given by free nucleons, as it is energetically favored to form nuclei. Hence, matter in β-equilibrium consists of nuclei immersed within an electron gas, which ensures overall charge neutrality. Naively, one would expect that the nucleus with the lowest overall mass per baryon would be the one to be found in β-equilibrium. Guided by our previous experience of the composition of charge neutral matter in β-equilibrium, we know that this is not true and that the least massive particle is not necessarily picked out. Hence, one needs to consider all possible nuclei to find the correct composition in β-equilibrium. The recipe is to find the combination of mass number A and charge Z of a nucleus that minimizes the total energy of the system for a given baryon number density by scanning through the whole possible chart of nuclei. This includes unknown ones up to the dripline of nuclei, where nuclei are so neutron-rich that they cannot bind any additional neutron anymore. In addition, the nuclei will form a regular lattice so as to minimize the Coulomb energy. This means that the lattice energy has to be included for the minimization of the total energy.

It turns out that the lattice energy is important for the composition of nuclei, but not for the EOS. At low densities, the nucleus ^{56}Fe is present in β-equilibrated matter, which is the one with the lowest mass per baryon.[2] With increasing baryon density, the electron density increases accordingly. It becomes increasingly likely that electrons transform protons to neutrons, as seen for the free gas case considered previously. The nuclei become increasingly neutron-rich with increasing baryon density until the neutron dripline for nuclei is reached with the nucleus having the highest neutron-to-proton ratio possible. The corresponding baryon number density is the so-called neutron-drip density of about $\rho \sim 4 \times 10^{11}\,\mathrm{g\,cm^{-3}}$. The neutron-drip density depends on the nuclear model used for determining the binding energy of the unknown neutron-rich nuclei, in particular the ones being located at the neutron dripline. Nuclei with a particular high binding energy will be favored in β-equilibrated matter. These favored nuclei are those with a closed shell of neutrons that have magic numbers for neutrons. The magic numbers for neutrons are $N = 2, 8, 20, 50, 82, 126$. According to model calculations, the ones most favorable for β-equilibrated matter are the neutron magic numbers $N = 50$ and 82.

The composition of β-equilibrated matter can be summarized as follows:

$\rho < 10^4\,\mathbf{g\,cm^{-3}}$**:** a gas of atoms, this is the ‘atmosphere’ of compact stars

$\rho = 10^4\,\mathbf{g\,cm^{-3}}$**:** a free gas of electrons and ^{56}Fe within a lattice structure forms

$\rho = 2 \times 10^6\,\mathbf{g\,cm^{-3}}$**:** electrons become relativistic

$\rho = 8 \times 10^6\,\mathbf{g\,cm^{-3}}$**:** a free gas of electrons and a lattice of nuclei beyond ^{56}Fe forms, the nuclei become increasingly neutron-rich with increasing density

$\rho = 4 \times 10^{11}\,\mathbf{g\,cm^{-3}}$**:** the neutron drip density, the limiting density for particle stable neutron-rich nuclei.

[2] We note in passing that the nucleus with the highest binding energy is ^{62}Ni. However, what counts is the total mass per baryon present in β-equilibrated matter. ^{56}Fe has a lower neutron fraction and the lower mass per baryon overcompensates the lower binding energy per baryon compared to ^{62}Ni.

3.4 Equation of State

In the following, we calculate the EOS for fermionic matter at vanishing temperature. This is the EOS for degenerate matter as it is found in compact stars. The EOS can be expressed in different forms. Most widely used is the relation of the pressure P with the energy density ε or the number density, which for zero temperature is simply:

$$P = P(\varepsilon) \tag{3.63}$$

or

$$P = P(n). \tag{3.64}$$

It is the additional input needed to close the equation of hydrostatic equilibrium and the one of mass conservation for describing compact star configurations.

In nuclear physics, one encounters another form of the EOS to describe nuclear matter, the relation of the binding energy per baryon as a function of baryon number density n_B:

$$\frac{E_b}{A} = \frac{E_b}{A}(n_B) = \frac{\varepsilon}{n_B} - m_N. \tag{3.65}$$

It is convention to use the mass number A in place of the baryon number B. The form of Eq. (3.65) can be transformed into the one of Eq. (3.63) by using thermodynamic relations. The energy per particle is a thermodynamic potential in the thermodynamic limit and a function of the number density and the entropy per particle. The derivative of the energy per particle with respect to the number density gives an expression for the pressure. Let us start with the following expression

$$\frac{\partial}{\partial n_B}\left(\frac{E}{A}\right) = \frac{\partial}{\partial n_B}\left(\frac{\varepsilon}{n_B}\right) = \frac{1}{n_B} \cdot \frac{\partial \varepsilon}{\partial n_B} - \frac{\varepsilon}{n_B^2}. \tag{3.66}$$

The derivative of the energy density with respect to the number density is just the chemical potential corresponding to the (conserved) number of particles. Using the Euler equation for zero entropy,

$$\varepsilon = -P + \mu_B \cdot n_B, \tag{3.67}$$

we arrive at

$$\frac{\partial}{\partial n_B}\left(\frac{E}{A}\right) = \frac{\mu_B}{n_B} - \frac{\varepsilon}{n_B^2} = \frac{P}{n_B^2}, \tag{3.68}$$

and the pressure can be expressed as

$$P = n_B^2 \cdot \frac{\partial (E/A)}{\partial n_B}. \tag{3.69}$$

The EOS is now in the form of Eq. (3.64) for zero temperature. We could have derived the result also directly by using Eq. (3.26). Note that Eq. (3.69) states that the slope of the energy per particle as a function of the number density is directly proportional to the pressure. Therefore, a minimum in the energy per particle corresponds to a vanishing pressure, that is, stable equilibrated matter.

We stress that the pressure given in the form of Eqs. (3.63) or (3.64) does not represent a thermodynamic potential and does not encode the full thermodynamic information, while Eq. (3.65) does. In fact, there appears an integration constant when calculating the energy density from the pressure in the form $P = P(n)$ using Eq. (3.69), which is undetermined. This would not be the case when the pressure is given as a function of the chemical potential, as in this case:

$$\varepsilon(\mu, T) = -P(\mu, T) + \mu \cdot n + T \cdot s, \tag{3.70}$$

with the thermodynamic relations

$$n = \frac{\partial P(\mu, T)}{\partial \mu} \qquad \text{and} \qquad s = \frac{\partial P(\mu, T)}{\partial T} \tag{3.71}$$

so that every thermodynamic quantity is fixed.

3.5 Properties of Free Fermi Gases

In the following we take a closer look at the EOS of degenerate matter at zero temperature as found in compact stars. We consider first a composite Fermi gas of electrons and nucleons. The pressure integral for the contribution from electrons is given by

$$P_e = \frac{g_e}{3(2\pi)^3} \int d^3k \, (k \cdot v_e), \tag{3.72}$$

with the velocity $v_e = k/E_e(k)$. The energy–momentum relation of electrons is that of free particles

$$E_e(k) = \sqrt{k^2 + m_e}, \tag{3.73}$$

with the electron mass m_e. Then the pressure integral reads

$$P_e = \frac{g_e}{6\pi^2} \int_0^{k_{F,e}} dk \frac{k^4}{E_e(k)}. \tag{3.74}$$

First we take the nonrelativistic limit $k_{F,e} \ll m_e$ by setting $v_e \approx k/m_e$ or equivalently $E_e \approx m_e$. The integration gives

$$P_e^{(\mathrm{nr})} = \frac{g_e}{6\pi^2} \int_0^{k_{F,e}} dk \frac{k^4}{m_e} = \frac{g_e}{30\pi^2} \frac{k_{F,e}^5}{m_e}. \tag{3.75}$$

The number density of electrons is given by

$$n_e = \frac{g_e}{(2\pi)^3} \int dk^3 = \frac{g_e}{2\pi^2} \int_0^{k_{F,e}} dk\, k^2 = \frac{g_e}{6\pi^2} k_{F,e}^3. \tag{3.76}$$

The nonrelativistic limit of the pressure in terms of the number density follows a power law

$$P_e^{(\mathrm{nr})} = \frac{1}{5} \left(\frac{6\pi^2}{g_e} \right)^{2/3} \frac{n_e^{5/3}}{m_e}, \tag{3.77}$$

with the exponent 5/3. As the pressure is inversely proportional to the fermion mass, the contribution from nucleons is suppressed by a factor m_e/m_N, where m_N is the nucleon mass, and from nuclei by $m_e/(A \cdot m_N)$ and can be ignored. Hence, the total pressure is to a good approximation just given by the contribution from electrons alone.

Charge neutrality demands that

$$n_e = n_p = Z \cdot n_A = \frac{Z}{A} n_N, \tag{3.78}$$

where Z and A are the charge and the mass number of the nucleus, respectively. The number density of the nucleus is denoted as n_A, the one for nucleons as n_N.

The general expression for the energy density is an integral over the energy

$$\varepsilon = \frac{g}{(2\pi)^3} \int d^3k\, E(k). \tag{3.79}$$

In the nonrelativistic limit, the energy is approximated by $E(k) \approx m$ and the energy density is given by

$$\varepsilon^{(\mathrm{nr})} = m \cdot \frac{g}{(2\pi)^3} \int d^3k = m \cdot n = \rho, \tag{3.80}$$

which is just the mass density ρ. Hence, the total energy density is dominated by the contribution from the nucleons

$$\varepsilon^{(\mathrm{nr})} = \rho_A = m_A \cdot n_A \approx \rho_N = m_N \cdot n_N. \tag{3.81}$$

Here, m_A is the mass of the nucleus, which includes effects from the binding energy. However, the typical binding energy of nuclei of $E_b/A \approx 8\,\mathrm{MeV}$ is much smaller than the nucleon mass, so that $m_A \approx A \cdot m_N$ is a good approximation. The total pressure can be expressed in terms of the mass density as

$$P^{(\mathrm{nr})} = \frac{1}{5} \left(\frac{6\pi^2}{g_e} \right)^{2/3} \frac{1}{m_e} \left(\frac{Z \cdot \rho_N}{A \cdot m_N} \right)^{5/3} \tag{3.82}$$

and finally the EOS in the nonrelativistic limit

$$P^{(\mathrm{nr})} \propto \left(\frac{Z}{A}\right)^{5/3} \cdot \rho_N^{5/3} \tag{3.83}$$

is a power-law with the exponent 5/3.

Let us turn now to the degeneracy pressure in the relativistic limit for electrons, which is $k_F \gg m_e$, but still $k_F \ll m_N$, so nucleons are treated as being nonrelativistic. The pressure integral equation (3.74) can be approximated by setting the velocity of electrons to be $v_e \approx c = 1$ in natural units or the energy of electrons to be $E_e^{\mathrm{rel}}(k) \approx k$:

$$P_e^{(\mathrm{rel})} = \frac{g_e}{6\pi^2} \int_0^{k_{F,e}} dk\, k^3 = \frac{g_e}{24\pi^2} k_{F,e}^4. \tag{3.84}$$

The expression for the number density of electrons, Eq. (3.76), remains unchanged so that

$$P_e^{(\mathrm{rel})} = \frac{1}{4}\left(\frac{6\pi^2}{g_e}\right)^{1/3} n_e^{4/3}. \tag{3.85}$$

Hence, the pressure in the relativistic limit is a power law in terms of the electron number density with the exponent 4/3. As we assume that the nucleons are still nonrelativistic, their contribution to the pressure can be ignored and their contribution to the energy density is still the dominant one:

$$\varepsilon^{(\mathrm{rel})} = \rho_A \approx \rho_N = m_N \cdot n_N. \tag{3.86}$$

Hence, the EOS in the relativistic limit

$$P^{(\mathrm{rel})} = \frac{1}{4}\left(\frac{6\pi^2}{g_e}\right)^{1/3} \left(\frac{Z \cdot \rho_N}{A \cdot m_N}\right)^{4/3} \tag{3.87}$$

is a power law

$$P^{(\mathrm{rel})} \propto \left(\frac{Z}{A}\right)^{4/3} \cdot \rho_N^{4/3}, \tag{3.88}$$

with the exponent 4/3.

Finally, we consider the EOS of a pure neutron gas, first for the nonrelativistic and then for the relativistic limit. The degeneracy pressure in the nonrelativistic limit ($k_F \ll m_n$) is given by

$$P_n^{(\mathrm{nr})} = \frac{g_n}{6\pi^2} \int_0^{k_{F,n}} dk \frac{k^4}{m_n} = \frac{g_n}{30\pi^2} \frac{k_{F,n}^5}{m_n}, \tag{3.89}$$

which is just the expression for electrons in the nonrelativistic limit, Eq. (3.75), by replacing the electron mass with the neutron mass m_n. Similarly, the expression of the number density of neutrons can be taken over from Eq. (3.76)

$$n_n = \frac{g_n}{6\pi^2} k_{F,n}^3. \tag{3.90}$$

The charge neutrality condition is fulfilled automatically. The energy density in the nonrelativistic limit is the mass density of neutrons

$$\varepsilon_n^{(\mathrm{nr})} = \rho_n = m_n \cdot n_n \tag{3.91}$$

so that the EOS

$$P_n^{(\mathrm{nr})} = \frac{1}{5}\left(\frac{6\pi^2}{g_n}\right)^{2/3} \frac{1}{m_n}\left(\frac{\rho_n}{m_n}\right)^{5/3} \tag{3.92}$$

is a power law

$$P_n^{(\mathrm{nr})} \propto \rho_n^{5/3}, \tag{3.93}$$

with an exponent $5/3$, such as in the case of a gas of electrons and nuclei considered previously.

Now, in the ultra-relativistic limit for neutrons $k_{F,n} \gg m_n$, the pressure reads

$$P_n^{(\mathrm{ur})} = \frac{g_n}{6\pi^2} \int_0^{k_{F,n}} dk\, k^3 = \frac{g_n}{24\pi^2} k_{F,n}^4. \tag{3.94}$$

The expression of the number density of neutrons, Eq. (3.90), remains unchanged and the pressure can be recast into

$$P_n^{(\mathrm{ur})} = \frac{1}{4}\left(\frac{6\pi^2}{g_n}\right)^{1/3} n_n^{4/3}. \tag{3.95}$$

However, the energy density changes in the ultra-relativistic limit. The general expression for the energy density is an integral over the neutron energy

$$\varepsilon_n = \frac{g_n}{(2\pi)^3} \int d^3k\, E_n(k), \tag{3.96}$$

which in the nonrelativistic limit is given by the mass density, as $E_n^{\mathrm{nr}}(k) \approx m_n$. However, in the ultra-relativistic limit, the energy density is given by

$$\varepsilon^{(\mathrm{ur})} = \frac{g_n}{(2\pi)^3} \int d^3k\, k, \tag{3.97}$$

as $E_n^{\mathrm{ur}}(k) \approx k$. Integration results in

$$\varepsilon^{(\mathrm{ur})} = \frac{g_n}{2\pi^2} \int_0^{k_{F,n}} dk\, k^3 = \frac{g_n}{8\pi^2} k_{F,n}^4. \tag{3.98}$$

Both, the pressure and the energy density are now a power law in the number density to the power of $4/3$. Comparing Eqs. (3.94) and (3.98), one finds that the EOS approaches in the ultra-relativistic limit

$$P^{(\mathrm{ur})} = \frac{1}{3}\varepsilon^{(\mathrm{ur})}, \tag{3.99}$$

which is just the EOS for a free gas of massless particles, as it should be in hindsight.

The EOSs found so far can be summarized as follows: there exists a general parameterization of the EOSs for free fermion gases in the form

$$P \propto n^{\Gamma} = \left(\frac{\rho}{m}\right)^{\Gamma}, \tag{3.100}$$

which is called a polytropic EOS or simply a polytrope. For the Tolman–Oppenheimer–Volkoff equation, one needs the relation between the pressure and the energy density, which one is tempted to write accordingly as

$$P \propto \varepsilon^{\gamma}. \tag{3.101}$$

We stress that the power coefficients Γ need not to be the same as γ, as we will show in the following.

In the nonrelativistic limit, the power coefficients are given by

$$P \propto n^{5/3} \propto \varepsilon^{5/3} \quad \Longrightarrow \quad \Gamma = \gamma = \frac{5}{3}, \tag{3.102}$$

in the relativistic limit (for a composite fermion gas) by

$$P \propto n^{4/3} \propto \varepsilon^{4/3} \quad \Longrightarrow \quad \Gamma = \gamma = \frac{4}{3}, \tag{3.103}$$

and in the ultra-relativistic limit (for a single fermion gas) by

$$P \propto n^{4/3} \text{ and } P = \frac{1}{3}\varepsilon \quad \Longrightarrow \quad \Gamma = 4/3 \text{ and } \gamma = 1. \tag{3.104}$$

Note that the results are generic for free Fermi gases, as outlined previously. Also note that ρ stands for the mass density of the nonrelativistic species, while ε denotes the more general energy density, which includes the kinetic energy.

The EOS of a free Fermi gas can be given analytically in a parametric form. The solution for the full integral for the pressure reads

$$P = \frac{g}{3(2\pi)^3} \int d^3k \, \frac{k^2}{E(k)} \tag{3.105}$$

$$= \frac{g}{6\pi^2} \int_0^{k_F} dk \, \frac{k^4}{\sqrt{k^2 + m^2}} \tag{3.106}$$

$$= \frac{g}{24\pi^2} \left\{ k_F E_f \left(k_F^2 - \frac{3}{2} m^2 \right) + \frac{3}{2} m^4 \cdot \ln \left[\frac{k_F + E_F}{m} \right] \right\}, \tag{3.107}$$

where $E_F = \sqrt{k_F^2 + m^2}$ is the Fermi energy and g the degeneracy factor. The solution for the full integral for the energy density reads

$$\varepsilon = \frac{g}{(2\pi)^3} \int d^3k \, E(k) \tag{3.108}$$

$$= \frac{g}{2\pi^2} \int_0^{k_F} dk \, k^2 \sqrt{k^2 + m^2} \tag{3.109}$$

$$= \frac{g}{8\pi^2} \left\{ k_F E_f \left(k_F^2 + \frac{1}{2} m^2 \right) - \frac{1}{2} m^4 \cdot \ln \left[\frac{k_F + E_F}{m} \right] \right\}. \tag{3.110}$$

For a given Fermi momentum k_F or a given number density n using the relation

$$n = \frac{g}{6\pi^2} k_F^3, \tag{3.111}$$

the pressure and the energy density can be computed parametrically. Note that the ultra-relativistic limit can be easily read off from Eqs. (3.107) and (3.110). Getting the nonrelativistic limit is less obvious and the reader is asked to derive it in the exercises.

3.6 Polytropes

We have seen that one parametrizes the low- and high-density behavior of a free gas of fermions by a polytropic EOS. In this section, we are going to explore in more detail the properties of polytropic EOSs, or polytropes for short. Polytropes are defined as a power law for the pressure P in terms of the number density n:

$$P = K \cdot n^\Gamma, \tag{3.112}$$

where K and Γ are constants. Note that the dimension of the constant K depends on Γ, whereas Γ is dimensionless. The definition of the constant polytropic power

Γ can be extended to one that depends on the number density by the logarithmic derivative

$$\Gamma = \frac{\partial \log P}{\partial \log n}, \tag{3.113}$$

which coincides with Eq. (3.112) for a density independent Γ. The generalized definition of Γ is useful to discuss, for example, the turnover from the nonrelativistic to the relativistic limit of Fermi gases, where Γ switches from 4/3 to 5/3 continously. We will stick to a constant Γ in the following discussion.

We know from the discussion of thermodynamics potentials that either the energy density is a thermodynamic potential and given in terms of the number density $\varepsilon = \varepsilon(n)$ or that the pressure is a thermodynamic potential in the bulk limit given in terms of the chemical potential $P = P(\mu)$. The form of a polytrope as defined in Eq. (3.112) is not given as a thermodynamic potential. Therefore, we cannot compute the missing thermodynamic quantities directly, which are ε and μ, from a polytrope. We resort to a trick by inverting the thermodynamic relation between the pressure and the energy density, see Eq. (3.69):

$$P = n^2 \cdot \frac{\mathrm{d}(\varepsilon/n)}{\mathrm{d}n}. \tag{3.114}$$

The energy density can then be rewritten in terms of the pressure as:

$$\frac{\mathrm{d}\,(\varepsilon/n)}{\mathrm{d}n} = \frac{P}{n^2} = K \cdot n^{\Gamma-2} \tag{3.115}$$

and integrated to give

$$\frac{\varepsilon}{n} = \frac{K}{\Gamma - 1} \cdot n^{\Gamma-1} + \text{const.} \tag{3.116}$$

There appears an integration constant, which needs to be fixed. For the low density limit the energy per particle E/N is just the particle mass

$$\frac{E}{N} = \frac{\varepsilon}{n} \to m \quad \text{for} \quad n \to 0, \tag{3.117}$$

so that the integration constant can be associated with the particle mass m for $\Gamma > 1$. The energy density for a polytrope is then given by

$$\varepsilon = m \cdot n + \frac{K}{\Gamma - 1} \cdot n^{\Gamma}. \tag{3.118}$$

Now we are in the position to calculate the chemical potential for polytropes:

$$\mu = \frac{d\varepsilon}{dn} = m + \frac{K \cdot \Gamma}{\Gamma - 1} \cdot n^{\Gamma-1}. \tag{3.119}$$

The reader is invited to check that the relations are thermodynamically consistent in the exercises.

Polytropes can feature an acausal behavior at high densities, which we will discuss now. The acausal behavior concerns the speed of sound in the medium, which can exceed the speed of light, leading to a violation of the principle of special relativity. The speed of sound (squared) in a medium is given by the partial derivative of the pressure with respect to the energy density at constant entropy

$$c_s^2 = \left.\frac{\partial P}{\partial \varepsilon}\right|_s . \tag{3.120}$$

For polytropes considered here, that is, at vanishing temperature, one can calculate that the speed of sound is

$$c_s^2 = \frac{dP}{dn} \cdot \frac{dn}{d\varepsilon} \to \Gamma - 1 \qquad \text{for } n \to \infty, \tag{3.121}$$

using the definition of a polytrope and Eq. (3.118). Hence, polytropes are getting acausal at high densities for $\Gamma > 2$. The limiting case $\Gamma = 2$ is interesting as it stands for the stiffest possible EOS, the one with the highest possible pressure for a given energy density, being compatible with causality. For $\Gamma = 2$, one finds that $P = \varepsilon$, so that $c_s^2 = 1$. Note that the speed of sound for a Fermi gas approaches $c_s = 1/\sqrt{3}$ in the high density limit, as one can read off from the ultra-relativistic limit, Eq. (3.99).

We close with some more words of caution. If polytropes are defined for some intermediate density regime, for example by approximating an EOS, then the integration constant can not be fixed with the particle mass but has to be adjusted to fit to the low density EOS used. Also, it is often assumed implicitly for polytropes that the number density can be expressed in terms of the mass density $n = \rho/m$ and the energy density is approximated by the mass density. We notice from Eq. (3.118) that this implies that the term originating from the kinetic energy of the particle is ignored so that one loses thermodynamic consistency. As such, one needs to check that this assumption holds to a sufficient degree for the density range considered. We have seen in Section 3.5 that this assumption does not hold for the ultra-relativistic limit for free Fermi gases.

Despite these caveats, polytropes are a useful tool for studying compact stars and we will encounter polytropes again and again when discussing the properties of white dwarfs and neutron stars in the chapters to follow.

Exercises

(3.1) The thermodynamic potential in the canonical ensemble is the free energy F, which is given by the Legendre transformation

$$F = F(V, T, N_i) \equiv U - T \cdot S. \tag{3.122}$$

The free energy is a function of the volume, the temperature, and the number of particles N_i and its total differential reads

$$\mathrm{d}F = -P \cdot \mathrm{d}V - S \cdot \mathrm{d}T + \sum_i \mu_i \cdot \mathrm{d}N_i. \tag{3.123}$$

Determine the free energy in the thermodynamic limit, that is, in the infinite volume limit $V \to \infty$. Derive the thermodynamic relations for the undetermined thermodynamic quantities (pressure, entropy density, and chemical potential) in the thermodynamic limit.

(3.2) In chemical equilibrium, the sum of the chemical potentials μ_i of particle i weighted by the change of the particle numbers dN_i vanishes

$$\sum_i \mu_i \cdot dN_i = 0.$$

Consider reactions in β-equilibrium, where reactions such as

$$p + e^- \to n + \nu_e$$

are in equilibrium. Show that the chemical potential of the electron and the positron are related to each other by

$$\mu_{e^-} = -\mu_{e^+}.$$

Can you show that this is a general rule for every particle and antiparticle? What is the chemical potential of the photon? Why? What is the chemical potential of a nucleus with mass number A and atomic number Z in β-equilibrium?

(3.3) Check the solutions for the full integrals of pressure and the energy density of a free Fermi gas, see Eqs. (3.107) and (3.110).

(3.4) Show that for a free Fermi gas, the chemical potential is equal to the Fermi energy $\mu = E_F$ by using the thermodynamic definition of μ.

(3.5) Determine the Fermi momentum for which the pressure of a Fermi gas switches from the nonrelativistic to the relativistic limit. Show that the expressions for the pressures of the two limits are equal for a Fermi momentum of $k_F = \frac{5}{4}m$.

(3.6) Check the thermodynamic consistency, that is, the Euler equation, of the EOS for a free Fermi gas. Rewrite the energy density ε in terms of the number density n and derive the pressure and the chemical potential by using thermodynamic relations. Do the same for the grand canonical ensemble by expressing the pressure as a function of the chemical potential $\mu = E_F$. Use first the nonrelativistic and relativistic limits and then the full expressions given in Eqs. (3.107) and (3.110).

(3.7) From an EOS of the form $\varepsilon = \varepsilon(n)$, derive the thermodynamic relation

$$\frac{\mathrm{d}\varepsilon}{\mathrm{d}n} = \frac{\varepsilon + P}{n}.$$

(3.8) Consider a massless Fermi gas. Convince yourself that the EOS is given by $P = \varepsilon/3$. For the EOS in the form $\varepsilon = \varepsilon(n)$, show that it corresponds to

$$\frac{\mathrm{d}\ln\varepsilon}{\mathrm{d}\ln n} = \frac{4}{3},$$

using the result from Exercise 3.7. Vice versa, show explicitly that this case corresponds to the ultra-relativistic limit for an EOS $P = \varepsilon/3$ by computing the pressure in terms of the number density n.

(3.9) The speed of sound is defined in Eq. (3.120). Show that the speed of sound can be expressed in general as

$$c_s^2 = \frac{P}{P + \varepsilon}\Gamma,$$

where Γ stands for the adiabatic index defined as

$$\Gamma = \frac{\mathrm{d}\ln P}{\mathrm{d}\ln n},$$

that is, it is related to the power Γ encountered for a polytropic EOS.

(3.10) Check the thermodynamic consistency, that is, the Euler equation, of the expressions derived for polytropes for the energy density, Eq. (3.118) and the chemical potential, Eq. (3.119). What happens for the case when $\Gamma = 1$?

(3.11) Show that the speed of sound for a polytrope in the high-density limit approaches $c_s^2 = \Gamma - 1$. What are the underlying assumptions on the value of Γ? Check the high-density limit of the speed of sound for the cases $\Gamma = 1$ and $\Gamma < 1$.

4
Compact Stars

4.1 Spheres in Hydrostatic Equilibrium

Let us consider a homogeneous sphere containing the total mass M within the total radius R. The mass density is denoted by $\rho(r)$. Integrating the mass density over a sphere with a radius r gives the mass $m_r(r)$ contained within the sphere of radius r:

$$m_r(r) = \int \mathrm{d}r'^3 \rho(r') = 4\pi \int_0^r \mathrm{d}r' \, r'^2 \rho'(r), \tag{4.1}$$

which is expressing the conservation of mass. Note, that $m_r(r)$ is the mass of the entire sphere of radius r and should not be confused with the mass at the radius r. The conservation of mass reads in differential form:

$$\frac{\mathrm{d}m_r(r)}{\mathrm{d}r} = 4\pi r^2 \rho(r). \tag{4.2}$$

The boundary conditions can be fixed by demanding that the mass at the center vanishes $m_r(r = 0) = 0$, while the mass at the surface is just given by the total mass $m_r(r = R) = M$ by definition.

In the following we consider the hydrostatic equilibrium within a homogeneous sphere. Hydrostatic equilibrium means that the pressure of matter onto an area A balances the gravitational force. The net force on matter vanishes locally. Let us assume that the sphere is in hydrostatic equilibrium up to radius r. Then imagine that one adds a small amount (or a shell) of matter of thickness $\mathrm{d}r$ within an arbitrary area A on top of the sphere. This adds an additional mass of

$$\mathrm{d}m_r(r) = A \cdot \rho(r)\mathrm{d}r \tag{4.3}$$

according to mass conservation. The additional pull of the gravitational force is given by

$$\mathrm{d}F_G(r) = -G\frac{m_r(r) \cdot \mathrm{d}m_r(r)}{r^2} = -G\frac{m_r(r) \cdot A \cdot \rho(r)\mathrm{d}r}{r^2}. \tag{4.4}$$

On the other hand, the pressure of the added matter over the area A also produces a force

$$dF_P(r) = A \cdot dP(r). \tag{4.5}$$

The additional pressure force has to balance the additional gravitational force to keep the hydrostatic equilibrium

$$dF_P(r) = A \cdot dP(r) = -G\frac{m_r(r) \cdot A \cdot \rho(r)dr}{r^2} = dF_G(r). \tag{4.6}$$

The area A cancels out so that one can consider hydrostatic equilibrium for an arbitrary area, as for example, a cylinder or the entire shell. It follows the equation of hydrostatic equilibrium

$$\frac{dP(r)}{dr} = -G\frac{m_r(r) \cdot \rho(r)}{r^2} \tag{4.7}$$

for the Newtonian case. The boundary conditions are that the pressure at the center is given by $P(0) = P_c$ and that the pressure vanishes at the surface $P(R) = 0$. Note, that the pressure has to vanish at the surface to ensure hydrostatic equilibrium.

4.1.1 Relativistic Hydrostatic Equilibrium

The Newtonian equation of hydrostatic equilibrium can be extended to general relativity. We have all necessary equations in Chapter 2 on general relativity. However, for the reader's convenience, we will summarize the condition of hydrostatic equilibrium in general relativity using partly a heuristic derivation.

In general relativity, the equation of hydrostatic equilibrium follows from the generalized relativistic form of the conservation of the energy–momentum tensor. The form of the energy–momentum tensor is assumed to be of the form of an ideal fluid, which means that there are no effects from dissipation, heat transport, or shear or bulk viscosity. The only quantities that enter the expression for the energy–momentum tensor are the pressure P and the energy density ε of the matter. From general relativity, only the metric tensor $g^{\mu\nu}$ and the four velocity u^μ can be additional ingredients for the expression of the energy–momentum tensor. Locally, the energy–momentum tensor has to be of the form

$$T^{\mu\nu} = \begin{pmatrix} \varepsilon & & & \\ & P & & \\ & & P & \\ & & & P \end{pmatrix}, \tag{4.8}$$

which is what we used to have in the Newtonian limit. Consider first the extension of this expression for its Lorentz-invariant form in flat space. For the relativistic form, the only matrices at hand are $\eta^{\mu\nu}$ and $u^\mu u^\nu$. It is then straightforward to see that the form of the energy–momentum tensor for an ideal fluid shall be assumed to be

$$T^{\mu\nu} = (\varepsilon + P) \cdot u^\mu u^\nu + P \cdot \eta^{\mu\nu} \tag{4.9}$$

in its general Lorentz-invariant form. In the local restframe, the four velocity is given by $u^\mu = (1,0,0,0)$ and one recovers Eq. (4.8). The relativistic form of the conservation of energy and momentum of the fluid is ensured by a vanishing covariant derivative

$$\partial_\mu T^{\mu\nu} = 0 \tag{4.10}$$

in flat space.

The extension to general relativity is now easy. One simply needs to replace the flat space metric $\eta^{\mu\nu}$ by the general metric $g^{\mu\nu}$ for curved spacetime

$$T^{\mu\nu} = (\varepsilon + P) \cdot u^\mu u^\nu + P \cdot g^{\mu\nu}. \tag{4.11}$$

The conservation of the energy–momentum tensor is now given by the corresponding covariant derivative form

$$\nabla_\mu T^{\mu\nu} = 0 \tag{4.12}$$

for general curved spacetime.

The effects from relativity and from the extension to curved spacetime give correction factors to hydrostatic equilibrium for each entity m_r, ρ, and radius r. The corrections have different origins. First, in general relativity, gravity couples to the total energy density not only the rest mass density. So, the mass density has to be replaced by the energy density. Second, gravity couples to the energy–momentum tensor, not only to the energy density. So, the pressure of matter will also enter as an additional correction. The pressure will appear as a correction to the mass density ρ and to the mass m_r contained within the sphere of radius r. Third, in curved spacetime there will be a correction from the metric tensor. For the static case in spherical symmetry, the corresponding metric is the Schwarzschild metric and the corresponding correction factor will be given by the Schwarzschild factor of the metric. One expects that the gravitational pull on matter will be stronger when including corrections from general relativity and the Schwarzschild factor shall enter, correspondingly modifying the radius r in the denominator. The final expression from general relativity for hydrostatic equilibrium for a static spherical spacetime, that is, for the Schwarzschild metric, is given by the Tolman–Oppenheimer–Volkoff (TOV) equation

$$\frac{dP}{dr} = -G\frac{m_r(r)\varepsilon(r)}{r^2}\left(1+\frac{P(r)}{\varepsilon(r)}\right)\left(1+\frac{4\pi r^3 P(r)}{m_r(r)}\right)\left(1-\frac{2Gm_r(r)}{r}\right)^{-1}, \tag{4.13}$$

see Eq. (2.149). In addition to replacing the mass density ρ with the energy density ε, there are three correction factors, for m_r, ε, and the radius r, in line with our arguments given earlier. The Newtonian limit can be easily recovered by setting $\varepsilon \to \rho$, $P = 0$, and ignoring the Schwarzschild factor in the denominator. Note that the three correction factors from general relativity all enhance the gravitational effects so that a larger pressure gradient is needed to ensure hydrostatic equilibrium compared to the Newtonian case.

The equation of the conservation of the mass in general relativity turns out to be

$$\frac{dm_r}{dr} = 4\pi r^2 \varepsilon(r), \tag{4.14}$$

see Eq. (2.137). One notes that replacing the mass density by the energy density seems to be sufficient to go from the Newtonian limit to the full expression in general relativity. Surprisingly, the Schwarzschild factor does not enter the expression. From the integral version of mass conservation, one expects that the Schwarzschild factor will enter via the integral measure of the curved spacetime of the Schwarzschild metric. In fact, it does not, so that m_r and thereby also the total mass of the sphere M is *smaller* than the covariant integral of the energy density over the whole sphere. This feature originates from the additional binding energy of matter in the presence of a gravitational field so that the mass seen from an outside observer is smaller compared to flat space.

4.1.2 Compact Stars of Noninteracting Neutrons

The TOV equations were solved numerically for the first time by Oppenheimer and Volkoff in their classic paper (Oppenheimer and Volkoff, 1939) on matter consisting of a free gas of neutrons, that is, noninteracting neutrons. The work of Oppenheimer and Volkoff was motivated by the work of Landau of 1932, who discussed for the first time the possible existence of dense stars looking like one giant nucleus (Landau, 1932). The interesting history surrounding the paper of Landau has been nicely described in Yakovlev et al. (2013).

For solving the TOV equation numerically, one needs to define the initial conditions of the two first-order differential equations, Eqs. (4.13) and (4.14). Usually one adopts initial values for the pressure and the mass at $r = 0$:

$$P(r=0) = P_c \qquad m_r(r=0) = 0. \tag{4.15}$$

For a given equation of state (EOS) of the form $P = P(\varepsilon)$, the two differential equations are then integrated from $r = 0$ until the pressure vanishes at the radius R of the sphere. This radius defines the radius of the compact star. The integrated mass at that radius R is denoted as the gravitational mass of the star $M = m_r(r = R)$. This procedure is the repeated for several values of the central pressure P_c, thereby generating a family of solutions of the TOV equations for a given fixed EOS.

There are a few general comments in order. First, the matter contained within radius r has to fulfill the condition that $r > 2Gm_r(r)$, as the Schwarzschild factor diverges for $r = 2Gm_r(r)$. This condition implies that the interior solution does not allow for the presence of a Schwarzschild black hole at any radius r. Second, it is clear from Eq. (4.13) that the pressure gradient has to be always negative for realistic forms of matter with $\varepsilon > 0$ and $P > 0$. A negative pressure gradient implies that the pressure is a continuously decreasing function of radius r, that is, the pressure cannot jump as a function of r. Physically, this is a consequence of hydrostatic equilibrium, of course. It also means, that if there exists a range in the EOS where the pressure does not change with energy density, the corresponding size of the shell of matter within that range of energy density will shrink to zero, as demanded by hydrostatic equilibrium. Hence, the energy density can jump as a function of radius r while the corresponding pressures before and after the jump in energy density are the same. Third, a compact star configuration with a well-defined total radius R implies that the pressure falls sufficiently fast as a function of energy density so that eventually the condition of a vanishing pressure at the surface of the compact star can be met. This condition hinges on the form of the EOS and it can happen that the pressure never vanishes for any radius r. Actually, there are known examples, also realistic ones, where this happens, in particular for matter at nonzero temperature. If the integration of the mass reaches a well-defined asymptotic limit, it implies that there is a halo of matter surrounding the compact star configuration. A radius R of that compact star configuration can then be defined at a nonvanishing minimum pressure delineating the regime of the dense core and the dilute halo. In other instances this might not be fulfilled. A well-known example is the spherical distribution of dark matter around galaxies where the mass increases linearly with radius r. Such a kind of solution will be discussed in connection with the Lane–Emden equation when discussing the properties of white dwarfs.

The EOS for a free gas of neutron matter generates compact star solutions with a well-defined radius R and a gravitational mass M. The numerical solution of the TOV equations for the EOS of a free gas of neutrons is depicted in Figure 4.1, which shows the gravitational mass M and the radius R of the neutron star as a function of the central energy density ε_c in units of the energy density of nuclear

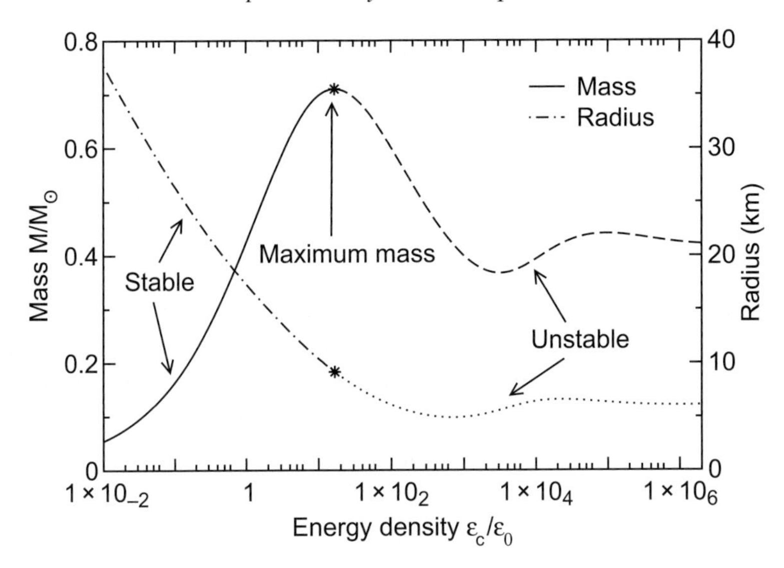

Figure 4.1 The mass and radius as a function of the central pressure for a compact star consisting of free neutron matter.

matter $\varepsilon_0 = 140\,\mathrm{MeV\,fm^{-3}}$ (see e.g., Section 7.3.1). The radius R decreases with increasing central energy density, reaches a minimum, and starts to oscillate around an asymptotic limit. The gravitational mass M increases with increasing central energy density, reaches a maximum, then a minimum, and starts to oscillate around an asymptotic limit. The first maximum in the gravitational mass is at a lower central energy density than the first minimum in the radius.

Figure 4.2 shows the gravitational mass M versus the radius R of a neutron star consisting of noninteracting neutrons. The sequence of solutions starts at small masses and large radii. For small central energy densities, the mass scales with the radius as $M \cdot R^3 =$ constant. With increasing central pressure, the mass increases while the radius decreases until a maximum in the mass is reached. Afterwards, the mass–radius relation spirals around a limiting value for asymptotically large central energy densities. This behavior for the radius and the mass is quite generic for compact stars. Most notably, there exists an upper bound on the mass of a compact star. For noninteracting neutron matter, the maximum mass is $M_{\mathrm{max}} = 0.71 M_\odot$, with a radius of $R = 9.1$ km.

It turns out that in the case studied here the solutions up to the maximum mass configuration are stable and that all other solutions beyond the maximum mass configuration are unstable. There are exceptions, though, under special conditions to be discussed in Chapter 9 on hybrid stars.

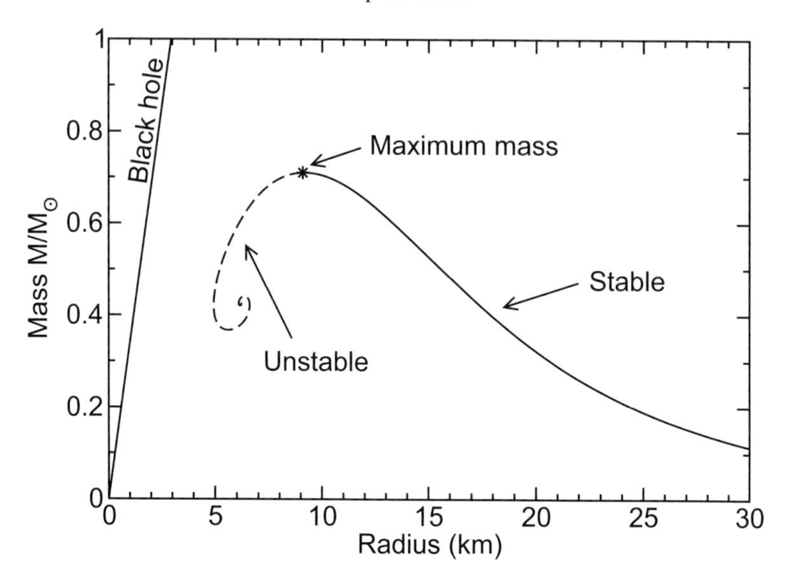

Figure 4.2 The mass–radius relation for a compact star consisting of free neutron matter. The dashed line denotes unstable configurations. The line labelled 'Black hole' shows the Schwarzschild radius.

4.2 Maximum Masses of Compact Stars

The existence of a maximum mass for neutron stars can be understood within Landau's argument for a maximum mass of compact stars (see Landau, 1932; Shapiro and Teukolsky, 1983). The argument can be extended to white dwarfs and to compact stars in general.

4.2.1 Landau's Argument for a Maximum Mass

Let us consider a static homogenous sphere of matter with a total mass M and a radius R consisting of free fermions of the mass m. Classically, the total energy $E(R)$ of a particle for radius R is the sum of the gravitational energy and the kinetic energy of the particle

$$E(R) = E_{\mathrm{G}} + E_{\mathrm{kin}} = -G\frac{Mm}{R} + E_{\mathrm{kin}}. \tag{4.16}$$

In general, the energy of the particle is given by the relativistic energy–momentum relation:

$$E(k) = \sqrt{k^2 + m^2}. \tag{4.17}$$

For a fluid of free fermions, the characteristic momentum is given by the Fermi momentum k_F, so we set $k = k_F$ in the following. The Fermi momentum is related to the number density as

$$n = \frac{g}{6\pi^2} k_f^3, \tag{4.18}$$

where g is the degeneracy factor. For neutrons, we have $g = (2s + 1) = 2$ from the spin degree of freedom. For simplicity, we set the number density equal to the average density $\bar{n}$ of the sphere, which is defined by

$$\bar{n} = \frac{N}{V} = \frac{3N}{4\pi R^3}, \tag{4.19}$$

where N stands for the number of fermions in the sphere. This gives the following relation for the Fermi momentum:

$$k_F = \left(\frac{6\pi^2}{g}\right)^{1/3} n^{1/3} = \left(\frac{9\pi}{2g}\right)^{1/3} \frac{N^{1/3}}{R}. \tag{4.20}$$

Assuming that the total mass of the sphere is just given by the sum of the masses of the fermions, one has

$$M = N \cdot m. \tag{4.21}$$

Let us consider now the ultra-relativistic limit for the kinetic energy of the fermion. Then, $E_{\text{kin}} = k_F$ and the total energy of the particle reads

$$E(R) = -G\frac{Mm}{R} + k_F. \tag{4.22}$$

Inserting the approximate expressions for the total mass M and the Fermi momentum one arrives at

$$E(R) = -G\frac{Nm^2}{R} + \left(\frac{9\pi}{2g}\right)^{1/3} \frac{N^{1/3}}{R}. \tag{4.23}$$

One notices that both terms scale as $1/R$. The total energy depends now only on the number of fermions in the sphere N for a fixed fermion mass m. For large values of N, the first term, the gravitational energy, dominates, for small values of N, the second term, the kinetic energy, dominates. In principle, there are three cases to be considered depending on the number of fermions N in the star:

$E(R) < 0$ **:** the system is unstable as the energy gets smaller and smaller for $R \to 0$, so that the system collapses eventually to a black hole

$E(R) > 0$ **:** the energy can be minimized by increasing the radius until the number density and correspondingly the Fermi momentum reaches the nonrelativistic region for a Fermi fluid ($k_F < m$). Then, $E_{kin} \propto k_F^2/m_f \propto R^{-2}$ and a stable minimum exists for a finite value of R

$E(R) = 0$: the system is marginally stable.

As the gravitational potential dominates for large values of N, giving a negative sign for the total energy, stable solutions exist only up to a maximum number of fermions $N_{\rm max}$ given by the marginally stable case. Hence, the limit of stability $E(R) = 0$ implies that

$$G N_{\rm max} m^2 = \left(\frac{9\pi}{2g}\right)^{1/3} N_{\rm max}^{1/3} \tag{4.24}$$

or

$$N_{\rm max} = \left(\frac{9\pi}{2g}\right)^{1/2} \left(\frac{m_{\rm P}}{m}\right)^3, \tag{4.25}$$

where we have set $G = 1/m_{\rm P}^2$ in natural units with $m_{\rm P}$ being the Planck mass. The maximum mass is then given by

$$M_{\rm max} = N_{\rm max} \cdot m = \left(\frac{9\pi}{2g}\right)^{1/2} \frac{m_{\rm P}^3}{m^2}. \tag{4.26}$$

The corresponding radius can be estimated by setting $k_F = m$, as this defines the onset of the ultra-relativistic limit for the fermions. Using Eq. (4.20), we get

$$R_{\rm crit} = \left(\frac{9\pi}{2g}\right)^{1/3} \frac{N_{\rm max}^{1/3}}{m} = \left(\frac{9\pi}{2g}\right)^{1/2} \frac{m_{\rm P}}{m^2}, \tag{4.27}$$

where we inserted $N_{\rm max}$ from Eq. (4.25). We define now the Landau mass $M_{\rm L}$ and the corresponding Landau radius $R_{\rm L}$ by

$$M_{\rm L} = \frac{m_{\rm P}^3}{m^2} \quad \text{and} \quad R_{\rm L} = \frac{m_{\rm P}}{m^2}. \tag{4.28}$$

So far we have not fixed the fermion mass and the expressions apply to any fermion star consisting of fermions of a mass m. For a neutron star with a fermion mass of the neutron, $m = m_n$, the actual values of the Landau mass and the Landau radius are

$$M_{\rm L} = \frac{m_{\rm P}^3}{m_n^2} = 1.848 M_\odot \qquad R_{\rm L} = \frac{m_{\rm P}}{m_n^2} = 2.729 \text{ km}. \tag{4.29}$$

We note that the values are not too far from the numerically exact ones for a neutron star consisting of free neutrons, which are $M_{\rm max} = 0.71 M_\odot$ and $R = 9.1$ km. The number of neutrons for the maximum mass configuration is about

$$N_{\rm max} \sim \frac{m_{\rm P}^3}{m_n^3} = \frac{M_{\rm L}}{m_n} \sim 2 \times 10^{57}. \tag{4.30}$$

Note that as the Landau mass is close to the mass of our Sun, the number of fermions in a neutron star is about the same as in the Sun.

4.2.2 Maximum Mass for White Dwarfs

Landau's argument can be readily extended to the case of white dwarfs. White dwarf material consists of two charged components, electrons and nuclei, which balance each other such that the total charge of the white dwarf vanishes. For each nucleus with a charge Z and mass number A, there are Z electrons around due to the charge neutrality condition. The number density of electrons has to be then, with n_A being the number density of nuclei with mass number A:

$$n_e = Z \cdot n_A = n_p, \tag{4.31}$$

which is just the number density of protons in the white dwarf. While electrons have a higher momentum than the nuclei due to their smaller mass, the more massive nuclei are responsible for the total mass of the white dwarf. The mass of nuclei is to good approximation given by $m_A \approx A \cdot m_p$. So the total energy for each nucleus reads

$$E(R) = -G\frac{Mm_A}{R} + Z \cdot k_{F,e}, \tag{4.32}$$

where $k_{F,e}$ is the Fermi momentum of the electrons. The total mass of the white dwarf is given by the sum of the masses of the nuclei or by the sum of the masses of the nucleons

$$M = N_N \cdot m_p = \frac{N_N}{A} \cdot m_A, \tag{4.33}$$

where we introduced the number of nucleons in the star N_N. Note that the number of protons balancing the charge of the electrons is given by $N_p = (Z/A) \cdot N_N = N_e$. The total energy for each nucleus now reads

$$E(R) = -G\frac{N_N A m_p^2}{R} + Z \cdot \left(\frac{9\pi}{2g_e}\right)^{1/3} \frac{N_e^{1/3}}{R}, \tag{4.34}$$

where we replaced the Fermi momentum of the electrons in terms of the number of electrons N_e with g_e being the degeneracy factor of the electrons. In terms of the number of nucleons, one finds

$$E(R) = -G\frac{N_N A m_p^2}{R} + Z \cdot \left(\frac{9\pi}{2g_e}\right)^{1/3} \frac{N_N^{1/3}}{R} \left(\frac{Z}{A}\right)^{1/3}. \tag{4.35}$$

The arguments for the stability of the white dwarf can be repeated analogously to the case of neutron stars. The critical condition for stability $E(R) = 0$ determines the maximum mass of white dwarfs and results in the condition for the maximum number of nucleons $N_{N,\max}$:

$$G N_{N,\max} A m_p^2 = Z \cdot \left(\frac{9\pi}{2g}\right)^{1/3} \left(\frac{Z}{A}\right)^{1/3} N_{N,\max}^{1/3}. \tag{4.36}$$

The maximum number of nucleons is then given by

$$N_{N,\max} = \left(\frac{9\pi}{2g_e}\right)^{1/2} \left(\frac{m_{\rm P}}{m_p}\right)^3 \left(\frac{Z}{A}\right)^2 . \tag{4.37}$$

There appears an additional factor $(Z/A)^2$ compared to the case of neutron stars, otherwise the expression is the same as the degeneracy factors are the same $g_e = g_n$. The corresponding maximum mass for white dwarfs is

$$M_{\max} = \left(\frac{9\pi}{2g_e}\right)^{1/2} \frac{m_{\rm P}^3}{m_p^2} \left(\frac{Z}{A}\right)^2 = 2.66 \cdot \frac{m_{\rm P}^3}{m_p^2} \left(\frac{Z}{A}\right)^2 . \tag{4.38}$$

Compare this expression to the numerically exact one, the Chandrasekhar mass, in natural units:

$$M_{\rm Chandra} = \left(\frac{2Z}{A}\right)^2 1.44 M_\odot = 3.12 \cdot \frac{m_{\rm P}^3}{m_p^2} \left(\frac{Z}{A}\right)^2 . \tag{4.39}$$

Except for the numerical prefactor, which is of the order of 1, the expressions are the same. Most interestingly, one recovers with Landau's argument the dependence of the maximum mass of white dwarfs on the charge-to-mass ratio (Z/A) of the nuclei and on the nucleon mass. Note that the electron mass does not enter the expression for the maximum mass of white dwarfs. Most strikingly, the expressions for the maximum mass of neutron stars and white dwarfs are similar and just differ by the charge-to-mass ratio squared. In summary

$$M_{\rm wd} \sim M_{\rm ns} \sim \frac{m_{\rm P}^3}{m_N^2} \sim 1 M_\odot, \tag{4.40}$$

as the maximum mass is determined in both cases by the nucleon mass. The critical radius for white dwarfs is given by $k_{F,e} = m_e$ so that

$$R_{\rm crit} = \left(\frac{9\pi}{2g_e}\right)^{1/3} \frac{N_{e,\max}^{1/3}}{m_e} \tag{4.41}$$

$$= \left(\frac{9\pi}{2g_e}\right)^{1/3} \frac{N_{N,\max}^{1/3}}{m_e} \left(\frac{Z}{A}\right)^{1/3} . \tag{4.42}$$

We realize that the electron mass m_e is entering one of our expressions for the first time. Using the expression for the maximum number of nucleons one arrives at

$$R_{\text{crit}} = \left(\frac{9\pi}{2g_e}\right)^{1/2} \frac{m_{\text{P}}}{m_p \cdot m_e} \left(\frac{Z}{A}\right), \tag{4.43}$$

so that the radius of a white dwarf is characteristically about

$$R_{\text{wd}} \sim \frac{m_{\text{P}}}{m_p \cdot m_e} \sim 5{,}000 \text{ km}, \tag{4.44}$$

which is about the size of Earth. Note, that the radius corresponding to the maximum mass of a white dwarf is related to the electron mass, or more generically to the mass of the fermion that is providing the pressure. The smaller the mass of this fermion, the larger the compact star. Comparing the expressions for the radius of the white dwarf with the one of a neutron star, see Eq. (4.27), one sees that they scale with the ratio of the proton to the electron mass. Modulo a factor of Z/A, the radius of a white dwarf is larger by the ratio of the proton to electron mass

$$\frac{R_{\text{wd}}}{R_{\text{ns}}} \sim \frac{m_p}{m_e} \sim 2{,}000, \tag{4.45}$$

that is, by about three orders of magnitude. Indeed, this is what one finds in more refined models of neutron stars and white dwarfs and from observations. Characteristically, the radius of a neutron star is about 10 km and that of white dwarfs is about 10,000 km.

4.3 Scaling Solutions for Compact Stars

Landau's argument for a maximum mass and the corresponding radius has been derived for arbitrary fermion masses. One notices that both the Landau mass and the Landau radius scale as the inverse fermion mass squared. In fact, this features originates from the scaling behavior of the TOV equations. Although the arguments in deriving the expressions for the Landau mass and radius seem to be simple minded, the result for the Landau's scaling solution are exactly the same for the full general relativistic treatment. The deeper reason behind this is the nearly magical power of scaling arguments, which are just based on dimensional reasoning. The scaling result can only depend on the dimensionful parameters of the problem in certain combinations that are allowed by the equations studied. In principle, Landau's derivation of the maximum mass and the radius are just that, looking for a reasonable combination of dimensionful parameters of the problem at hand to arrive at a scaling solution. We will delineate the procedure in the following in more detail and extend it to more general classes of EOSs.

We note in passing that scaling solutions are well known and studied in stellar evolution. A particular kind of scaling solution of the equations of hydrostatic equilibrium for stellar configurations will be adopted, for example, for the study of the Lane–Emden equations for white dwarfs, see Chapter 5.

4.3.1 Scaling of the TOV Equations

First let us introduce dimensionsless quantities for the EOS:

$$P = \varepsilon_0 \cdot P' \qquad \varepsilon = \varepsilon_0 \cdot \varepsilon' \tag{4.46}$$

which we denote by a prime. Here ε_0 stands for a constant with the dimension of an energy density (which is in natural units equivalent to the dimension of a pressure). The aim is to look for a solution of the TOV equations in terms of that energy density ε_0 alone. We rescale the other dimensionful quantities of the TOV equations, which are the radial coordinate r and the mass m_r with radius r such that the TOV equations are scale-free:

$$r = a \cdot r' \qquad m_r = b \cdot m'_r, \tag{4.47}$$

with dimensionful coefficients a and b to be determined. Again, the primed quantities denote the scale invariant quantities. The rescaled TOV equation in terms of the dimensionless primed quantities reads

$$\frac{\varepsilon_0 \cdot dP'}{a \cdot dr'} = -G\frac{bm'_r\varepsilon_0\varepsilon'}{a^2 \cdot r'^2}\left(1 + \frac{\varepsilon_0 \cdot P'}{\varepsilon_0 \cdot \varepsilon'}\right)\left(1 + \frac{4\pi a^3 \cdot r'^3\varepsilon_0 P'}{b \cdot m'_r}\right)\left(1 - \frac{2Gb \cdot m'_r}{a \cdot r'}\right)^{-1}. \tag{4.48}$$

Three remaining dimensionful quantities are present, the coefficients a and b and the gravitational constant G. One sees that the TOV equations are becoming independent from those remaining dimensionful quantities under the conditions that

$$G \cdot b = a \qquad a^3 \cdot \varepsilon_0 = b. \tag{4.49}$$

Note that this holds true also for the correction terms from general relativity, including the Schwarzschild factor of the metric. The conditions from Eq. (4.49) can be recast to determine the unknown coefficients

$$a = (G \cdot \varepsilon_0)^{-1/2} \qquad b = (G^3 \cdot \varepsilon_0)^{-1/2}. \tag{4.50}$$

Then the dimensionless TOV equation reads

$$\frac{dP'}{dr'} = -\frac{m'_r\varepsilon'}{r'^2}\left(1 + \frac{P'}{\varepsilon'}\right)\left(1 + \frac{4\pi r'^3 P'}{m'_r}\right)\left(1 - \frac{2m'_r}{r'}\right)^{-1} \tag{4.51}$$

without any dimensionful quantity appearing in the expression. One can convince oneself that the same scaling conditions for the radius and the mass in Eq. (4.50) transform the equation of mass conservation to a fully dimensionless form:

$$\frac{dm'_r}{dr'} = 4\pi r'^2 \varepsilon'. \tag{4.52}$$

For a given EOS in the form $P' = P'(\varepsilon')$, the dimensionless TOV equations can be solved numerically in terms of a dimensionless mass–radius relation, energy density profile, and so on. The physical quantities are recovered by fixing the scaling energy density ε_0. That means that each quantity of the solution scales with the factor ε_0 to some power. In particular, any mass and radius rescales as given by Eqs. (4.47) and (4.50):

$$m_r = \frac{m'_r}{(G^3\varepsilon_0)^{1/2}} \qquad r = \frac{r'}{(G\varepsilon_0)^{1/2}}, \tag{4.53}$$

which includes also the maximum mass $M_{\rm max}$ and its radius $R_{\rm crit}$, which have to scale as $1/\sqrt{\varepsilon_0}$ also

$$M_{\rm max} = \frac{M'}{(G^3\varepsilon_0)^{1/2}} \qquad R_{\rm crit} = \frac{R'}{(G\varepsilon_0)^{1/2}}, \tag{4.54}$$

or in terms of the Planck mass $m_{\rm P}$:

$$M_{\rm max} = M' \cdot \frac{m_{\rm P}^3}{\varepsilon_0^{1/2}} \qquad R_{\rm crit} = R' \cdot \frac{m_{\rm P}}{\varepsilon_0^{1/2}}. \tag{4.55}$$

The entire mass–radius relation has to scale accordingly. Recall that the primed quantities are dimensionless numbers to be determined numerically by solving the TOV equations.

The scaling relations of Eq. (4.55) have several important applications. The recipe is as follows: solve the dimensionless TOV equations for a given type of dimensionless EOS and then rescale the results with $\sqrt{\varepsilon_0}$ to arrive at physical values. Let us give some examples in the following.

4.3.2 EOS for a Free (Massive) Fermi Gas

For a free gas of fermions, the only scale that appears is the fermion mass. We rescale the EOS by the fermion mass simply by $P' = P/m_f^4$ and $\varepsilon' = \varepsilon/m_f^4$, which implies that the scaling constant is $\varepsilon_0 = m_f^4$. Then, the maximum mass and the corresponding radius from Eq. (4.55) are given by

$$M_{\text{max}} = M' \cdot (G^3 \cdot m_f^4)^{-1/2} = M' \cdot \frac{m_P^3}{m_f^2} \tag{4.56}$$

$$R_{\text{crit}} = R' \cdot (G \cdot m_f^4)^{-1/2} = R' \cdot \frac{m_P}{m_f^2}, \tag{4.57}$$

where we recognize the Landau mass and Landau radius as solutions. So, we have shown that the Landau mass and radius are solutions of the full general relativistic equations for compact stars, the TOV equations, and not just for the classical limit. The dimensionless prefactors M' and R' can be computed either by solving the dimensionless TOV equations with the EOS for a free Fermi gas or by using the known numerical values for a free gas of neutrons with an appropriate rescaling. Numerically one finds that these prefactors are given by $M' = 0.384$ and $R' = 3.367$. Let us look at the compactness of a star, which we defined as the ratio of the gravitational mass to its radius

$$C = \frac{GM}{R}. \tag{4.58}$$

Note that the compactness of the star is dimensionless and measures how close the star's radius is to that of a black hole of the same mass. A compact star of free fermions has the maximum compactness at the maximum mass and is given by the ratios of the dimensionless prefactors of Eqs. (4.56) and (4.57)

$$C_{\text{max}} = \frac{M'}{R'} = 0.11, \tag{4.59}$$

which is independent on the fermion mass.

4.3.3 EOS for a Relativistic Fermi Gas with a Vacuum Term

The EOS of relativistic particles

$$P = \frac{1}{3}\varepsilon \tag{4.60}$$

is similar to a polytrope with a power of $\gamma = 1$. As we will see, this EOS produces unstable configurations in the Newtonian case, that is, for the Lane–Emden equation, see Chapter 5. Moreover, the EOS does not contain any dimensionful quantity so there is nothing to define a dimensionful scaling constant ε_0. However, the EOS can be modified to be able to arrive at stable solutions for the TOV equations. The trick is to introduce a dimensionful constant to the EOS in the form

$$P = \frac{1}{3}(\varepsilon - \varepsilon_0), \tag{4.61}$$

where ε_0 is an offset of the energy density. The EOS has now the special property that the pressure vanishes at a nonvanishing energy density ε_0. This feature is absent in the EOSs studied so far. The EOS of fermion gases or of neutron matter have a vanishing pressure at a vanishing energy density, even for the case of interacting fermions that is to be studied later.

In general, microscopic theories of matter based on a quantum field theoretical description have a vanishing energy density in the vacuum by definition. More precisely, one can add a constant to the Lagrangian density and it would not change the equations of motion of the quantum fields so that one can fix the energy density of the vacuum to be 0. For gravity an additional constant in the Lagrangian density counts and has an impact on the equations of motion. In fact, for gravity this additional constant is nothing less than what is known as the cosmological constant.

There are known cases of microscopic models, where the form of equation equation (4.61) appears. One of the most famous ones is that of the MIT bag model, which we will discuss in more detail in Chapter 8 on quark stars. The MIT bag model considers a gas of free relativistic quarks that are confined within a bag stabilized by some pressure B from the outside. The quantity B is assumed to be a constant and usually dubbed the MIT bag constant. The total pressure in the bag is then

$$P = P_{\text{free}} - B. \tag{4.62}$$

Thermodynamic consistency demands that the energy density has to be

$$\varepsilon = \varepsilon_{\text{free}} + B, \tag{4.63}$$

which the reader is invited to check as an exercise. Note that the pressure from the outside is the pressure of the vacuum, that is, it corresponds to nonvanishing vacuum contributions. Also, as seen from Eq. (4.63), the vacuum pressure corresponds to an energy density with a negative sign. The nonvanishing vacuum energy density can be attributed to nonvanishing vacuum expectation values of the quantum chromodynamics (QCD) vacuum, in particular to the quark and gluon condensates. We note in passing that similar reasoning would also apply to the electroweak theory and the nonvanishing vacuum expectation value of the Higgs field.

The EOS of the MIT bag model can be recast to the form of Eq. (4.61). As we know that for a relativistic massless gas of particles,

$$P_{\text{free}} = \frac{1}{3}\varepsilon_{\text{free}} \tag{4.64}$$

the equations can be combined to the final form

$$P = \frac{1}{3}(\varepsilon - 4B), \tag{4.65}$$

which is of the form of Eq. (4.61) with $\varepsilon_0 = 4B$. From our scaling analysis, we can write down immediately how the maximum mass and the corresponding radius depend on the value of the MIT bag constant B:

$$M_{\rm max} = M' \cdot (G^3 \cdot \varepsilon_0)^{-1/2} = 2.01\, M_\odot \cdot \frac{(145\,\text{MeV})^2}{\sqrt{B}} \tag{4.66}$$

$$R_{\rm crit} = R' \cdot (G \cdot \varepsilon_0)^{-1/2} = 10.9\,\text{km} \cdot \frac{(145\,\text{MeV})^2}{\sqrt{B}}, \tag{4.67}$$

where $B^{1/4} = 145\,\text{MeV}$ is the standard MIT bag model value and the numerical prefactors are taken from the known solution for that value (see e.g., Baym and Chin, 1976; Witten, 1984). The mass–radius relation of the kind of EOSs that have a vanishing pressure at a nonvanishing energy density are special and correspond to the class of selfbound stars. Selfbound stars have a vanishing pressure at the surface of the star by the properties of matter and do not need gravity to be stabilized – hence, the name selfbound. A maximum mass configuration exists, though, when the gravitational energy reaches the scale given by ε_0, destabilizing the compact star. The maximum compactness for the EOS of the form of Eq. (4.65) is given by

$$C_{\rm max} = \frac{GM_{\rm max}}{R_{\rm crit}} = \frac{M'}{R'} = 0.271 \tag{4.68}$$

and is independent of the choice of the MIT bag constant B or the scaling energy density ε_0. In comparison to Eq. (4.59), these selfbound stars are considerably more compact than compact stars made of free fermions.

4.3.4 Limiting EOS from Causality

Zel'dovich has shown (Zel'dovich, 1961) that the stiffest possible EOS is not of the form of an ultra-relativistic gas of particles, see Eq. (4.60), but of the form

$$P = \varepsilon. \tag{4.69}$$

Indeed, by looking at the speed of sound, one realizes that for this EOS, the speed of sound squared is

$$c_s^2 = \frac{\partial P}{\partial \varepsilon} = 1, \tag{4.70}$$

so that the speed of sound is equal to the speed of light, the maximum value allowed by causality. Therefore, one can call the Zel'dovich EOS, Eq. (4.69), also the limiting causal EOS, as it gives the maximum pressure for a given energy density allowed by causality. The Zel'dovich EOS can be realized in microscopic models, if the dominant term of the energy density scales as

$$\varepsilon = c \cdot n^2, \tag{4.71}$$

where n stands for the number density and c is a constant with dimensions $(1/\text{mass})^2$. For neutron stars, n would be the baryon number density. Thermodynamic consistency then fixes the pressure to be

$$P = c \cdot n^2, \tag{4.72}$$

with the same prefactor so that one recovers the Zel'dovich equation, Eq. (4.69). We will encounter the Zel'dovich EOS as the high-density EOS for an interacting gas of fermions in Section 4.4.

If we look at the Zel'dovich EOS, Eq. (4.69), we realize that no mass scale appears and, moreover, that the EOS will lead to unstable solutions for the TOV equations, as the polytropic index is like the case for the ultra-relativistic gas discussed in Section 4.3.3. We know now how to arrive at stable solutions for the TOV equation, we simply introduce a constant to the EOS. It is convenient to introduce the following extended form of the Zel'dovich EOS

$$P = P_f + (\varepsilon - \varepsilon_f), \tag{4.73}$$

with a nonvanishing pressure P_f for an energy density ε_f. Note that for $P_f = 0$ we would recover an EOS for selfbound stars. Here we have now a different application in mind. Assume that we know the low-density EOS up to some pressure P_f with the corresponding energy density $\varepsilon_f = \varepsilon_f(P_f)$ for neutron stars. Then, the stiffest possible EOS for higher densities has to be the Zel'dovich EOS of the form given in Eq. (4.73). As the EOS provides the maximum pressure allowed by causality, the corresponding maximum mass is the highest possible mass and gives an upper limit to the maximum mass allowed by causality.

The neutron matter EOS is known to about the saturation density of normal nuclear matter with an energy density of $\varepsilon_{\text{nm}} = 140\,\text{MeV fm}^{-3}$. The pressure at that energy density is about two orders of magnitude smaller, so it can be safely ignored. Then, the limiting EOS contains only one dimensionful quantity and we can proceed with our scaling analysis. The maximum mass then has to scale with the fiducial energy density ε_f as

$$M_{\text{max}} = M' \cdot (G^3 \cdot \varepsilon_f)^{-1/2} = 4.2 M_\odot \cdot \sqrt{\frac{\varepsilon_{\text{nm}}}{\varepsilon_f}}, \tag{4.74}$$

where $\varepsilon_{\rm nm}$ is the ground-state energy density of nuclear matter. This expression for the limiting maximum mass of a neutron star was derived by Rhoades and Ruffini (1974) and is called the Rhoades–Ruffini mass limit.

It turns out that the numerical factor is rather insensitive to the choice of the low-density nuclear EOSs as the pressure of the neutron star matter EOS is so small. However, in their original paper, Rhoades and Ruffini chose twice the saturation density as their fiducial energy density setting $\varepsilon_f = 2\varepsilon_{\rm nm}$. Looking at Eq. (4.74), one arrives at a limiting maximum mass of a neutron star of $3M_\odot$, a value often generically quoted in the literature as the maximum mass of a neutron star allowed by causality. As we do not know at present the neutron star matter EOS at $2\varepsilon_{\rm nm}$, that value of the maximum mass should be taken with corresponding care. The maximum mass of a neutron star allowed by causality alone with our present knowledge of the neutron star matter EOS up to the energy density at saturation density is $4.2M_\odot$.

4.3.5 Selfbound Linear EOS

The Zel'dovich EOS in the form of Eq. (4.73) with vanishing fiducial pressure $P_f = 0$ and the MIT bag model EOS, Eq. (4.65), can be combined to a special class of EOSs

$$P = s \cdot (\varepsilon - \varepsilon_0)\,, \tag{4.75}$$

with a dimensionless constant s. As s is equivalent to the speed of sound, $s = c_s^2$, in the medium

$$c_s^2 = \frac{\partial P}{\partial \varepsilon} = s, \tag{4.76}$$

it is bounded by $0 < s \leq 1$. For $s = 1/3$, we recover the MIT bag model, for $s = 1$, the Zel'dovich EOS as the limiting causal EOS. EOSs of the type of Eq. (4.75) are the simplest type of EOS, having a nonvanishing energy density at vanishing pressure. The solutions of the TOV equations will constitute classes of so-called selfbound stars, which are bound by themselves as the pressure vanishes at the surface of the star, even when gravity is absent. Note that ordinary stars including compact stars are stabilized by the nonvanishing pressure gradient of matter being counterbalanced by the gravitational force.

In all cases for the selfbound EOS of Eq. (4.75), the solution of the TOV equation will exhibit that the maximum mass and its radius will scale with $1/\varepsilon_0^{1/2}$, as shown earlier. The numerical prefactors, however, will now depend on the parameter s. For $s = 1/3$, one can use Eq. (4.66) and rewrite it in the form

$$M_{\max} = 2.57\, M_\odot \cdot \left(\frac{\varepsilon_{\text{nm}}}{\varepsilon_0}\right)^{1/2} \qquad R_{\text{crit}} = 14.0\,\text{km} \cdot \left(\frac{\varepsilon_{\text{nm}}}{\varepsilon_0}\right)^{1/2}, \tag{4.77}$$

for $s = 2/3$, one finds numerically

$$M_{\max} = 3.64\, M_\odot \cdot \left(\frac{\varepsilon_{\text{nm}}}{\varepsilon_0}\right)^{1/2} \qquad R_{\text{crit}} = 16.4\,\text{km} \cdot \left(\frac{\varepsilon_{\text{nm}}}{\varepsilon_0}\right)^{1/2}, \tag{4.78}$$

and for the limiting causal EOS $s = 1$

$$M_{\max} = 4.23\, M_\odot \cdot \left(\frac{\varepsilon_{\text{nm}}}{\varepsilon_0}\right)^{1/2} \qquad R_{\text{crit}} = 17.6\,\text{km} \cdot \left(\frac{\varepsilon_{\text{nm}}}{\varepsilon_0}\right)^{1/2}. \tag{4.79}$$

Here we have computed the dimensionful numerical prefactors for the energy density of saturated nuclear matter $\varepsilon_{\text{nm}} = 140\,\text{MeV fm}^{-3}$. We see that the equation for the maximum mass for the case $s = 1$ coincides with the one for the Rhoades–Ruffini mass limit presented in Eq. (4.74). We stress that the relative changes of the numerical prefactors for the maximum mass and the corresponding radius are different and do not scale with the parameter s. In fact, the mass–radius relations for all three cases are different and cannot be mapped into each other by rescaling.

Note that the compactness, the ratio of the maximum mass, and the radius are independent of the chosen value of the scale, ε_0, which drops out. Hence, the dimensionless form of the EOS alone defines the compactness of the whole mass–radius relation, including the maximum mass configuration. In most cases, this defines also a maximum compactness for all stable solutions to the TOV equations, as the maximum compactness usually coincides with the maximum mass. For the cases considered here, one finds that the maximum mass configuration gives the maximum compactness of

$$C_{\max} = \frac{GM_{\max}}{R_{\text{crit}}} = \frac{M'}{R'} = \begin{cases} 0.271 & \text{for } s = 1/3 \\ 0.328 & \text{for } s = 2/3\,, \\ 0.354 & \text{for } s = 1 \end{cases} \tag{4.80}$$

where M' and R' are the dimensionless numerical prefactors for the mass and radius, as defined in Eq. (4.55). The compactness is related to the redshift factor $1 + z$ via

$$1 + z = \frac{1}{\sqrt{1 - \frac{2GM}{R}}} \tag{4.81}$$

and the maximum redshift can be determined to be

$$z_{\max} = \frac{1}{\sqrt{1 - \frac{2GM_{\max}}{R_{\text{crit}}}}} - 1 = \frac{1}{\sqrt{1 - \frac{2M'}{R'}}} - 1. \tag{4.82}$$

The numerical values are as follows

$$z_{\max} = \begin{cases} 0.478 & \text{for } s = 1/3 \\ 0.705 & \text{for } s = 2/3 \\ 0.851 & \text{for } s = 1 \end{cases}, \tag{4.83}$$

so that the maximum redshift stays in all cases below 1 and is at maximum $z_{\max} = 0.851$. As the case $s = 1$ constitutes the stiffest possible EOS, it leads to the most compact star configurations allowed by causality. It was shown by Lindblom that this maximum redshift applies also for the case of neutron stars (Lindblom, 1984).

The central energy density for the maximum mass configuration is the maximum central energy density possible for the given EOS. For the cases considered here, one gets the following numerical values:

$$\frac{\varepsilon_{\max}}{\varepsilon_0} = \begin{cases} 4.81 & \text{for } s = 1/3 \\ 3.54 & \text{for } s = 2/3 \\ 3.03 & \text{for } s = 1 \end{cases}. \tag{4.84}$$

Note that those numbers just give the relative ratio of the central energy density to the surface energy density for the maximum mass configuration. It is instructive to quote the maximum central energy densities for the maximum mass of $M_{\max} = 2M_{\odot}$, which are

$$\frac{\varepsilon_{\max}}{\varepsilon_{\mathrm{nm}}} = \begin{cases} 7.91 & \text{for } s = 1/3 \\ 11.8 & \text{for } s = 2/3 \\ 13.5 & \text{for } s = 1 \end{cases} \tag{4.85}$$

in terms of the energy density of nuclear matter $\varepsilon_{\mathrm{nm}}$. So, for a maximum mass of $M_{\max} = 2M_{\odot}$ and the causal EOS with $c_s^2 = 1$, the maximum possible energy density is $\varepsilon_{\max} = 13.5\varepsilon_{\mathrm{nm}}$, which constitutes the ultimate energy density of observable cold matter (Lattimer and Prakash, 2005). Note that the maximum energy density changes inversely with the maximum mass. So a higher maximum mass results in a lower ultimate energy density. In fact, the maximum mass scales with $\varepsilon_0^{-1/2}$ and, obviously, the maximum energy density with ε_0. So the product of the maximum mass squared times the maximum energy density is independent of ε_0, stated first explicitly by Lattimer and Prakash (2011). In other terms, the maximum energy density scales inversely with the maximum mass squared and to embarass them I will denote the scaling relation for the ultimate energy density as the Lattimer–Prakash relation:

$$\varepsilon_{\mathrm{ultimate}} = 13.5\varepsilon_{\mathrm{nm}} \left(\frac{2M_{\odot}}{M_{\max}} \right)^2 . \tag{4.86}$$

If one finds that the maximum mass is $M_{\max} = 3M_{\odot}$, for example, the ultimate energy density from causality will be only $\varepsilon_{\mathrm{ultimate}} = 6\varepsilon_{\mathrm{nm}}$.

4.4 Interacting Fermions

So far we considered free ideal gases of fermions. Interactions between the fermions have been neglected for considering the maximum masses of compact star configurations and their corresponding radius. In the following, we discuss the possible impact of interactions on the mass–radius relation and the maximum mass of compact stars.

We assume that the contribution from interactions is proportional to the number density of fermions n to the power of γ. Motivated by this, we add an interaction term to the EOS of the form

$$\varepsilon_{\text{int}} = c \cdot n^{\gamma}, \tag{4.87}$$

with c being a constant that determines the strength of the interaction. The corresponding contribution to the pressure has to be then of the form

$$P_{int} = -\left.\frac{\partial E_{\text{int}}}{\partial V}\right|_{N,T=0} = n^2 \frac{(\partial \varepsilon_{\text{int}}/n)}{\partial n} = (\gamma - 1) \cdot c \cdot n^{\gamma} = (\gamma - 1) \cdot \varepsilon_{\text{int}}. \tag{4.88}$$

The speed of sound is then given by

$$c_s^2 = \frac{\partial P}{\partial \varepsilon} = \gamma - 1 \tag{4.89}$$

if the EOS is dominated by the interaction terms. Then, the overall EOS will be acausal for $\gamma > 2$. For two-body interactions between fermions, $\gamma = 2$ seems to be a reasonable value. In fact, such a kind of interaction term can be motivated from relativistic mean-field models where the interaction between fermions is mediated by the exchange of vector mesons (see e.g., Glendenning, 2000).

Let us consider now an EOS for interacting fermions, including the kinetic terms for the pressure and energy density. First have a look at the nonrelativistic limit:

$$\varepsilon^{(\text{nr})} = m \cdot n + c \cdot n^2 \qquad p^{(\text{nr})} = \frac{1}{5m^{8/3}} \left(\frac{6\pi^2}{g}\right)^{2/3} \cdot n^{5/3} + c \cdot n^2. \tag{4.90}$$

For low densities, the first terms dominate

$$\varepsilon^{(\text{nr})} \approx m \cdot n \qquad p^{(\text{nr})} \approx \frac{1}{5m^{8/3}} \left(\frac{6\pi^2}{g}\right)^{2/3} \cdot n^{5/3} \propto \varepsilon^{5/3} \tag{4.91}$$

and we recover the low-density limit of a noninteracting Fermi gas, a polytrope with a power of $5/3$. At some intermediate density the terms from the interactions will start to be dominant. With increasing density, the pressure will be dominated by the interaction term, as the kinetic term of the nonrelativistic pressure is much smaller compared to the mass term of the energy density. The energy density can still be dominated by the mass term for some intermediate density so that

$$\varepsilon \approx m \cdot n \qquad p \approx c \cdot n^2 \propto \varepsilon^2 \tag{4.92}$$

and we recover a polytrope with a power of $\Gamma = 2$. Finally, in the ultra-relativistic limit, the interaction terms dominate the pressure and the energy density. Hence,

$$\varepsilon \approx c \cdot n^2 \qquad p \approx c \cdot n^2 \approx \varepsilon \tag{4.93}$$

and the EOS approaches the limiting causal EOS with a power of one in terms of the energy density ε.

From the discussion of the general solution of compact stars for polytropic EOSs, see Section 5.2, we will see later that a polytrope with a power of $\Gamma = 5/3$ gives a mass–radius relation of $M \cdot R^3 = \text{const.}$, a polytrope with a power of $\Gamma = 2$ gives a mass–radius relation with a constant radius, and a polytrope with a power of $\Gamma = 1$ gives unstable solutions. Pretty generically, we conclude that a compact star consisting of interacting fermions will have a mass–radius relation that looks like that of noninteracting fermions at large radii, that is, at low densities, that is independent of the mass for low radii, that is, at moderate densities, and that reaches a maximum mass at high densities. Beyond the density corresponding to the maximum mass, the solutions are unstable.

It is even possible to make some generic statements about the maximum mass for the case of interacting fermions. First, let us introduce a dimensionless measure of the interaction strength. By dimensionless analysis, the coefficient c of the interaction term has the units of mass^{-2}. Let us denote that mass scale as the interaction mass scale m_I and the mass of the fermion as m_f. We can rewrite the contribution from the interaction to the pressure by

$$P_{\text{int}} = c \cdot n^2 = \frac{1}{m_I^2} \cdot n^2. \tag{4.94}$$

By rescaling all the terms with the Fermi mass, one arrives at

$$\frac{P_{\text{int}}}{m_f^4} = \frac{m_f^2}{m_I^2} \cdot \left(\frac{n}{m_f^3}\right)^2 = y^2 \cdot \left(\frac{n}{m_f^3}\right)^2, \tag{4.95}$$

with the dimensionless interaction strength $y = m_f/m_I$.

Figure 4.3 shows the mass–radius relation of compact stars with interacting fermions for different interaction strengths y. Note that the plot is shown in units of the Landau mass and the Landau radius, so that all quantities are dimensionless and valid for arbitrary fermion masses. One sees that for small values of the interaction strength $y < 1$, the mass–radius relation scales as $M \propto R^{-3}$ for large radii and looks like the one for the noninteracting case. For large values of the interaction

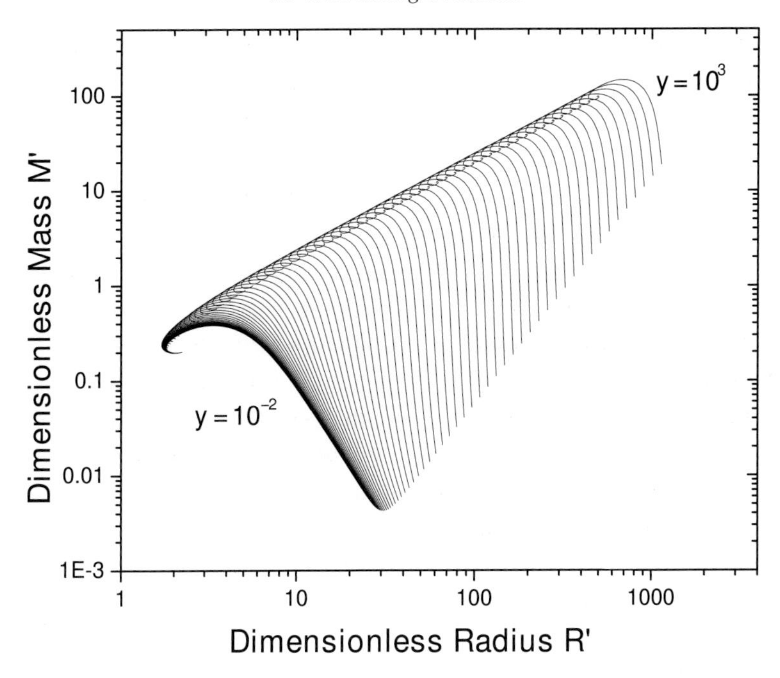

Figure 4.3 The mass–radius relation for compact stars consisting of interacting fermions for different interaction strengths y (in 20 equal steps for each decade) given in units of the Landau mass and Landau radius. Reprinted with permission from Narain et al. (2006). Copyright (2006) by the American Physical Society

strength $y > 1$, the mass–radius relation becomes independent of the mass with a constant radius. The maximum mass and the radius are growing linearly with the interaction strength y. In summary, for the weakly interacting case, ($y < 1$): $M \cdot R^3 = \text{const.}$ and $M_{\text{max}} = \text{const.}$, while for the strongly interacting case, ($y > 1$): $R \approx \text{const.}$ and $M_{\text{max}} \propto y$.

The different behavior of the mass–radius relation for the weakly and the strongly interacting cases can be understood in simple terms. For the weakly interacting case, the fermions are becoming relativistic before the interaction terms start to dominate the pressure, so that the maximum mass is reached before interactions are important. One recovers the case for noninteracting fermions. For the strongly interacting case, the interaction terms are dominating the pressure and energy density before the fermions are becoming relativistic. The maximum mass is then reached when the interaction terms are starting to become dominant for ε, that is, when the contribution from the mass term and the interaction term for the energy density are about equal

$$m_f \cdot n_c \approx c \cdot n_c^2 \tag{4.96}$$

and the corresponding critical density n_c can be estimated to scale with the interaction strength y as

$$n_c \sim m_f/c = m_f \cdot m_I^2 = \frac{m_f^3}{y^2}. \tag{4.97}$$

For comparison, the critical density for the free case corresponds to $n_c \sim m_f^3$, as $n \sim k_F^3$ and the critical Fermi momentum is just given by the onset of having relativistic fermions $k_{F,c} = m_f$. We realize now that for $y < 1$, the instability stemming from relativistic fermions is happening at a critical density, which is lower than the one stemming from the interaction terms. Hence, the mass–radius relation looks like that of a compact star with noninteracting fermions. For $y > 1$, however, the instability stemming from the interaction terms is happening at a density lower than the one stemming from relativistic fermions.

The critical values for the pressure and the energy density for $y > 1$ are scaling as

$$\varepsilon_c \sim m_f \cdot n_c \sim m_f^2 \cdot m_I^2 \sim \frac{m_f^4}{y^2} \qquad P_c \sim \frac{1}{m_I^2} \cdot n_c^2 \sim \frac{1}{m_I^2} \cdot \frac{m_f^6}{y^4} \sim \frac{m_f^4}{y^2}. \tag{4.98}$$

We know that for the noninteracting case, the critical energy density and the pressure scales as m_f^4 and the maximum mass and the corresponding radius scales as given by the Landau mass and radius, that is, proportional to $1/m_f^2$. As the critical energy density and pressure for the strongly interacting case scales now as m_f^4/y^2, in analogy the maximum mass and the corresponding radius have to scale as

$$M_L^{\text{int}} = y \cdot \frac{m_\text{P}^3}{m_f^2} = y \cdot M_L \qquad R_L^{\text{int}} = y \cdot \frac{m_\text{P}}{m_f^2} = y \cdot R_L \qquad (y > 1), \tag{4.99}$$

hence, linear in the interaction strength y. Numerically one finds the following expression for arbitrary values of y:

$$M_{\text{max, int}} = (0.384 + 0.165 \cdot y) \cdot M_L \tag{4.100}$$

and

$$R_{\text{crit, int}} = (3.367 + 0.797 \cdot y) \cdot R_L. \tag{4.101}$$

The compactness of a star is given by the ratio of the radius to its Schwarzschild radius. The maximum compactness is given by the compactness of the maximum mass configuration. For weakly interacting fermions $y \ll 1$, one recovers the maximum compactness of a free fermion star. Fermion stars with strongly interaction fermions can be more compact. In summary,

$$C_{\text{max}} = \frac{GM_{\text{max}}}{R_{\text{crit}}} = \begin{cases} 0.11 & \text{for } y \ll 1 \\ 0.21 & \text{for } y \gg 1 \end{cases} \tag{4.102}$$

for our choice of the interaction terms in the EOS. We see that interactions can make a neutron star more compact by a factor of two compared the noninteracting case. Of course, a different ansatz for the interaction between the fermions could result in an even more compact configurations.

Finally, let us put the interaction strength y in perspective with the known forces. In weak interactions of the standard model, the interaction strength is given by Fermi's constant, which is in natural units given by

$$G_F = 1.1663787(6) \times 10^{-5}\,\text{GeV}^{-2} = (292.806\,\text{GeV})^{-2} \tag{4.103}$$

and the corresponding interaction mass scale would be $m_I \sim 300\,\text{GeV}$. For strong interactions of the standard model (QCD), the typical mass scale is given by $\Lambda_{\text{QCD}} \sim 200\,\text{MeV}$, as known from perturbative QCD and the running of the strong coupling constant α_s (see Chapter 8). In the nonperturbative regime, QCD can be described by chiral effective field theory, where the interaction strength is controlled by the pion decay constant $f_\pi = 92\,\text{MeV}$. In fact, interaction terms to first order in the chiral expansion for the pion–nucleon interaction depend on $1/f_\pi^2$. So, for the strong interaction, the typical mass scale would be roughly $m_I \sim 100\,\text{MeV}$. For a neutron star, where the mass of the fermion (the neutron) is about 1 GeV, one would expect sizable corrections from interactions, as $y \sim m_f/m_I = m_n/f_\pi \sim 10$. The maximum mass of a neutron star, including effects from strong interactions would then be much larger than the one for the free case. In fact, this is indeed seen in refined models of neutron stars, see Chapter 7. The numerical expression for the maximum mass, Eq. (4.100), gives then a maximum mass of a neutron star of about $3.8M_\odot$ compared to $0.71M_\odot$ for the free case.

One can even include gravity in these considerations. The corresponding mass scale is, of course, Newton's constant or, equivalently, the Planck mass, which is in natural units:

$$G = \frac{1}{m_\text{P}^2} = (1.220890(13) \times 10^{19}\,\text{GeV})^{-2}. \tag{4.104}$$

As any reasonable fermion mass is well below the Planck scale, one would recover in any reasonable case the weakly interacting case $y < 1$ for any fermion star where the fermions are only interacting on scales of the gravitational interaction, which is equivalent to the noninteracting case for compact star configurations.

4.5 Boson Stars

So far we have considered compact stars made of fermions where the Fermi pressure counterbalances the gravitational pull of matter. In this section we are discussing the case when compact stars consist of bosons instead of fermions.

Boson stars are normally considered as a solution to the combined Einstein equations with the Klein–Gordon equation of a scalar boson. As we will see, the essential features of these solutions can be well understood in terms the of TOV equations and the EOS of bosonic matter, which we will pursue in the following.

Bosons at zero temperature are in a Bose condensate, where all particles are occupying the lowest energy state. For a free gas of bosons, the lowest energy state has zero momentum, so there is no pressure generated by the bosonic particles. Naturally, one can ask oneselves if boson stars can be stable, if there is no pressure at all. Here comes quantum mechanics to the rescue. The Heisenberg uncertainty principle demands that there is an uncertainty in the momentum of particles, which is also valid for bosons. A massive particle has a Compton wavelength, which corresponds to a nonvanishing momentum of the particle, even if it is in the lowest energy state.

We can use Landau's argument for a maximum mass also for the case of bosons. Let us consider a sphere of bosons and have a look at a relativistic boson sitting at the radius R of that sphere. The total energy of that boson is the sum of gravitational energy and kinetic energy

$$E(R) = E_g + E_{\text{kin}} = -G\frac{m_b M}{R} + E_{\text{kin}}, \tag{4.105}$$

with the total mass of the boson star $M = N \cdot m_b$, where N is the number of bosons in the compact star and m_b the mass of the boson. For a relativistic boson, the energy is equal to the momentum. The uncertainty principle relates the momentum of the boson to the size of the quantum system so that

$$E_{\text{kin}} = k_b \sim \frac{1}{R}, \tag{4.106}$$

where R is the radius of the boson star. Note that this implies that the boson star is something like a macroscopic quantum ball. So, the kinetic energy and the gravitational energy of the boson scale, with $1/R$ as in the case for fermions:

$$E(R) = -G\frac{N \cdot m_b^2}{R} + \frac{1}{R}. \tag{4.107}$$

In the nonrelativistic limit, the kinetic energy of the bosons is proportional to $E_{\text{kin}} = k_b^2/m_b \sim 1/R^2$, as in the case for fermions. One can repeat the stability analysis for bosons in an analogy to the one of fermions. For a positive total energy, the compact star will expand to minimize the total energy. Eventually, the boson becomes nonrelativistic, the kinetic energy scales as $1/R^2$, generating a stable minimum at a nonvanishing value of the radius R. For a negative total energy, the compact star will shrink to minimize the total energy, eventually collapsing to a black hole. Hence, there exists a maximum number of bosons $N_{\max}$, which is given by the limiting case of stability $E(R) = 0$. This condition translates to

$$G \cdot N_{\max} \cdot m_b^2 = 1, \tag{4.108}$$

so that the maximum number of bosons is given by

$$N_{\max} = \frac{1}{G \cdot m_b^2} = \frac{m_P^2}{m_b^2}. \tag{4.109}$$

The maximum mass of a compact star of free bosons is then

$$M_{\max} = N_{\max} \cdot m_b = \frac{m_P^2}{m_b}, \tag{4.110}$$

which differs by a factor m_P/m_b compared to the Landau maximum mass for free fermions, Eq. (4.28). Usually, $m_P \gg m_b$, so that the maximum mass of a boson star is orders of magnitude smaller compared to the maximum mass of a fermion star with the same particle mass. The critical radius can be estimated from the onset of relativity for the boson's kinetic energy from Eq. (4.106) to be

$$R_{\text{crit}} = \frac{1}{m_b}, \tag{4.111}$$

which is just the Compton wavelength of the boson. The critical radius differs by the same factor m_P/m_b from the case for free fermions. Hence, boson stars are orders of magnitude smaller compared to fermion stars that consist of particles with the same particle mass. As the critical radius of a boson star is determined by the Compton wavelength, which is a quantum mechanical quantity, the radius is microscopically small and boson stars can be considered to be quantum balls.

Interactions between the boson will change the picture drastically. Let us proceed by introducing an interaction term to the energy density, which scales as the number density n squared, similar to the case we studied before for fermions:

$$\varepsilon = m_b \cdot n + c \cdot n^2, \tag{4.112}$$

where c is a constant of dimension $(1/\text{mass})^2$, specifying the strength of the interaction. We ignore the effect from the quantum zero-point energy, which we consider to be negligible compared to the interaction energy. A posteriori we will see that this is indeed justified. The pressure can be calculated by using Eq. (3.69) to be

$$P = c \cdot n^2. \tag{4.113}$$

For low number densities, the mass term in the energy density dominates and one recovers an EOS of the form $P \sim \varepsilon^2$, which is a polytrope with a power law with the power of $\gamma = 2$. We know that this type of EOS leads to stable compact star configurations with a constant radius in the mass–radius diagram. At high densities, the energy density is dominated by the interaction term and we recover an EOS of the form $P = \varepsilon$, which is a power law with $\gamma = 1$. We have seen that this EOS leads to unstable compact star configurations. The limiting case is where the mass term and the interaction term in the expression of the energy density are of equal magnitude, giving us a criteria for determining the maximum mass of a compact star with interacting bosons. The critical number density is accordingly determined via

$$m_b \cdot n_c = c \cdot n_c^2 \to n_c = \frac{m_b}{c}, \tag{4.114}$$

giving a critical energy density of about

$$\varepsilon_c \sim m_b \cdot n_c = \frac{m_b^2}{c} = m_b^2 \cdot m_I^2 = \frac{m_b^4}{y^2}. \tag{4.115}$$

Here we introduced the interaction mass scale m_I defined via the relation $c = 1/m_I^2$ and the interaction strength $y = m_b/m_I$ in an analogy with the case of interacting fermions. Now we are in the position to use the scaling properties of the TOV equations, Eq. (4.55). The maximum mass and the corresponding radius of a compact star with interacting bosons is then given by choosing the scaling energy density to be $\varepsilon_0 = m_b^4/y^2$, which results in

$$M_{\text{max}} = M' \cdot y \cdot \frac{m_{\text{P}}^3}{m_b^2} = M' \cdot y \cdot M_L \tag{4.116}$$

and

$$R_{\text{crit}} = R' \cdot y \cdot \frac{m_{\text{P}}}{m_b^2} = R' \cdot y \cdot R_L. \tag{4.117}$$

The numerical values for the scaling coefficients are $M' = 0.164$ and $R' = 0.763$. Surprisingly, we recover the Landau mass and radius for boson stars for an interaction strength of $y = 1$. This means that boson stars have a similar maximum mass and a radius as its fermionic counterparts with the same particle mass if interactions are included in the EOS. The dependence of the maximum mass and the corresponding radius on the interaction strength y is linear, as in the case for fermion stars with an interaction strength of $y > 1$, see Eqs. (4.100) and (4.101). Also, the numerical prefactors are very close to the fermionic case. Interestingly, if one sets the interaction strength by choosing the interaction mass scale to the Planck scale, $m_I = m_{\text{P}}$, one arrives at the expression for the maximum mass and the corresponding radius for the noninteracting case, Eqs. (4.110) and (4.111), derived by using Landau's argument.

Figure 4.4 shows the mass–radius relation for boson stars for different interaction strengths y. The mass and the radius are given in dimensionless units, that is, in

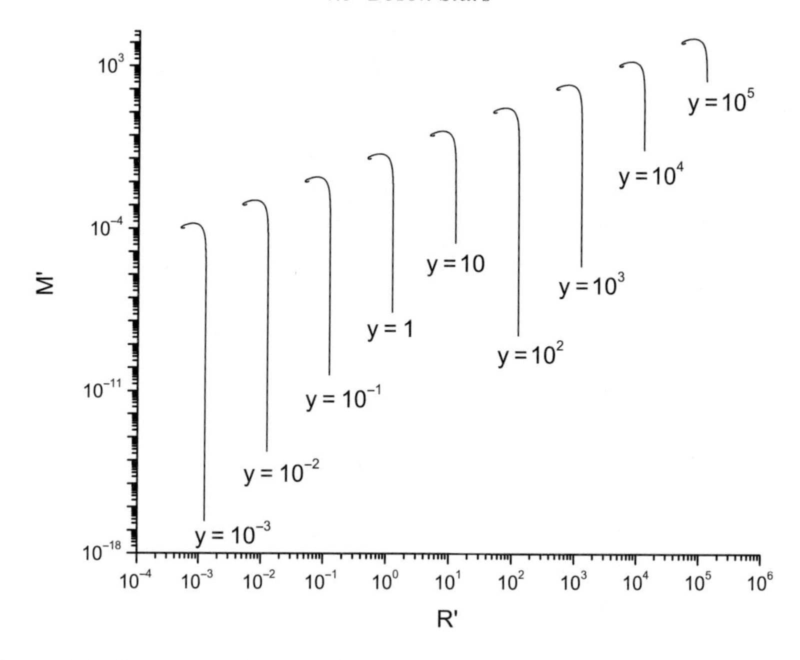

Figure 4.4 The mass–radius relation for compact stars consisting of interacting bosons for different interaction strengths y given in units of the Landau mass and Landau radius. Reprinted with permission from Agnihotri et al. (2009). Copyright (2009) by the American Physical Society

units of the Landau mass M_L and the Landau radius R_L. One sees that the mass–radius relation exhibits a constant radius over a large range of compact star masses. The maximum mass as well as the characteristic radius scale linearly with the interaction scale y. Note that the plot has a double logarithmic scale.

The maximum compactness of boson stars is given by the compactness of the maximum mass configuration

$$C_{\mathrm{max}} = \frac{M'}{R'} = 0.21 \tag{4.118}$$

and is independent of the boson mass and the interaction strength y. Hence, this value would be also applicable for noninteracting boson stars by choosing the interaction mass scale to be the Planck scale, as discussed earlier. The maximum compactness of boson stars is about the same compared to the case of strongly interacting fermion stars, see Eq. (4.102).

The history of boson stars is quite interesting. Wheeler was the first to study self-gravitating spheres for the combined Einstein–Maxwell equations, that is, compact stars of photons, which he dubbed geons (Wheeler, 1955) but the solutions turned out to be unstable. Derrick showed that there exists no stable time-independent solution of a nonlinear equation of motion of a scalar field (Derrick, 1964), also known as Derrick's theorem. Later, Kaup followed by a work of Ruffini and Bonazzola found stable solutions of the Einstein equation together with the Klein–Gordon equations for a fundamental scalar field (Kaup, 1968; Ruffini and Bonazzola, 1969). Derrick's theorem does not apply, as time-dependent solutions were considered for the scalar field. The resulting boson stars were of microscopic sizes and prevented from collapse to a black hole just by Heisenberg's uncertainty principle. This is the case we considered for noninteracting bosons earlier and our estimates for the maximum mass and the radius of the boson star do actually apply. Colpi, Shapiro, and Wasserman introduced a scalar field with self-interactions for the study of boson stars (Colpi et al., 1986). They found that the mass as well as the radius of a boson star with scalar selfinteractions increases drastically. The resulting relations of the maximum mass and the corresponding radius scale with the Landau mass and radius times the square-root of interaction strength parameter, which they showed to be consistent with a scaling analysis, in an exact analogy to the scaling analysis we discussed earlier. Another class of boson stars were introduced by Friedberg, Lee, and Pang for nontopological soliton fields (Friedberg et al., 1987), where the property of the soliton ensured the stability of the boson star.

In the standard model of particle physics, there exists no known stable massive boson. Bosons in the standard model are unstable on the timescale of electroweak interactions, such as the W, Z, and Higgs boson. The only stable bosons, the gluons and the photons, have a vanishing mass and likely form an unstable boson star. There are massive bosons known in strong interaction physics, as the scalar and vector mesons. These mesons are unstable, too, and most of them decay on the timescale of strong interactions. The most stable mesons are pseudoscalar mesons, which decay on the timescale of weak interactions. The charged pions have a mass of $m_\pi \sim 140\,\text{MeV}$ and a lifetime of $\tau \sim 3 \times 10^{-8}$ s for the K_L^0. The kaons have a mass of about $m_K \sim 500\,\text{MeV}$ and lifetimes of up to $\tau \sim 5 \times 10^{-8}$ s. These timescale are still orders of magnitudes too small to be of astrophysical interest for boson stars. There might be stable bosons beyond the standard model, as for example, the hypothetical axion, a pseudoscalar particle that is considered a candidate for cold dark matter or other unknown fundamental scalar fields.

Exercises

(4.1) Estimate the maximum mass and the corresponding radius for hypothetical compact stars made of the following fermions via the Landau mass $M_L = m_{\text{P}}^3/m_f^2$ and the Landau radius $R_L = m_{\text{P}}/m_f^2$:

(a) for a neutrino star with $m_\nu = 1\,\text{eV}$
(b) for a neutralino star with $m_{\nu_s} = 100\,\text{GeV}$.

How does the maximum mass and the corresponding radius change when including interactions? Use the interaction mass scale from QCD and from weak interactions.

(4.2) Show that the mass–radius relation for compact stars made of a free Fermi gas with the mass m and the degeneracy factor g scales as $1/(\sqrt{g}m^2)$. What would be the maximum mass of gas of free nucleons, if protons are uncharged, that is, for a degeneracy factor of $g = 4$ instead of $g = 2$ for the case of free neutrons?

(4.3) Look at an interaction-dominated gas of nucleons of the form $\varepsilon = n^2/m_I^2$. Show that the pressure has the form $p = \varepsilon = n^2/m_I^2$ by using thermodynamic relations.

(4.4) Show that a pressure of the form $P(\mu) = a \cdot \mu^2 - c$, where a and c are constants, results in an EOS of the form $P = \varepsilon - 2c$ independent of the value of a.

(4.5) Derive the thermodynamic relation for the speed of sound squared

$$c_s^2 = \frac{\mathrm{d}P}{\mathrm{d}\varepsilon} = \frac{\mathrm{d}\ln\mu}{\mathrm{d}\ln n} \tag{4.119}$$

using the chain rule and thermodynamic relations for vanishing temperature. Solve the differential equation for a constant speed of sound $s = c_s^2$ and show that in general

$$P(\mu) = a \cdot \mu^{(s+1)/s} - c, \tag{4.120}$$

where a and c are constants. Check that the solution results in an EOS of the form $P = s \cdot (\varepsilon - \varepsilon_{\text{vac}})$, with a vacuum energy $\varepsilon_{\text{vac}} = (1 + 1/s) \cdot c$. Conclude how the corresponding mass–radius relation has to scale with parameters a and c.

5
White Dwarfs

In this chapter we discuss the properties of white dwarfs including a more general discussion of equations of state (EOSs) in the form of polytropes and the resulting properties of compact star configurations. Let us start with a brief review of the history of white dwarfs.

5.1 A Brief History of White Dwarfs

White dwarfs where the first type of compact stars to be observed. It happened by serendipity that the first white dwarf was seen in a telescope. The story goes that Alvan Clark tested his father's new 18-inch refracting telescope in 1862. He pointed it to the brightest star in the sky, Sirius in the constellation Canis Major, which means 'great dog'. Therefore, Sirius is also called the 'dog star'. However, what seemed to be one star for the naked eye or for a small telescope turned out to be double star system in the new powerful telescope of Alvan Clark's father. Besides the bright main star Sirius A, there is a small companion star close to it, Sirius B, which is now called the Pup. In follow-up observations, the mass has been determined to be close to the mass of the Sun and the radius to be similar to the size of Earth, which leads to an average density orders of magnitude larger than that of ordinary stars. In 1915, the spectrum of Sirius B was measured and found to be 'white', implying a surface temperature similar to the considerably larger companion star Sirius A. The extreme compactness of Sirius B could be confirmed by a measurement of the gravitational redshift of spectral lines originating from the surface of the star in 1925. The modern values for Sirius B are $M = 1.018 \pm 0.011 M_\odot$, $R = 0.00864 \pm 0.00012 R_\odot$, and $T_{\text{eff}} = 25.193 \pm 37$ K from observations with the Hubble Space Telescope (Barstow et al., 2005; Bond et al., 2017a). As of writing this book, Sirius B is the closest and brightest known white dwarf.

The theoretical explanation for the unusual properties of this white dwarf came a year later, when Dirac formulated the Fermi–Dirac statistics, the basis for calculating degenerate matter, and Fowler applied it to describe white dwarfs (Fowler, 1926). Ultimately, Chandrasekhar took into account effects from special relativity and derived a maximum mass for white dwarfs (Chandrasekhar, 1931a, 1931b). We will delineate his calculation for the maximum mass of white dwarfs in the following.

5.2 Mass–Radius Relation for Polytropes

First, to understand the concept of a maximum mass for white dwarfs, we consider the hydrostatic equilibrium of spheres with matter described by a polytrope. Recall that a polytrope is defined as

$$P = K \cdot n^{\Gamma}, \tag{5.1}$$

where P is the pressure, n is the number density, Γ is the adiabatic index, and K is a constant. A polytrope can be also defined as

$$P = K' \cdot \rho^{\Gamma}, \tag{5.2}$$

where $\rho = m \cdot n$ is the mass density with the mass of the particle m. The definition is equal to the first one by setting the constant $K' = K/m^{\Gamma}$.

A sphere in hydrostatic equilibrium balances the gravitational pull with the pressure of matter pushing outward. Then the pressure gradient has to fulfill the differential equation

$$\frac{\mathrm{d}P(r)}{\mathrm{d}r} = -G\frac{m_r(r) \cdot \rho(r)}{r^2}, \tag{5.3}$$

which is the equation of hydrostatic equilibrium derived in Section 4.1. Recall that the quantity $m_r(r)$ is the mass contained within sphere of radius r:

$$m_r(r) = \int \mathrm{d}r'^3 \rho(r') = 4\pi \int_0^r \mathrm{d}r' \, r'^2 \rho(r'). \tag{5.4}$$

The total mass of the sphere M is then given by

$$M = m_r(R) = 4\pi \int_0^R \mathrm{d}r' \, r'^2 \rho(r'), \tag{5.5}$$

where R is the radius of the sphere.

Let us make a simple consideration to find a crude solution to the differential equation of hydrostatic equilibrium equation (5.3). We assume that the pressure gradient is approximately given by the differences of the pressure in the core at $r = 0$, the central pressure $P_c = P(0)$, and that at the surface at

$r = R$, which vanishes $P(R) = 0$. For the mass density $\rho(r)$, we adopt a constant average mass density

$$\bar{\rho} = \frac{3M}{4\pi R^3}. \tag{5.6}$$

Then by setting $r = R$ on the right-hand side of Eq. (5.3), we get the approximate relation

$$\frac{dP}{dr} \approx \frac{\Delta P}{\Delta R} \approx \frac{P(0) - P(R)}{-R} = -\frac{P_c}{R} \propto -G\frac{M\bar{\rho}}{R^2}. \tag{5.7}$$

Using the expression for the average mass density, $\bar{\rho}$ it follows

$$P_c \propto G\frac{M}{R} \cdot \frac{M}{R^3} = G\frac{M^2}{R^4}. \tag{5.8}$$

This relation is well known in stellar astrophysics for relating the central pressure to the mass and radius for ordinary stars in hydrostatic equilibrium.

Next we use the EOS for polytropes in the form $P \propto \rho^\Gamma$ and insert it to get rid of the central pressure:

$$P_c = K' \cdot \bar{\rho}^\Gamma \propto K' \cdot \left(\frac{M}{R^3}\right)^\Gamma \propto G\frac{M^2}{R^4}. \tag{5.9}$$

The left-hand side stands for the pressure gradient of matter and is proportional to $R^{-3\Gamma}$. The right-hand side stands for the pull of gravity and is proportional to R^4. The gravitational side will win for decreasing radii if $\Gamma < 4/3$. Hence, matter described by a polytrope with $\Gamma < 4/3$ will be unstable and will collapse to a black hole.

Rearranging Eq. (5.9), one arrives at an equation relating the mass and the radius of the sphere in hydrostatic equilibrium for polytropic EOSs

$$M^{2-\Gamma} \cdot R^{3\Gamma-4} \propto \frac{K'}{G} = \text{const.} \tag{5.10}$$

This equation has several interesting features that we are going to discuss now for a given polytropic power Γ.

- $\Gamma < 4/3$: The sphere is unstable and will collapse to a black hole, as outlined earlier.
- $\Gamma = 4/3$: The limiting case of stability. The mass–radius relation reduces to a constant mass

  $$M = \text{const.},$$

  which is independent of the radius R. For $\Gamma = 4/3$, the factor for the polytropic EOS is proportional to $m^{-4/3}$, the mass of the particle that for white dwarfs is the nucleon mass, see Eq. (3.87). Setting the gravitational constant $G = 1/M_P^2$ Eq. (5.10) gives a constant mass of

$$M \propto \left(\frac{K'}{G}\right)^{3/2} \propto \frac{M_P^3}{m^2},$$

which is nothing less than the famous Chandrasekhar mass.

- $\Gamma = 5/3$: This corresponds to a polytrope describing the nonrelativistic limit of a Fermi gas. The mass–radius relation results in

$$M \cdot R^3 = \text{const.}$$

Note that the mass of the star increases with *decreasing* radii.

- $\Gamma = 2$: We encountered such a polytrope when discussing an interaction dominated EOS, see Eq. (4.92). The mass–radius relation is simply

$$R = \text{const.},$$

that is, a constant radius independent of the mass of the star.

- $\Gamma = 2 + \delta$: Here δ is a positive quantity, $\delta > 0$. The mass–radius relation changes its character as now

$$M \cdot R^{-(2/\delta+3)} = \text{const.},$$

so that the mass of the star increases with *increasing* radius. One can turn the statement around: An increasing mass of the star with the radius indicates that the pressure increases more rapidly than with the density squared.

- $\Gamma \to \infty$: This is the limit for an incompressible fluid with a constant mean mass density $\bar{\rho} \propto M/R^3$. That the mean mass density is constant can be also seen from the mass–radius relation, which is the limit of sending $\delta \to \infty$ in the previous case:

$$M \cdot R^{-3} = \text{const.}$$

The mass of the star increases with the radius cubed. This case corresponds to spheres that are stable by themselves, gravity is not needed to ensure stability. Such stars are called selfbound stars.

5.3 Lane–Emden Equation

Let us take a look at the full solution to hydrostatic equilibrium using Eq. (5.3) in spherical symmetry. Combining with the equation for mass conservation, Eq. (5.4), one arrives at the following combined integro-differential equation:

$$\frac{\mathrm{d}P(r)}{\mathrm{d}r} = -G\frac{\rho(r)}{r^2} \cdot 4\pi \int_0^r \mathrm{d}r'\, r'^2 \rho(r'). \tag{5.11}$$

In order to have just a differential equation, we arrange terms to the left and take the derivative with respect to the radius r of both sides:

$$\frac{\mathrm{d}}{\mathrm{d}r}\left(\frac{r^2}{\rho(r)}\frac{\mathrm{d}P(r)}{\mathrm{d}r}\right) = -G\cdot 4\pi r^2 \rho(r). \tag{5.12}$$

The equation now looks very similar to a Poisson equation in spherical symmetry

$$\frac{1}{r^2}\frac{\mathrm{d}}{\mathrm{d}r}\left(r^2\frac{\mathrm{d}\phi(r)}{\mathrm{d}r}\right) = -4\pi G\cdot\rho(r) \tag{5.13}$$

by setting for the potential the gravitational potential ϕ:

$$\frac{\mathrm{d}\phi(r)}{\mathrm{d}r} = -G\frac{m_r(r)}{r^2} = \frac{1}{\rho(r)}\frac{\mathrm{d}P(r)}{\mathrm{d}r}, \tag{5.14}$$

which is just the equation of hydrostatic equilibrium written in a different form. In the following we focus on polytropic EOSs. We want to recast Eq. (5.12) in a dimensionless form in order to study scaling solutions, as done in Section 4.3. Let us introduce a dimensionless quantity for the radius ξ first

$$r = \xi\cdot r_c, \tag{5.15}$$

where the dimensionful quantity r_c is a characteristic scale for the radius. Using the new dimensionless radial coordinate and a polytropic EOS, one arrives at

$$\frac{K'\Gamma}{r_c^2\xi^2}\cdot\frac{\mathrm{d}}{\mathrm{d}\xi}\left(\xi^2\rho(\xi)^{\Gamma-2}\frac{\mathrm{d}\rho(\xi)}{\mathrm{d}\xi}\right) = -4\pi G\cdot\rho(\xi). \tag{5.16}$$

Now we introduce a dimensionless quantity for the density θ such that the expression in the brackets simplifies

$$\rho(\xi) = \theta(\xi)^{1/(\Gamma-1)}\cdot\rho_c = \theta(\xi)^n\cdot\rho_c, \tag{5.17}$$

where ρ_c is a characteristic mass density scale. Here we introduce for convenience the polytropic index n defined as:

$$n = \frac{1}{\Gamma-1} \quad\rightarrow\quad \Gamma = 1+\frac{1}{n}. \tag{5.18}$$

Then,

$$\begin{aligned}\rho(\xi)^{\Gamma-2}\frac{\mathrm{d}\rho(\xi)}{\mathrm{d}\xi} &= \rho(\xi)^{1/n-1}n\rho_c\,\theta(\xi)^{n-1}\frac{\mathrm{d}\theta(\xi)}{\mathrm{d}\xi}\\ &= \theta(\xi)^{1-n}\rho_c^{1/n-1}n\rho_c\,\theta(\xi)^{n-1}\frac{\mathrm{d}\theta(\xi)}{\mathrm{d}\xi}\\ &= n\rho_c^{1/n}\frac{\mathrm{d}\theta(\xi)}{\mathrm{d}\xi}.\end{aligned} \tag{5.19}$$

The equation for hydrostatic equilibrium for polytropes reads now

$$\frac{K'(n+1)\rho_c^{1/n-1}}{4\pi G r_c^2 \xi^2} \cdot \frac{\mathrm{d}}{\mathrm{d}\xi}\left(\xi^2 \frac{\mathrm{d}\theta(\xi)}{\mathrm{d}\xi}\right) = -\theta(\xi)^n. \tag{5.20}$$

One sees that one can get rid of all dimensionful quantities by defining the characteristic radius to be

$$r_c = \left(\frac{K'(n+1)\rho_c^{1/n-1}}{4\pi G}\right)^{1/2}. \tag{5.21}$$

Finally, we arrive at the Lane–Emden equation

$$\frac{1}{\xi^2} \cdot \frac{\mathrm{d}}{\mathrm{d}\xi}\left(\xi^2 \frac{\mathrm{d}\theta(\xi)}{\mathrm{d}\xi}\right) = -\theta(\xi)^n, \tag{5.22}$$

which is dimensionless and depends only on the polytropic index n. It is a second-order differential equation, which can be solved for the scaling function $\theta(\xi)$ by specifying two boundary conditions. First, the density shall vanish at the surface of the sphere

$$\theta_n(\xi = \xi_n) = 0, \tag{5.23}$$

where ξ_n is the dimensionless radius of the sphere for a given polytropic index n. Second the derivative of the scaling function at the center shall vanish

$$\theta'(\xi = 0) = \left.\frac{\mathrm{d}\theta(\xi)}{\mathrm{d}\xi}\right|_{\xi=0} = 0. \tag{5.24}$$

This choice can be motivated by looking at the equation of hydrostatic equilibrium, Eq. (5.3), and taking the limit $r \to 0$. Then one sees that the pressure gradient vanishes at the center so that for polytropes the mass density gradient has to vanish too. In addition, we normalize the scaling function such that the characteristic mass density is the one at the center of the sphere, that is, we set $\theta(0) = 1$.

There are three analytic solutions known to the Lane–Emden equation, for $n = 0, 1$, and 5. For the case $n = 0$, the solution is

$$\theta_0(\xi) = 1 - \frac{\xi^2}{6} \qquad \xi_0 = \sqrt{6}. \tag{5.25}$$

We note that this case corresponds to the case $\Gamma \to \infty$. Hence, the solution describes a sphere with an incompressible fluid and a constant mass density. The solution for the case $n = 1$ reads

$$\theta_1(\xi) = \frac{\sin\xi}{\xi} \qquad \xi_1 = \pi, \tag{5.26}$$

which is the case $\Gamma = 2$. This is the case for an interaction-dominated EOS, where the pressure is proportional to the density squared (see the discussion in Section 4.4). The case $n = 5$ has the solution

$$\theta_5(\xi) = \left(\frac{1}{1+\xi^2/3}\right)^{1/2} \qquad \xi_5 \to \infty, \tag{5.27}$$

where the radius of the sphere is infinite. However, the mass of the sphere is finite. This last case is a limiting case, as for $n > 5$, also the mass increases with the radius without any bound.

5.4 Chandrasekhar Mass

We turn our discussion now to finding solutions to the Lane–Emden equation that are applicable for white dwarfs. White dwarf matter consists of a lattice of nuclei surrounded by a degenerate gas of electrons balancing the charge of the nuclei. We recall the discussion of free fermion gases of Section 3.5, where we found that a degenerate gas can be described by polytropic EOSs. In the low density limit, the degenerate gas is nonrelativistic and has a polytropic EOS with a power of $\Gamma = 5/3$. In the high density limit, the degenerate gas is relativistic and the polytrope turns to one with a power of $\Gamma = 4/3$. Hence, for describing white dwarf matter, the polytropic indices corresponding to a degenerate gas of fermions in the nonrelativistic ($\Gamma = 5/3 \to n = 3/2$) and in the relativistic limit ($\Gamma = 4/3 \to n = 3$) are the interesting ones. However, those cases have to be considered by solving the Lane–Emden equation numerically, as no analytic solution is known. Fortunately, the numerical solutions for the Lane–Emden equation for various values of the polytropic index n are well known and have been tabularized, even well before the advent of computers (see e.g., Chandrasekhar, 1939). The solutions can be given in terms of the radius ξ_n and the derivative of the scaling function at the surface $\theta'(\xi_n) = (d\theta(\xi)/d\xi)|_{\xi_n}$. The dimensionful physical quantities are determined by fixing the coefficient of the polytrope K', which sets the overall scale of the physical solution (besides the gravitational constant G). The solutions for the total mass M and the radius R of the sphere can be written down analytically and are parametrically given in terms of the central mass density ρ_c. From the definition of ξ and r_c, the radius of the sphere R is then

$$R = \xi_n \cdot r_c = \xi_n \cdot \left(\frac{K'(n+1)}{4\pi G}\right)^{1/2} \cdot \rho_c^{(1-n)/2n}. \tag{5.28}$$

Using the definition of θ and ρ_c, the total mass of the sphere can be expressed as an integral over ξ:

$$M = 4\pi \int_0^R \mathrm{d}r\ r^2 \rho(r) \tag{5.29}$$

$$= 4\pi r_n^3 \rho_c \int_0^{\xi_n} \mathrm{d}\xi\ \xi^2 \theta(\xi)^n. \tag{5.30}$$

The integral can be solved by using the Lane–Emden equation, Eq. (5.22):

$$M = -4\pi r_n^3 \rho_c \int_0^{\xi_n} \mathrm{d}\xi\ \frac{\mathrm{d}}{\mathrm{d}\xi}\left(\xi^2 \frac{\mathrm{d}\theta}{\mathrm{d}\xi}\right) \tag{5.31}$$

$$= -4\pi r_n^3 \rho_c \cdot \xi_n^2 \left.\frac{\mathrm{d}\theta}{\mathrm{d}\xi}\right|_{\xi_n} \tag{5.32}$$

$$= -4\pi \left(\frac{K'(n+1)}{4\pi G}\right)^{3/2} \xi_n^2 \cdot \theta'(\xi_n) \cdot \rho_c^{(3-n)/2n}. \tag{5.33}$$

Eqs. (5.28) and (5.33) determine the mass–radius relations of stars with a polytropic EOS of index n in terms of the central mass density ρ_c. One can eliminate the central mass density ρ_c to arrive at a mass–radius relation of:

$$M^{n-1} \cdot R^{3-n} = -\frac{1}{4\pi}\left(\frac{K'(n+1)}{G}\right)^n \xi_n^{n+1}\left(\theta'(\xi_n)\right)^{n-1} = \text{const.} \tag{5.34}$$

In terms of the power of the polytropes Γ, the mass–radius relation reads

$$M^{2-\Gamma} \cdot R^{3\Gamma-4} = -\left(\frac{1}{4\pi}\right)^{\Gamma-1} \frac{K'(n+1)}{G} \xi_n^\Gamma \left(\theta'(\xi_n)\right)^{2-\Gamma} = \text{const.}, \tag{5.35}$$

which is exactly the same mass–radius relation found by using simple dimensional arguments for the equation of hydrostatic equilibrium, Eq. (5.10). But now we are in the position to give the exact numbers for the constant appearing on the right-hand side of the mass–radius relation. Of particular interest is the case $n = 3$ or $\Gamma = 4/3$, where the dependence on the radius R drops out. The mass–radius relation results in a constant mass given by

$$M = -\frac{1}{(4\pi)^{1/2}}\left(\frac{4K'}{G}\right)^{3/2} \xi_3^2\ \theta'(\xi_3) = \text{const.} \tag{5.36}$$

For white dwarfs, we know the coefficient in front of the polytrope in the relativistic limit, see Eq. (3.87):

$$K' = \frac{1}{4}\left(\frac{6\pi^2}{g_e}\right)^{1/3}\left(\frac{Z}{A \cdot m_N}\right)^{4/3}. \tag{5.37}$$

Table 5.1 *The solutions to the Lane–Emden equation for some choices of a given polytropic index n.*

n	Γ	ξ_n	$-\xi_n^2\theta'(\xi_n)$
0	∞	$\sqrt{6}$	$2\sqrt{6}$
1	2	π	π
3/2	5/3	3.65375	2.71406
3	4/3	6.89685	2.01824
5	6/5	∞	$\sqrt{3}$

Setting for the degeneracy factor of the electron $g_e = 2$ for its two spin degrees of freedom, the limiting mass is given by

$$M_{\text{Chandra}} = -\frac{\sqrt{3\pi}}{2}\xi_3^2\,\theta'(\xi_3)\left(\frac{Z}{A\cdot m_N}\right)^2\left(\frac{\hbar c}{G}\right)^{3/2}, \tag{5.38}$$

where we inserted back the physical constants $\hbar$ and c. This mass is the expression found by Chandrasekhar for the maximum mass of a white dwarf (Chandrasekhar, 1931b). It is interesting to note that the Chandrasekhar mass depends on three physical constants of nature: the gravitational constant G, the velocity of light c, and the Planck constant $\hbar$ combining, gravity, special relativity, and quantum mechanics, respectively. The last term in the expression of the Chandrasekhar mass is just the Planck mass cubed, so that the Chandrasekhar mass reads

$$M_{\text{Chandra}} = -\frac{\sqrt{3\pi}}{2}\xi_3^2\,\theta'(\xi_3)\left(\frac{Z}{A}\right)^2\frac{m_{\text{P}}^3}{m_N^2}. \tag{5.39}$$

Besides the numerical prefactor, the dependencies on the physical quantities are exactly the same as found when discussing the maximum mass of white dwarfs using Landau's argument, see Eq. (4.38). The value of the Chandrasekhar mass can be calculated by using the numerical values of the solution of the Lane–Emden equation for $n = 3$ from Table 5.1. As the white dwarfs consists of a lattice of nuclei surrounded by a gas of electrons, it is more appropriate to use the mass per mass number of the nucleus present in a white dwarf than the nucleon mass, which is smaller due to the binding energy of the nucleus. From stellar evolution theory, it is reasonable to assume that a white dwarfs consists of ^{4}He, ^{12}C,[1] or ^{16}O. Let us therefore take as a reasonable mass scale the atomic mass unit m_u, which is defined as the mass per atomic number of ^{12}C: $m_u = m(^{12}\text{C})/12 = 931.49\,\text{MeV}$. The numerical value of the Chandrasekhar mass can then be conveniently expressed in the form

[1] The white dwarf would be like a diamond in the sky.

$$M_{\text{Chandra}} = 1.456 M_\odot \left(\frac{2Z}{A}\right)^2 \left(\frac{m_u}{m_A/A}\right)^2, \tag{5.40}$$

where m_A/A is the mass per atomic mass number A of the nucleus present in the white dwarf (using the proton mass instead of m_u results in a numerical prefactor of $1.435 M_\odot$). Note that for all the nuclei listed, the charge-to-mass ratio is always $Z/A = 1/2$. In fact, for light nuclei the most stable isotope has always that charge-to-mass number ratio. It is different for the most stable nucleus ^{56}Fe, which has a ratio of $Z/A = 26/56 = 0.464$, so the limiting mass for a white dwarf consisting of iron would be considerably smaller, $M_{\text{Chandra}}(^{56}\text{Fe}) = 1.258 M_\odot$. However, such a white dwarf cannot be formed in standard stellar evolution theory. If a white dwarf accretes matter and collapses instead of exploding in a supernova of type Ia, a hypothetical iron white dwarf could be formed. This mechanism is called an accretion-induced collapse. An iron white dwarf has not been observed so far.

5.5 Coulomb Corrections

As nuclei are charged and their charges are compensated by electrons, there are separate charges present in white dwarfs locally. The most stable configuration in order to minimize the Coulomb energy is a solid lattice structure. We look at a Wigner–Seitz cell with a radius r_0 around a nucleus and estimate the contribution from Coulomb interactions to the EOS. The charge due to the nucleus with the mass number A is given by its proton number or charge Z. The number of electrons has to balance that charge so that the number density of electrons n_e times the volume of the sphere of radius r_0 has to be equal to the charge of the nucleus

$$Z = \frac{4\pi r_0^3}{3} \cdot n_e. \tag{5.41}$$

The Coulomb self energy of a charged sphere due to the electrons alone is given by the integral

$$E_{ee} = \int_0^{r_0} \mathrm{d}q \frac{q(r)}{r} = \frac{3}{5} \frac{Z^2 e^2}{r_0}, \tag{5.42}$$

where we used that

$$q(r) = Ze \frac{r^3}{r_0^3} \tag{5.43}$$

is the charge within the radius r of the sphere with a radius r_0. The interaction between the electrons and the ion gives a contribution of the form

$$E_{\text{ion}} = \int_o^{r_0} \mathrm{d}q \frac{Ze}{r} = -\frac{3}{2} \frac{Z^2 e^2}{r_0}, \tag{5.44}$$

where the charge of the ion $q = Ze$ has been inserted instead of the charge of the electrons within the radius r. The sum of the two contributions combine to

$$E_{\text{Coul}} = E_{ee} + E_{\text{ion}} = -\frac{9}{10}\frac{Z^2e^2}{r_0}. \tag{5.45}$$

The prefactor $9/10$ is very close to the numerical value for a body-centered cubic lattice, which is 0.90981. The Coulomb energy from the lattice per electron can be expressed in terms of the electron number density as

$$\frac{E_{\text{Coul}}}{Z} = -\frac{9}{10}Z^{2/3}e^2\left(\frac{4\pi}{3}n_e\right)^{1/3}, \tag{5.46}$$

where we replaced the radius of the Wigner–Seitz cell r_0 using the charge neutrality condition, Eq. (5.41). The pressure can be calculated by using the thermodynamic relation, see Eq. (3.69),

$$P_{\text{Coul}} = n_e^2\frac{d(E_{\text{Coul}}/Z)}{dn_e} = -\frac{3}{10}Z^{2/3}e^2\left(\frac{4\pi}{3}\right)^{1/3} n_e^{4/3}. \tag{5.47}$$

We note that the contribution from the lattice to the pressure has the form of a polytrope with $\Gamma = 4/3$. We also note that the pressure is negative, as is the contribution from the lattice to the energy density. The EOS from the Coulomb contribution reads

$$P_{\text{Coul}} = \frac{1}{3}\varepsilon_{\text{Coul}}, \tag{5.48}$$

like that of an ultrarelativistic free gas of particles. We can compare the Coulomb contribution to the pressure to that of the Fermi pressure of electrons. In the relativistic limit, the contribution of the Fermi pressure grows also as $n^{4/3}$, see Eq. (3.85). However, the Coulomb contribution is suppressed compared to the Fermi pressure by one to two orders of magnitudes as

$$Z^{2/3}e^2 = \alpha \cdot Z^{2/3} = 0.01 \text{ (for He) to } 0.03 \text{ (for O)}, \tag{5.49}$$

where $\alpha = e^2 \approx 1/137$ is the fine structure constant (in natural units). There is, however, a critical density at low densities, where the Coulomb contribution can be of equal magnitude compared to the Fermi pressure. In the nonrelativistic limit, the Fermi pressure is proportional to $n^{5/3}/m_e$. The two contributions are balancing each other for an electron density of:

$$n_{e,\text{crit}} = \frac{Z^2\alpha^3}{2\pi^3}m_e^3, \tag{5.50}$$

which gives a critical mass density of

$$\rho_{\text{crit}} = m_u \frac{A}{Z} \cdot \frac{Z^2 \alpha^3}{2\pi^3} m_e^3 = (5 - 70) \cdot \text{g cm}^{-3} \tag{5.51}$$

for helium to oxygen nuclei. For a white dwarf, these are extremely low densities that are more common for planets. The total pressure vanishes at these critical densities, so that the compact star is bound by itself. It consists of a solid structure with a constant density. The mass–radius relation will then change to $M \propto R^3$ for such low densities – for planets.

5.6 Structure of White Dwarfs

We are now in the position to discuss the generic structure of white dwarfs. Let us start from the surface of the white dwarf and then continue toward the center by increasing the mass density. The outermost layer consists of atoms, the electrons are still bound to the atomic nuclei. This is the so-called atmosphere of white dwarfs. With increasing density, the electron density rises and so does the electron Fermi momentum and the electron Fermi energy, that is, the kinetic energy of the electrons. When the electron kinetic energy equals the atomic binding energy of the electron, all electrons can be stripped from the nucleus. The electrons will form a Fermi gas, the atomic nuclei a Coulomb lattice. This condition marks the end of the atmosphere and the beginning of the solid interior of the white dwarf. We can estimate the critical mass density up to which the solid crust turns into the atmosphere for a white dwarf. Let us consider nuclei with mass number A and charge Z being immersed in a gas of electrons. The electron number density for a free Fermi gas is given by

$$n_e = \frac{k_f^3}{3\pi^2}, \tag{5.52}$$

where k_F is the Fermi momentum of the electrons. By charge neutrality, the number density of nucleons is given by

$$n_N = \frac{A}{Z} n_p = \frac{A}{Z} n_e, \tag{5.53}$$

so that the mass density can be written as

$$\rho = m_u \cdot n_N = m_u \frac{A}{Z} \frac{k_F^3}{3\pi^2}, \tag{5.54}$$

where m_u is the atomic mass unit. The electron binding energy for hydrogen is 13.6 eV. For heavier systems, the binding energy of the last bound electron is the relevant one for a fully ionized system. The binding energy of atoms with just one electron scales as the charge of the atomic nucleus squared. The critical condition for the transition from the atmosphere to the solid crusts correspond to set the Fermi energy equal to the binding energy of the last electron:

$$E_F = \frac{k_F^2}{2m_e} = E_b(Z) = 13.6\ \text{eV} \cdot Z^2. \tag{5.55}$$

For carbon with $Z = 6$, the corresponding binding energy is 490 eV, for oxygen with $Z = 8$, it is 870 eV, and for iron with $Z = 26$, 9.19 keV. The critical mass density then reads

$$\rho_c = \frac{2^{3/2}}{\pi^2} A \cdot Z^2 m_u m_e^{3/2} E_b(Z = 1)^{3/2} = 0.378\ \text{g} \cdot \text{cm}^{-3} A \cdot Z^2, \tag{5.56}$$

which is $\rho_c = 163\,\text{g}\,\text{cm}^{-3}$ for ^{12}C, $\rho_c = 387\,\text{g}\,\text{cm}^{-3}$ for ^{16}O, and $\rho_c = 1.43 \times 10^4\,\text{g}\,\text{cm}^{-3}$ for ^{56}Fe.

For higher densities, the electrons will become eventually relativistic. In the relativistic limit, the mass density can be expressed as a function of the Fermi energy of the electrons by setting $E_F = k_F = m_e$ in Eq. (5.54):

$$\rho_{\text{rel}} = m_u \cdot n_N = \frac{m_u A}{3\pi^2 Z} m_e^3 = 9.74 \times 10^5\ \text{g} \cdot \text{cm}^{-3} \cdot \frac{A}{Z}. \tag{5.57}$$

The critical density for the onset of relativistic electrons depends then on the charge-to-mass ratio of the nuclei in the white dwarf. The critical density is then $\rho_{\text{rel}} = 1.95 \times 10^6\,\text{g}\,\text{cm}^{-3}$ for nuclei ^{4}He, ^{12}C, and ^{16}O, and $\rho_{\text{rel}} = 2.10 \times 10^6\,\text{g}\,\text{cm}^{-3}$ for ^{56}Fe. These critical densities are important for the stability of the white dwarf as we know. Note that the composition has not changed up to this density.

We can estimate that when the electron density becomes so large, one can expect to start having a change in the composition. For high electron densities, it will become eventually energetically more favorable to transform electrons with protons to neutrons, when the kinetic energy of electrons reaches the mass difference of neutrons and protons. The change in the composition corresponds then to a transformation of nuclei to more neutron-rich nuclei. The critical condition is reached when the Fermi energy of electrons equals $E_F = m_n - m_p = 1.29\,\text{MeV}$. The corresponding critical mass density corresponds to

$$\rho_n = \frac{m_u \cdot A}{3\pi^2 Z}(m_n - m_p)^3 = 1.57 \times 10^7\ \text{g} \cdot \text{cm}^{-3} \cdot \frac{A}{Z}. \tag{5.58}$$

The critical neutronization density is then $\rho_n = 3.13 \times 10^7\,\text{g}\,\text{cm}^{-3}$ for nuclei ^{4}He, ^{12}C, and ^{16}O, and $\rho_n = 3.37 \times 10^7\,\text{g}\,\text{cm}^{-3}$ for ^{56}Fe, in both cases an order of magnitude higher than the critical density for relativistic electrons. Hence, only the most massive white dwarfs can be expected to have a change of the composition in their cores. Note that while the average mass density of a white dwarf is about

$$\bar{\rho} = \frac{3M}{4\pi R^3} = 3.82 \times 10^6 \text{ g cm}^{-3} \left(\frac{M}{M_\odot}\right)\left(\frac{5 \times 10^3 \text{ km}}{R}\right)^3 \tag{5.59}$$

for typical values of $M = 1M_\odot$ and $R = 5{,}000\,\text{km}$, the central mass density can be much larger. In realistic EOSs, the central mass density of the most massive white dwarf can reach values of several 10^9 g cm^{-3}, as we will see in Chapter 7 (see e.g., Fig. 7.9).

One can also estimate the thickness of the different layers of a white dwarf without actually performing a numerical simulation. Take a look again at the equation of hydrostatic equilibrium

$$\frac{\mathrm{d}P}{\mathrm{d}r} = -G\frac{m_r \rho}{r^2}, \tag{5.60}$$

where $m_r = 4\pi r^2 \rho$ is the mass contained within a sphere of radius r. At the inner boundary of a small layer of a star, say here for the atmosphere of a white dwarf, the mass m_r is approximately the total mass of the star $m_r \approx M$ and the radius the total radius of the star $r \approx R$. The pressure gradient can be estimated by the pressure at the inner boundary of the layer $\Delta P = -P_c$ over the thickness of the layer ΔR:

$$\frac{\mathrm{d}P}{\mathrm{d}r} \approx -\frac{P_c}{\Delta R} \approx -G\frac{M\rho_c}{R^2}, \tag{5.61}$$

where ρ_c is the mass density at the inner boundary of the layer. The relative thickness of the layer relative to the radius of the star can then be estimated by the expression

$$\frac{\Delta R}{R} \approx \frac{R}{GM}\frac{P_c}{\rho_c}. \tag{5.62}$$

The thickness of the outermost layer of a star in hydrostatic equilibrium is given by the inverse of the compactness of the star and the ratio of the pressure to the energy density at the inner boundary of the layer. Let us now insert the EOS in the form of a nonrelativistic electron gas with a degeneracy factor of $g_e = 2$. The pressure of electrons reads

$$P^{(\text{nr})} = \frac{1}{15\pi^2}\frac{k_F^5}{m_e}, \tag{5.63}$$

where k_F is the Fermi momentum of the electrons. The mass density is given by

$$\rho = m_u \cdot n_N = m_u \cdot n_e \frac{A}{Z} = m_u \cdot \frac{k_F^3}{3\pi^2} \cdot \frac{A}{Z}, \tag{5.64}$$

where we used the condition of charge neutrality to replace the number density of nucleons n_n with the number density of electrons n_e. The ratio of the pressure to the energy density turns out to be simply

$$\frac{P^{(\mathrm{nr})}}{\rho} = \frac{k_F^2}{5m_e m_u} \cdot \frac{A}{Z} = \frac{2E_F}{5m_u} \cdot \frac{A}{Z}, \tag{5.65}$$

where E_F is the Fermi energy of the electrons. So we arrive at the remarkable result that the thickness of the outer layer is basically determined by the ratio of the electron Fermi energy over the nucleon mass times the inverse compactness of the star:

$$\frac{\Delta R}{R} \approx \frac{R}{GM} \cdot \frac{2E_F}{5m_u} \cdot \frac{A}{Z} \tag{5.66}$$

modulo the charge-to-mass ratio of the nuclei. For a typical compactness for white dwarfs of $C = GM/R \approx 10^{-4}$ and a Fermi energy of $E_F = 13.6\,\mathrm{eV}\cdot Z^2$, we estimate a thickness of the atmosphere of a white dwarf of

$$\frac{\Delta R}{R} \approx 5.84 \times 10^{-5} \left(\frac{10^{-4}}{C}\right) \cdot Z \cdot A. \tag{5.67}$$

A hydrogen layer would have a thickness of about 300 m for a white dwarf with a radius of $R = 5{,}000\,\mathrm{km}$. For oxygen, one finds a thickness of about 40 km, for iron, a thickness of more than 400 km.

We can use the formula also for neutron stars just by changing the value of the compactness from $C \approx 10^{-4}$ to $C \approx 0.2$. Then, the thickness of an atmosphere layer of a neutron star will be only

$$\frac{\Delta R}{R} \approx 2.92 \times 10^{-8} \left(\frac{0.2}{C}\right) \cdot Z \cdot A, \tag{5.68}$$

that is, from a tiny 0.3 mm layer for hydrogen and 4 cm one for oxygen up to about 40 cm for iron for a 10-km neutron star.

5.7 Thermal Effects for White Dwarfs

So far we have considered white dwarf matter at vanishing temperature. We assumed that white dwarf matter is well described by a degenerate gas of electrons within a lattice of nuclei. However, white dwarfs are the relics of ordinary stars that have core temperatures of several million kelvin. In order to be able to treat white dwarf matter by ignoring effects from a nonvanishing temperature, one needs to quantify a critical condition, where one can expect that effects from degeneracy are dominating thermal effects.

The key point is to compare the corresponding energy scales, that is, the kinetic energy from the Fermi energy for a degenerate Fermi gas with the thermal energy. In the following we consider a free gas of fermions that is appropriate for treating white dwarf matter. As we will see, for neutron star matter this is not the case anymore.

We will use natural units in the following also for temperatures with the convention that the Boltzmann constant is set to $k_B = 1$. Temperatures are then expressed as energies, in MeV, and the conversion of the units from kelvin to MeV is given by

$$k_B = 8.6173303(50) \times 10^{-5}\ \mathrm{eV\ K}^{-1}, \tag{5.69}$$

as taken from the particle data group (Tanabashi et al., 2018). The thermal energy in natural units is then simply

$$E_{\mathrm{th}} = T. \tag{5.70}$$

An energy scale of $E = 1$ MeV corresponds to a temperature of

$$T = 1.16 \times 10^{10}\ \mathrm{K} \cdot \left(\frac{E_{\mathrm{th}}}{1\ \mathrm{MeV}} \right), \tag{5.71}$$

which is huge even on astrophysical scales. The core temperature of our Sun is about $T = 1.5 \times 10^7$ K, which corresponds to an energy scale of about $E_{\mathrm{th}} = 1.3$ keV. Measured surface temperatures of white dwarfs are typically about 10,000 K corresponding to an energy scale of about 1 eV. So, for white dwarfs we can safely assume that the thermal energy is well below the electron mass of 0.511 MeV so that electrons can be treated nonrelativistically. Actually, there are only a few of astrophysical systems where one knows that the scale of the temperature exceeds the electron mass, which are, for example, core-collapse supernovae, proto-neutron stars, and neutron star mergers.

So let us consider the kinetic energy from the Fermi energy E_F of a free gas of electrons in the nonrelativistic limit

$$E_F^{(\mathrm{kin})} = \sqrt{k_F^2 + m_e^2} - m_e \approx \frac{k_F^2}{2m_e}, \tag{5.72}$$

which just reflects the energy–momentum relation of nonrelativistic particles. The Fermi momentum k_F is related to the number density n_e for a degenerate electron gas by

$$n_e = \frac{g_e}{(2\pi)^3} \int \mathrm{d}^3k = \frac{g_e}{6\pi^2} k_F^3, \tag{5.73}$$

that is, the integral over a sphere in momentum space up to the Fermi momentum k_F. For electrons, $g_e = 2$, which we adopt in the following. The Fermi momentum is then related to the number density by

$$k_F = \left(3\pi^2 n_e\right)^{1/3}. \tag{5.74}$$

So we can express the kinetic energy in terms of the number density as

$$E_F^{(\text{kin})} = \frac{(3\pi^2)^{2/3}}{2m_e} \cdot n_e^{2/3} = \frac{(3\pi^2)^{2/3}}{2m_e m_u^{2/3}} \cdot \left(\frac{Z}{A}\right)^{2/3} \cdot \rho^{2/3}, \tag{5.75}$$

where we replaced the number density of electrons with the mass density $\rho = (A/Z)m_u n_e$. By comparing with the thermal energy, we arrive at the critical condition for degeneracy:

$$\rho_{\text{crit}} = \frac{2^{3/2} m_u m_e^{3/2}}{3\pi^2} \left(\frac{A}{Z}\right) \cdot T_{\text{crit}}^{3/2} = 0.0121 \text{ g cm}^{-3} \left(\frac{A}{2Z}\right) \cdot \left(\frac{T}{10^4 \text{ K}}\right)^{3/2}, \tag{5.76}$$

where the thermal energy equals the kinetic energy of a degenerate gas of electrons. We see that thermal effects can be safely ignored for a density larger than $1.21 \times 10^{-2}\,\text{g cm}^{-3}$ at the surface of a white dwarf for a surface temperature of $T = 10{,}000\,\text{K}$. As we have seen before, that density is well within the atmosphere of a white dwarf, so that the electron gas can be very well approximated by a degenerate Fermi gas at vanishing temperature inside white dwarfs.

There is another way to estimate the critical condition for degeneracy, by looking at the pressure for nondegenerate gas in comparison to that for a degenerate one. The pressure of a nondegenerate electron gas is that of an ideal gas

$$P_{nd} = n_e \cdot T. \tag{5.77}$$

For a nonrelativistic degenerate electron gas, the pressure is given by Eq. (5.63)

$$P^{(\text{nr})} = \frac{1}{15\pi^2} \frac{k_F^5}{m_e} = \frac{(3\pi^2)^{2/3}}{5m_e} n_e^{5/3}, \tag{5.78}$$

where we used the relation for the Fermi momentum $k_F = (3\pi^2 n_e)^{1/3}$. The two pressures are equal for a critical temperature of

$$T_{\text{crit}} = \frac{(3\pi^2)^{2/3}}{5m_e} \cdot n_e^{2/3} = \frac{(3\pi^2)^{2/3}}{5m_e m_u^{2/3}} \cdot \left(\frac{Z}{A}\right)^{2/3} \cdot \rho_{\text{crit}}^{2/3} \tag{5.79}$$

or a critical mass density of

$$\rho_{\text{crit}} = \frac{5^{3/2} m_u m_e^{3/2}}{3\pi^2} \left(\frac{A}{Z}\right) \cdot T_{\text{crit}}^{3/2} = 0.0478 \text{ g cm}^{-3} \left(\frac{A}{2Z}\right) \cdot \left(\frac{T_{\text{crit}}}{10^4 \text{ K}}\right)^{3/2}. \tag{5.80}$$

We realize that the critical density for equal pressures has the same dependence on the temperature as seen before in Eq. (5.76), that is, it scales with the temperature to the power of 3/2. Moreover, the difference in the prefactor is just given by $(5/2)^{3/2} = 3.95$, about a factor of four. So the Fermi energy equals the thermal energy at slightly lower density compared to the density where the degeneracy pressure equals that for a nondegenerate ideal gas.

There is, however, an effect from the temperature on the structure of white dwarfs, which is related to the Coulomb lattice in the interior of the white dwarf. When the temperature reaches the binding energy of the Coulomb lattice, the Coulomb lattice will melt. Up to a certain density, nuclei will not form a lattice but roam freely in the hot electron gas surrounding them. This layer of a melted Coulomb lattice between the atmosphere and the solid inner part of the white dwarf is aptly called the 'ocean' of the white dwarf.

Let us estimate the critical density in dependence on the temperature for the appearance of the ocean in a white dwarf. Thereby we are looking at the ratio of the Coulomb energy of the lattice with a nuclear charge Z to the thermal energy T. We will call that ratio Γ:

$$\Gamma = \frac{Z^2 e^2}{r_{\text{cell}} \cdot T}, \tag{5.81}$$

where r_{cell} is the radius of one cell of the Coulomb lattice around the central charge Z (the Wigner–Seitz cell). The ratio Γ is a plasma parameter and called the Coulomb interaction parameter. Surprisingly, the value of Γ needed to melt the Coulomb lattices seems to be quite generic. Monte Carlo fluid simulations for Coulomb lattice have shown that a classic one-component plasma lattice melts for $\Gamma_{\text{m}} = 175.0 \pm 0.4$ (Potekhin and Chabrier, 2000). For a white dwarf, there are two components, nuclei and electrons, not only one. However, for the Coulomb lattice in a white dwarf, the critical value turns out to be not much different for carbon, where $\Gamma_{\text{m}}(^{12}\text{C}) = 178.4 \pm 0.2$, and smaller for a mixture of carbon and oxygen (Horowitz et al., 2010). We will adopt $\Gamma_{\text{m}} = 178$ for our estimates for white dwarfs. One can relate the mass density from the size of the cell to the radius r_{cell}:

$$\rho = A \cdot m_u \cdot n_{\text{cell}} = A \cdot m_u \cdot \frac{3}{4\pi r_{\text{cell}}^3} \tag{5.82}$$

or in reverse

$$r_{\text{cell}} = \left(\frac{3 A m_u}{4\pi\rho} \right)^{1/3}. \tag{5.83}$$

The melting temperature is then given by

$$T_m = \frac{e^2 Z^2}{\Gamma_m} \left(\frac{4\pi\rho}{3Am_u} \right)^{1/3} \tag{5.84}$$

or in numbers

$$T_m = 2.01 \times 10^4 \text{ K} \cdot \left(\frac{Z}{6} \right)^{5/3} \cdot \left(\frac{2Z}{A} \right)^{1/3} \left(\frac{\rho}{1 \text{ g cm}^{-3}} \right)^{1/3} . \tag{5.85}$$

This temperature is in the range of the observed surface temperature of white dwarfs, which can reach values of 100,000 K, so that, indeed, there can exist a small liquid ocean for hot white dwarfs.

5.8 Observation of White Dwarfs

In this subsection we discuss the global properties of white dwarfs, in particular their masses and radii. White dwarfs have been also studied for their cooling behavior. From their cooling age, one has deduced the age of the globular clusters M4 of about 12.7 Gyrs and the age of our galaxy to be about 7 Gyrs (Hansen et al., 2002). The oldest white dwarfs have cooled so much that they are not visible anymore. They are also sometimes called black dwarfs. Still, black dwarfs can be studied by their gravitational interactions, for example, in a binary system.

Before we start we need to discuss some important results for white dwarfs from stellar evolution theory. As discussed in the Introduction, white dwarfs are the stellar endpoint of stars with masses between 0.08 $M_\odot$ and 8 $M_\odot$. For $0.08 M_\odot < M < 0.4 M_\odot$, the star will only burn hydrogen to helium. The final white dwarf will be a helium white dwarf. For $0.4 M_\odot < M < 8 M_\odot$, the star will burn hydrogen to helium and then helium to mainly carbon and oxygen. The final white dwarf will be a carbon-oxygen white dwarf. Note, that the mass loss in the evolution of massive stars is so huge that even a $8 M_\odot$ star will end in a white dwarf with a mass not exceeding the Chandrasekhar mass limit. The relic white dwarf is just the carbon-oxygen core of the final star configuration, being just the amount of carbon and oxygen being produced in the lifetime of the star. The outer shells of lighter elements, helium and hydrogen, have been ejected into the interstellar medium. As the white dwarf is initially quite hot and shining bright in ultraviolet light, it illuminates those shells thrown out in space, resulting in those beautiful sights of a so-called planetary nebula. The name is misleading and is of historical origin, as the nebula resemble fuzzy planets in low-resolution telescopes. A planetary nebula has nothing to do with planets, it is a telltale of a newborn white dwarf.

One can make a rough estimate of the timescales it takes to form a helium white dwarf and a carbon-oxygen white dwarf from an ordinary star that is burning hydrogen to helium. Empirically, one has found a relation between the luminosity of a star L, the energy emitted per unit time, and the mass of a star M:

$$\frac{L}{L_\odot} = \left(\frac{M}{M_\odot}\right)^\alpha, \tag{5.86}$$

which is a power law with an exponent of about $\alpha \sim 4$. Here, $L_\odot$ denotes the luminosity of our Sun. A certain fraction of the star is converted from hydrogen to helium, typically 10%, within the life of the star. As mass is converted to energy, the total energy released in the lifetime of the star is proportional to its initial mass. As the luminosity is the energy emitted per unit time, the lifetime of the star τ can then be estimated by the ratio of the mass over the luminosity of the star:

$$\tau \sim \frac{M}{L} \propto M^{1-\alpha} \propto M^{-3}. \tag{5.87}$$

Hence, a star with twice the mass of our Sun will live about eight times shorter, a star with half the mass of our Sun will live about eight times longer than our Sun, respectively. The present age of our Sun is about 4.5 Gyrs and it is estimated that it will shine for another 4.5 Gyrs before becoming a white dwarf. As the age of the universe is about 13.8 Gyrs, all stars with a mass of $0.4M_\odot$ and less will be still burning helium only and no helium white dwarf could have been formed so far. We conclude that basically all presently existing white dwarfs in the universe can only be carbon-oxygen white dwarfs.

Observationally, one defines different spectral types of white dwarfs by their spectral lines. White dwarfs with hydrogen lines are classified as DA white dwarfs, having in particular strong Balmer hydrogen lines. This spectral type constitute by far the most common spectral type of white dwarfs, about 80%. DB white dwarfs have helium lines, DC white dwarfs show a continuous spectrum. For other spectral types see Table 5.2. For determining the radius of white dwarfs, one usually constraints the sample to DA white dwarfs, so that one has in all cases to model just a hydrogen atmosphere.

As of 2017, there are more than 30,000 white dwarfs known and listed in the Montreal White Dwarf Database (see Dufour et al., 2017). Most of them, about 80%, are DA white dwarfs. Typical magnetic fields are in the range of a few 10^7 G to several 10^8 G in some exceptional cases. The contribution of the observed surface temperatures T_{eff} peaks at about 10,000 K with most temperatures lying in the

Table 5.2 *Spectral types of white dwarfs.*

Spectral type	DA	DB	DC	DQ	DZ
Spectral lines	Hydrogen	Helium	None	Carbon	Metals

range of a few thousand kelvin to about 30,000 K, but also temperatures of up to 200,000 K are listed. Reported surface gravities are usually given on a logarithmic scale (in cgs units). Typical values are between $\log(g) = 7$ and 9, but the distribution peaks sharply at $\log(g) = 8$. The distribution of masses peaks sharply at about $0.6M_\odot$, with typical masses ranging from $0.2M_\odot$ to about $1.0M_\odot$. The masses given in the database are not necessarily directly measured but derived quantities. They are calculated by using the theoretical mass–radius relation, so the data cannot be used to provide a test of it. However, in a few cases this is possible, and we are going to discuss those ones here.

There are several ways to measure the mass and radius of white dwarfs. Let us summarize the possible observables that are depending on those two quantities. First of all, there is the luminosity of the white dwarf. A black body with a given temperature T emits energy weighted by the Planck distribution function

$$f(k) = \frac{1}{\exp(k/T) - 1}, \tag{5.88}$$

which is just the Bose–Einstein distribution function for photons. Here k is the wavenumber of the photon. Integrating the energy over the Planck distribution function gives the total energy density (note that $E(k) = k$ for massless particles)

$$\varepsilon = \frac{g}{(2\pi)^3} \int d^3k \; E(k) \cdot f(k) = \frac{\pi^2}{15} T^4, \tag{5.89}$$

with the degeneracy factor for photons $g = 2$ for their two possible spin states. So the energy density rises with the temperature to the fourth power. Consider now the flux of energy emitted of the black body per steradian from a sphere. At each point on the surface, the radiation will be isotropically emitted on a half-sphere. The radiation emitted per steradian $\mathrm{d}\Omega$ with respect to the normal of the surface has to be weighted by the cosine of the angle θ. The normalized integral of the energy emitted over the half-sphere gives the flux emitted per unit time and unit area

$$F_{\rm wd} = \frac{1}{4\pi} \int \mathrm{d}\Omega \; \cos(\theta) \cdot \varepsilon = \frac{1}{4\pi} \int_0^{2\pi} \mathrm{d}\phi \int_0^{\pi/2} \mathrm{d}\theta \cos(\theta) \cdot \varepsilon = \frac{1}{4}\varepsilon. \tag{5.90}$$

So the proper averaging over the emission angles gives a factor $1/4$. Finally, the total energy emitted per unit time integrated over the whole surface of the sphere with radius R results in the Stefan–Boltzmann law for the luminosity

$$L = 4\pi R^2 \cdot F_{\rm wd} = 4\pi R^2 \cdot \sigma T^4, \tag{5.91}$$

where $\sigma = \pi^2/60$ is the Stefan–Boltzmann constant (in natural units). By determining the luminosity of the white dwarf and its effective temperature, the Stefan–Boltzmann law gives the radius of the white dwarf. However, the luminosity cannot be measured from the emission of the white dwarf alone. What is measured on

Earth is the flux F, the energy emitted per unit time and unit area. Picture yourself a sphere around the white dwarf with the radius being the distance of the white dwarf to Earth. As the flux integrated over the surface of this sphere is just the luminosity of the white dwarf as is the flux integrated over the surface of the white dwarf, the flux measured on Earth will be reduced by the relative ratio of the area of the two spheres, that is, by the ratio of the radius R of the white dwarf and the distance D squared:

$$F_{\text{Earth}} = \frac{R^2}{D^2} F_{\text{wd}} = \frac{R^2}{D^2} \cdot \sigma T^4. \tag{5.92}$$

In order to get the radius of the white dwarf, one needs to know the distance to the white dwarf D. For white dwarfs within 1 kpc or so, one can measure the distance by parallax, as done with the satellite mission Hipparcos (1989–1993) and Gaia (since 2013) to high precision.

A more refined fit to the spectra of white dwarfs can be done by fitting the spectra to a model of the atmosphere of white dwarf. Models of the spectra of white dwarfs include in addition to the effective temperature T_{eff} the surface gravity $g = GM/R^2$ as an input parameter. These parameters are fitted to the form of the hydrogen Balmer spectral line profiles by model atmospheres. Another way to determine the radius of a white dwarf is by a measurement of the gravitational redshift of a spectral line $z = GM/R$. In both cases, if one knows the mass of white dwarf from, for example, a measurement of its Keplerian orbit around a companion star, the radius can be fixed without knowing the distance to the white dwarf.

Such a measurement of the mass can be achieved for visual binaries, where the white dwarf is so close that one can observe directly the companion star. Such close-by white dwarfs with visual companion stars are the canonical white dwarfs Sirius B, Procyon B, Stein 2051 B, and 40 Eridani B. The masses of Procyon B (Bond et al., 2015) and Sirius B (Bond et al., 2017a) could be determined with precise measurements by the Hubble Space Telescope, the mass of 40 Eridani B with new high-precision astrometric measurements (Mason et al., 2017). Under special circumstances, one can determine the mass of the white dwarf by gravitational lensing, if the white dwarf passes in front of a background star. In fact, the mass of the nearby white dwarf Stein 2051 B has been determined by this method, see Sahu et al. (2017).

The measured values of the surface temperature, the mass, and the radius of these close-by white dwarfs are summarized in Table 5.3. The corresponding mass–radius diagram is shown in Figure 5.1 and compared to theoretical predictions for the mass–radius relation of white dwarfs. One notices different lines for the mass–radius relation. We know from our discussion that the Chandrasekhar maximum mass depends sensitively on the charge-to-mass ratio of the nucleus being present

Table 5.3 *Measured white dwarf properties for nearby visual binary systems (taken from Bond et al., 2017b). © AAS. Reproduced with permission.*

	T_{eff} (K)	Radius ($R_\odot$)	Mass ($M_\odot$)
Procyon B	$7,740 \pm 50$	0.01232 ± 0.00032	0.592 ± 0.006
Sirius B	$25,369 \pm 46$	0.008098 ± 0.000046	1.018 ± 0.011
Stein 2051 B	$7,122 \pm 181$	0.0114 ± 0.0004	0.675 ± 0.051
40 Eridani B	$17,200 \pm 110$	0.01308 ± 0.00020	0.573 ± 0.018

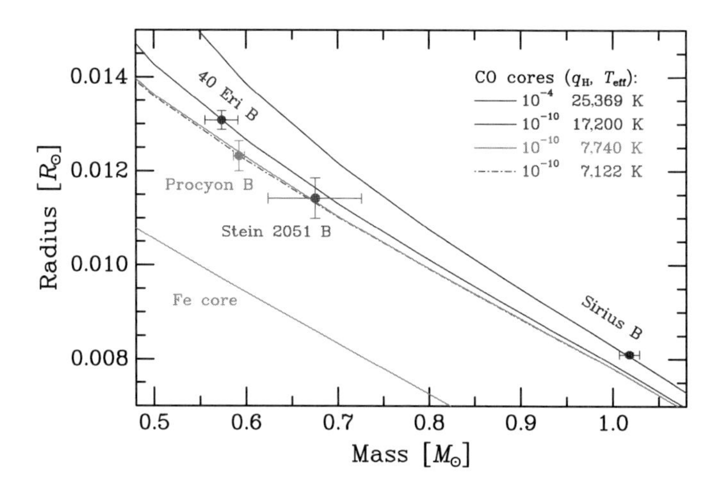

Figure 5.1 The mass–radius relation of white dwarfs (taken from Bond et al., 2017b). © AAS. Reproduced with permission

in the white dwarf. A white dwarf with an iron core has a smaller mass than one with a carbon-oxygen core. However, as argued, stellar evolution theory predicts that at present basically all white dwarf are carbon-oxygen white dwarfs. Figure 5.1 shows the mass–radius relation for iron and for carbon-oxygen white dwarfs. Indeed, none of the measured data points are found at the mass–radius relation line for iron white dwarfs but are lying on the line for carbon-oxygen white dwarfs. We notice also that the mass and radius of 40 Eridani B and Sirius B is above the standard carbon-oxygen white dwarf mass–radius relation. The relatively high surface temperatures of 40 Eridani B of about 17,000 K and of Sirius B of about 25,000 K are enough to have a marked impact on their radii. The radius of hot white dwarfs will be bloated up by such a high temperature. The mass–radius relation for white dwarfs with the measured surface temperature of the four white dwarfs under consideration are also shown in the figure for comparison. Taking into account these thermal effects is essential to explain the mass and radius of 40 Eridani B and of Sirius B, while effects from the lower temperatures of the other two white dwarfs are too small to be seen. In the legend of Figure 5.1 appears another parameter: the mass fraction of the H atmosphere relative to the mass of the white dwarf $q_H = M_{\rm H}/M_{\rm wd}$. Two different values are considered: a thick hydrogen atmosphere with $q_{\rm H} = 10^{-4}$ and a thin hydrogen atmosphere with $q_{\rm H} = 10^{-10}$. Sirius B is modeled with a thin hydrogen atmosphere, the other white dwarfs with a thick hydrogen atmosphere. Sirius B is special as it has a helium atmosphere. However, a small amount of hydrogen is present in the atmosphere, which is sufficient to produce the hydrogen Balmer spectral lines being used to derive its mass and radius. The model atmosphere has to be adjusted to take this into account, as the line profiles are sensitive to the composition of the atmosphere. All in all, one sees that the measurements of the mass and radius fit perfectly well the theoretical predictions for the mass–radius relation of carbon-oxygen white dwarfs.

We close the discussion of mass–radius measurement with a warning. The match of the measured mass and radius of white dwarfs with theoretical predictions for carbon-oxygen white dwarfs has not always been that perfect. Older measurements for, for example, Procyon B led to the conclusion that it might be an iron white dwarf. Only more recent refined measurements with the Hubble Space Telescope could resolve that issue, see Provencal et al. (2002). One learns that it is difficult to arrive at reliable masses and radii of white dwarfs and one should be cautious. We will keep this in mind when discussing the mass–radius relation for neutron stars later on.

Exercises

(5.1) Use the approximate relation for hydrostatic equilibrium, Eq. (5.8), to estimate the central pressure for the Sun, the Earth, a typical white dwarf, and a typical neutron star (use the data of Table 1.1).

(5.2) Show that the characteristic radius r_c as defined in Eq. (5.21) introduced to arrive at the Lane–Emden equation corresponds to the Landau radius up to a prefactor of $\mathcal{O}(1)$. Show also that the boundary condition, Eq. (5.24), can be motivated by looking at the equation of hydrostatic equilibrium and taking the limit $r \to 0$.

(5.3) Check that the Lane–Emden equation, Eq. (5.22), has the three known analytic solutions for $n = 0, 1$, and 5, as given in the text. Check also the values of the Table 5.1 for solutions needed to determine the mass and radius of the compact star, that is, the radius ξ_n and the derivative of the density $\theta'(\xi = \xi_n)$ where the density vanishes.

(5.4) Consider a Wigner–Seitz cell of radius r_0 around a nucleus with charge Z filled with a gas of electrons ensuring charge neutrality. Derive the intermediate steps of the calculation of the Coulomb energy for the cell by looking at the contributions from the electron–electron and electron–ion interactions (see Eq. (5.45)).

(5.5) What is the limit on the net charge for astrophysical spheres? Look at the forces acting on a charged particle at the surface of the charged sphere. Then look at the total energy of a charged sphere in comparison to the gravitational binding energy. Compare the two estimates. What is your conclusion?

6
Pulsars

6.1 The Discovery of Pulsars

... here was I trying to get a Ph.D. out of a new technique, and some silly lot of little green men had to choose my aerial and my frequency to communicate with us.

This is how Jocelyn Bell-Burnell, born Bell, recalls her serendipitous discovery of pulsars in 1967 in a way appropriate for the after-dinner speech at the Eight Texas Symposium on Relativistic Astrophysics in Boston[1] in 1976 (Bell-Burnell, 1977). She was a PhD student of Anthony Hewish at that time and her thesis project was the study of scintillations in the interplanetary medium from the emissions from quasars. After setting up the radio antenna, it was her sole responsibility to run the radio telescope and analyze the data by hand, which was pouring out as a paper chart that summed up to several kilometers. She discovered a regular scruff in the radio data that seemed to come from some place outside the solar system. The signal was so regular and precise that it suggested first that extraterrestrials were sending messages, dubbed little green men by her. However, she discovered three more scruffs of this type from different regions of the sky, hinting at a natural phenomenon. The discovery of these pulsating radio sources, pulsars in short, was published in 1968 (Hewish et al., 1968). The word pulsar for pulsating star was invented by the journalist Anthony Michaelis.

6.2 Pulsars Are Rotation-Powered Neutron Stars

The regular pulses of pulsars have the following three basic features:

(1) the pulse periods are short and range from milliseconds to seconds

[1] The first two meetings were held in Texas and the title was kept for all subsequent meetings, even though they are now held all over the world.

(2) the pulses are extremely regular
(3) the pulse period increases with time.

In the discovery paper, white dwarfs and neutron stars were mentioned as possible sources. The first pulsar discovered had a period of 1.3 s. However, the duration of emission could be constrained to be much less than a second, to 0.016 s. If the source is emitting coherently from a given area, the size of this area cannot be larger than the distance light can travel within the duration of emission. Hence, the emitting source cannot be larger than $\delta r = c \cdot \delta t \approx 4{,}800\,\text{km}$. Therefore, the emitting source involves compact sources, either compact stars or black holes. Black holes need to couple to ordinary matter to emit radiation, so they alone cannot be responsible for pulsars. In addition, such highly accurate patterns do not belong to the realm of plasma physics, more to celestial mechanics.

One can think of three different emission scenarios from these compact sources: a compact binary system, an oscillating compact star, or a rotating compact star. Let us go through the possible scenarios step by step.

White dwarfs have radii of a few thousand kilometers. Therefore, binary systems of white dwarfs will be larger than the limit in size given and cannot be the sources of pulsars. Binary neutron stars would be allowed. However, as binary neutron stars are getting closer to their companion star by emitting gravitational waves, the period will decrease with time in contradiction to the observation. Binary compact stars as a source for pulsars are therefore ruled out.

Let us consider oscillations of compact stars now. The typical oscillation period of stars Π can be estimated by assuming a standing wave with the wavelength given by the radius R of the sphere: $\lambda \sim R$. The dispersion relation for a sound wave is given by the speed of sound c_s as

$$c_s = \lambda \cdot \nu \sim \frac{R}{\Pi}, \tag{6.1}$$

where ν is the frequency and Π the period of the sound wave. We can estimate the speed of sound from the equation of state in the form $c_s^2 = P_c/\bar{\rho}$, where P_c is the central pressure and $\bar{\rho} \sim M/R^3$ the average density of the star. The central pressure can be estimated by the equation for hydrostatic equilibrium, Eq. (5.8):

$$P_c \sim \frac{GM^2}{R^4}. \tag{6.2}$$

Putting everything together, we find that the speed of sound squared is given by the compactness of the star C:

$$c_s^2 \sim \frac{GM}{R} = C \tag{6.3}$$

and the oscillation period of a standing sound wave can be estimated by the simple expression

$$\Pi \sim \frac{R}{C^{1/2}} \sim (G\bar{\rho})^{-1/2} \sim 1\,\mathrm{h} \cdot \left(\frac{1\ \mathrm{g\ cm^{-3}}}{\bar{\rho}}\right)^{1/2} . \tag{6.4}$$

The dependance of the period on the inverse square root of the average density is commonly used in stellar astrophysics. For the average density of our Sun of $\bar{\rho} = 1.4\,\mathrm{g\,cm^{-3}}$, one gets a period of about 1 hour. This can be compared to the characteristic 5-minute oscillations seen for our Sun, so our estimate is a little bit too high for solar oscillations. However, the solar oscillations are not resulting in a dramatic change of the luminosity. The prominent example of brightly oscillating stars are Cepheids, which are so bright that they have been used to determine distances in the universe by their universal period–luminosity relation. Polaris belongs to the class of classic cepheids, for example. For one cepheid, one knows the mass and the radius precisely as it is in an eclipsing binary system (Pietrzynski et al., 2010). The measured mass is $4.14 \pm 0.05 M_\odot$ and the radius $32.4 \pm 1.5 R_\odot$. Our simple estimate for the pulsation period results in a period of about 4 days, which fits nicely the observed period of $\Pi = 3.80\,\mathrm{days}$. For a typical average density of white dwarfs of $\bar{\rho} = 10^6\ \mathrm{g\ cm^{-3}}$, one arrives at a typical oscillation period of a few seconds. For a typical neutron star with an average density of about $\bar{\rho} = 10^{14}\,\mathrm{g\,cm^{-3}}$, the oscillation period is less than 1 ms. While these periods are in the range of the ones seen for pulsars, the observed wide range of pulse periods observed rules out just one characteristic frequency (modulo overtones) from oscillations.

Finally, we consider rotating compact stars. The maximum angular frequency ω can be estimated by the balance of the centrifugal force with the gravitational one for a particle of mass m sitting at the equator of a sphere:

$$m \cdot R \cdot \omega_{\mathrm{max}}^2 = \frac{GMm}{R^2} \quad \text{or} \quad \omega_{\mathrm{max}} = \left(\frac{GM}{R^3}\right)^{1/2} . \tag{6.5}$$

Hence, the minimum period of rotation $P_{\mathrm{min}} = 2\pi/\omega_{\mathrm{max}}$ is given by

$$P_{\mathrm{min}} = 2\pi \left(\frac{R^3}{GM}\right)^{1/2} = \frac{2\pi R}{C^{1/2}} = 0.47 \times 10^{-3}\ \mathrm{s} \cdot \left(\frac{R}{10\ \mathrm{km}}\right) \left(\frac{0.2}{C}\right)^{1/2} , \tag{6.6}$$

where C is the compactness of the star. Note that the minimum period has the same parametric dependence as the one for oscillations, Eq. (6.4). So, white dwarfs would have a minimum period that was too high to explain periods of a few milliseconds, while rotating neutron stars can rotate with such low periods. Hence, pulsars can be explained by rotating neutron stars. Indeed, Pacini pointed out shortly before the discovery of pulsars that rotating neutron stars in combination with a high magnetic field can produce electromagnetic radiation (Pacini, 1967). The model of a rotation-powered neutron star as the source of a pulsar received widespread acceptance with the paper of Gold, which appeared shortly after the discovery (Gold, 1968). The lighthouse picture of a pulsar with a radiation in the form of rotating beacon still stands today.

6.3 Properties of Pulsars

So, we know today that pulsars are rotation-powered neutron stars. They emit regular pulses in radio waves. The Australian Telescope National Facility (ATNF) lists more than 2,700 pulsars in their online database as of 2019 (Manchester et al., 2005). The period of their pulses ranges from 10 ms to 10 s. Pulsars are extremely precise, the change in their period is only $\dot{P} = 10^{-10}$ to 10^{-21}! In comparison, modern atomic clocks have reached a precision of 10^{-16}, so that some pulsars are sending out pulses that are more accurate than our standard of time on Earth. In fact, an ensemble of pulsars has been used to define a standard of time with a precision comparable to the uncertainties of the international atomic time (Hobbs et al., 2012).

Pulsars are labeled with the prefix PSR, standing for pulsating source of radio emission, followed by their celestial coordinates using the right ascension and the declination with respect to the spring equinox. As the point of the spring equinox on the sky is slowly changing with time due to the precession of Earth's rotation axis, the coordinates need to be fixed to a certain standard of time, the so-called standard epoch. Fixing it to the year 1950, the standard epoch B1950.0 is denoted by a 'B', for the year 2000, the standard epoch J2000.0 is denoted by a 'J'. It was decided that the switch from the older standard epoch to the newer one was to happen in 1984. Therefore, the prefix tells us, whether the pulsar was discovered before 1984 (the ones with a 'B' as prefix) or 1984 or afterward (the ones with a 'J' as prefix).

The fastest pulsar known is the pulsar J1748-2446ad, spinning at 716 Hz (Hessels et al., 2006). So, the prefix 'J' tells us that the pulsar is one of the newly discovered pulsars. The next four digits give the right ascension in hours and minutes followed by the declination angle. The declination angle has a minus sign, so the pulsar is located in the southern hemisphere. The two letters appearing at the end are special, as there have been several pulsars with the same coordinates discovered, necessitating the need for a refined labeling. The reason that there are so many pulsars known with that same celestial coordinates is that the coordinates point to the globular cluster Terzan 5.

Globular clusters are known to harbor many compact stars, white dwarfs as well as neutron stars. They consists of several million stars packed within a radius of just 10–100 pc. They are the oldest structures in our galaxy, even older than the galactic disk. No wonder that globular clusters harbor so many compact stars, as the ordinary stars had enough time to reach the final end point of stellar evolution, thereby creating white dwarfs and neutron stars. The globular cluster Terzan 5 is the fifth entry of a list of globular clusters discovered by the astronomer Agop Terzan. Terzan 5 has 37 known pulsars. 47 Tucanae is another globular cluster in our galaxy. It is located in the constellation Tucana in the southern hemisphere. The number in front of the constellation usually denotes a star. When 47 Tucanae was discovered, the resolution of the available telescope was not enough to resolve the supposed star to its true nature. 47 Tucanae contains 25 known pulsars.

Several discovered pulsars have a companion and are called binary pulsars. The companion has been seen to be a main sequence star, a planet, a white dwarf, or even another neutron star. In one case, one was able to detect all the pulses of the companion neutron star: the system of the pulsars PSR J0737−3039A and PSR J0737−3039B constitue the only known double pulsar system (Lyne et al., 2004; Kramer et al., 2006). Measuring the pulses of both neutron stars allowed for the most stringent test of general relativity in the strong field limit, see (Kramer and Wex, 2009), and the following discussion. The best known binary pulsar is the so-called Hulse–Taylor pulsar, or PSR B1913+16. Its discovery and the subsequent indirect determination of the emission of gravitational waves from the highly relativistic binary lead to the awarding of the Nobel prize in 1993 to Russell Hulse and Joseph Taylor.

There exists even triple pulsar systems, that is, pulsars with two companions. One of them is the pulsar PSR B1257+12, where two planets orbit the pulsars (Wolszczan and Frail, 1992; Wolszczan, 1994). At the time of its discovery in 1992, it was the first detection of exoplanets, that is, of planets outside the solar system. The other known triple pulsar systems are PSR J0337+1715, where the pulsar has two white dwarfs as companions (Ransom et al., 2014), and PSR B1620-26, where the pulsar has a white dwarf and a planet as companions (Thorsett et al., 1999).

Pulsars have been also detected in the gamma-ray band by periodic emission of gamma-ray pulses. The most prominent one is the Geminga pulsar, which is only seen as a gamma-ray pulsar but not in the radio wave band (the expression Geminga means ‘not there’ in the Milanese dialect).

6.4 The Zoo of Neutron Stars

While the majority of observed neutron stars has been found in the form of the classic pulsar, there are several other classes of astronomical objects that are also associated with neutron stars.

Magnetars: Strictly speaking, this is a class of neutron stars that have huge magnetic fields. Observationally, they are associated with the two classes of astronomical objects listed here. Both classes have in common that they have long spin periods in the range of a few seconds and large spin-down rates. They emit X-rays continuously. No companion has been found.

Anomalous X-ray pulsars (AXPs): Pulsating X-ray sources in the soft X-ray range with an unusually high X-ray luminosity.

Soft gamma-ray repeaters (SGRs): Give short intense bursts in the hard X-ray and soft gamma-ray range that are repeating.

While these two classes were treated separately initially, it is now widely accepted that AXPs and SGRs are magnetars that draw their energy output not from the rotational energy but from the huge surface magnetic fields of the order of 10^{14}–10^{15} Gauss.

Central compact objects (CCOs): These are bright X-ray spots in the center of supernova remnants. They show thermal-like soft X-ray emissions.

Isolated neutron stars (INSs): Seven thermal sources were discovered by the ROSAT X-ray satellite and dubbed the magnificent seven (M7). They are all close by and just a few hundred thousand years old. The closest one is RX J1856-3754 (here the prefix RX stands for a source measured by the ROSAT X-ray satellite), with a distance of only $D = 123^{+11}_{-15}$ pc (Walter et al., 2010). The spectrum of RX J 1856-3754 is a perfect black body, with an effective temperature of $T = 63$ keV (Burwitz et al., 2003).

Rotating radio-transients (RRATs): They release short sporadic bursts of radio waves. Long periods of a few seconds with large spin-down rates have been detected.

Fast radio bursts (FRBs): These are millisecond events of radio emission with a high dispersion measure, hinting at an extragalactic source. The first one detected was found by reexamining archival data and became known as the Lorimer event (Lorimer et al., 2007).

X-ray binaries: These are accreting systems involving the emission of X-rays. The companion of the neutron stars is either a massive star (high-mass X-ray binary [HMXB]) with masses around $20M_\odot$ or a star with a mass around $1M_\odot$ (low-mass X-ray binary [LMXB]). Some of them emit regular pulses and are therefore called (accretion-powered) X-ray binary pulsars. Some of the LMXBs show a sudden blast of X-ray emission and are called X-ray bursters.

This spectral classification does not have strict dividing lines. On the contrary, isolated neutron stars, for example, show slight pulsations, which relates to a strong magnetic field and a possible relation to magnetars. The long periods and the large spin-down rates of isolated neutron stars and RRATS might hint at a connection between these two classes. Also, RRATs and FRBs show similar features in their radio emission, the dividing criterion might be just their galactic and extragalactic origin, respectively.

6.5 The Dipole Model of Pulsars

In this section we discuss a simple model of the pulsed emission from rotation-powered neutron stars. Let us consider a rotating sphere with a magnetic field in the form of a dipole. The axis of rotation and the magnetic poles are misaligned. The emission of electromagnetic radiation is assumed to originate from the magnetic north and south poles of the sphere along the lines of the magnetic field.

6.5.1 Rotational Energy Loss

First, let us discuss the energy stored in rotational energy of a neutron star. The rotational energy is given by

$$E_{\rm rot} = \frac{1}{2} I \cdot \Omega^2, \tag{6.7}$$

where I is the moment of inertia and Ω the angular frequency. The moment of inertia is the second moment of the mass density $\rho(r)$ relative to the rotation axis. For a spherically symmetric sphere one gets using spherical coordinates,

$$I = \int {\rm d}V\, r^2 \cdot \rho(r) = \frac{8\pi}{3} \int {\rm d}r\, r^4 \cdot \rho(r), \tag{6.8}$$

where r is the distance to the center of the sphere. You are asked to verify this expression in the exercises. For a homogenous sphere with a constant mass density ρ_0, the moment of inertia can be calculated to be

$$I = \frac{8\pi}{15} \rho_0 R^5 = \frac{2}{5} M \cdot R^2 = f \cdot M \cdot R^2, \tag{6.9}$$

where R is the radius and M the total mass of the sphere. The prefactor f is called the moment of inertia prefactor, which is $f = 0.4$ for a constant mass density, that is, an incompressible fluid, and will be different for a nonuniform radial density profile (e.g., $f = 0.3307$ for our planet Earth due to its denser core, see Williams, 1994). As the density increases toward the center of a star (or planet), the value of f will be smaller in reality. For a neutron star with a mass of $M = 1.4M_\odot$ and a radius of $R = 10\,\mathrm{km}$, one finds a moment of inertia of

$$I = 1.11 \times 10^{38}\,\mathrm{kg\,m^2} \left(\frac{M}{1.4M_\odot}\right)\left(\frac{R}{10\,\mathrm{km}}\right)^2, \tag{6.10}$$

assuming $f = 0.4$, that is, a constant mass density.

The energy loss of rotational energy is just the time derivative of the rotational energy

$$\frac{\mathrm{d}E_{\mathrm{rot}}}{\mathrm{d}t} = I\Omega\dot{\Omega} = I\frac{4\pi^2}{P^3}\cdot \dot{P}, \tag{6.11}$$

assuming that the moment of inertia does not change. Here we have set $P = 2\pi/\Omega$, where P is the period of the neutron star. The Crab pulsar has a frequency of $\nu = 29.946\,923(1)\,\mathrm{Hz}$ with a time derivative of $\dot{\nu} = -3.775\,35(2) \times 10^{-10}$ (Lyne et al., 2015). We use $P = 1/\nu$ and $\dot{P} = -\dot{\nu}/\nu^2$ and adopt the previous value for the moment of inertia to get an energy loss for the Crab pulsar of

$$\dot{E}_{\mathrm{rot}} = 4.97 \times 10^{38}\,\mathrm{erg\,s^{-1}} \left(\frac{0.0334\,\mathrm{s}}{P}\right)^3 \left(\frac{\dot{P}}{4.21 \times 10^{-13}}\right). \tag{6.12}$$

For comparison, the luminosity of the Sun is $L_\odot = 3.828 \times 10^{33}\,\mathrm{erg\,s^{-1}}$, so the energy loss of the Crab pulsar is about five orders of magnitude larger than that of the Sun.

It was known even before the discovery of the Crab pulsar that the Crab nebula shows an accelerated expansion that could not be explained by the radiation pressure of the nebula alone (Baade, 1942). There is an extra energy source that drives the accelerated expansion of the Crab nebula. The extra energy seen in the Crab nebula is estimated to be about $10^{38}\,\mathrm{erg\,s^{-1}}$ (Davidson and Fesen, 1985), which fits perfectly the energy loss of the Crab pulsar derived earlier. Hence, the rotational energy released by the Crab pulsars powers the accelerated expansion of the Crab nebula. This is a nice confirmation of the picture that pulsars are indeed rotation-powered neutron stars.

6.5.2 Electromagnetic Emission

Let us consider now the electromagnetic emission from a rotating neutron star with a magnetic field. We consider the simplest magnetic field configuration, that is, that of a dipole with a magnetic moment $\boldsymbol{m}$. The magnetic field for a dipole has the following form in spherical coordinates

$$\boldsymbol{B} = \frac{3\boldsymbol{e}_r(\boldsymbol{m}\cdot\boldsymbol{e}_r) - \boldsymbol{m}}{r^3} = \frac{|\boldsymbol{m}|}{r^3}\left(2\cos\theta\cdot\boldsymbol{e}_r + \sin\theta\cdot\boldsymbol{e}_\theta\right), \tag{6.13}$$

where $\boldsymbol{m}$ is the magnetic dipole moment, $\boldsymbol{e}_r$ is a unit vector pointing in the direction of $\boldsymbol{r}$, and $\boldsymbol{e}_\theta$ is the corresponding unit vector for the angle, θ. At the equator of the sphere, $r = R$ and $\boldsymbol{e}_r$ is perpendicular to the magnetic dipole moment $\boldsymbol{m}$, so the magnitude of the magnetic field on the surface of the star is given by

$$B_{\text{surf}} = \frac{|\boldsymbol{m}|}{R^3}. \tag{6.14}$$

Note that the magnitude of the magnetic field at the poles will be a factor two higher.

The electromagnetic emission for an accelerated charge q with the velocity v is proportional to the acceleration squared. The emission pattern is that of a dipole, peaked along the forward and backward directions and proportional to $\sin^2\theta$. Hence, the energy emitted per unit time, the power P, per steradian is of the form

$$\frac{\mathrm{d}P_{\text{dipole}}}{\mathrm{d}\Omega} = \frac{q^2}{4\pi}\dot{v}^2 \sin^2\theta, \tag{6.15}$$

where we assume small (nonrelativistic) velocities, $v \ll c$. Here a dot indicates the derivative with respect to time. Integrating over the angle gives Larmor's formula for the energy loss at nonrelativistic velocities

$$\frac{\mathrm{d}E_{\text{dipole}}}{\mathrm{d}t} = -P_{\text{dipole}} = -\frac{2}{3}q^2\dot{v}^2. \tag{6.16}$$

For a rotating dipole we use, the dipole moment can be written as $\boldsymbol{d} = q \cdot \boldsymbol{r}$, so that the energy emitted per unit time is proportional to the second time derivative of the magnitude of the dipole moment squared:

$$\frac{\mathrm{d}E_{\text{dipole}}}{\mathrm{d}t} = -\frac{2}{3}|\ddot{\boldsymbol{d}}|^2. \tag{6.17}$$

Lamor's formula can be applied to an electric as well as to a magnetic dipole, so that the rate of electromagnetic emission for a rotating magnetic dipole is

$$\frac{\mathrm{d}E_{\text{dipole}}}{\mathrm{d}t} = -\frac{2}{3}|\ddot{\boldsymbol{m}}|^2, \tag{6.18}$$

that is, proportional to the second time derivative of the magnetic dipole moment squared. If the magnetic dipole moment is not pointing along the axis of rotation, it will be time-dependent and tap electromagnetic energy from the rotating neutron star. Let α be the angle between the axis of rotation, which is parallel to the z-axis, and the magnetic dipole moment. Then the magnetic dipole moment reads

$$\boldsymbol{m} = B_{\text{surf}}R^3 \cdot (\sin\alpha\cos\Omega t, \sin\alpha\sin\Omega t, \cos\alpha), \tag{6.19}$$

where we replaced the magnitude of the magnetic dipole moment with that of the magnetic field on the surface using Eq. (6.14). The total emitted power, that is, energy per time, is then given by

$$\frac{\mathrm{d}E_{\text{dipole}}}{\mathrm{d}t} = -\frac{2}{3} B_{\text{surf}}^2 R^6 \Omega^4 \sin^2 \alpha. \tag{6.20}$$

The important thing to note here is that the energy loss due to the magnetic dipole emission is proportional to the angular frequency Ω to the fourth power. Equating the rotational energy loss, Eq. (6.11), with the energy loss from a magnetic dipole gives

$$\dot{E}_{\text{rot}} = I\Omega\dot{\Omega} = -\frac{2}{3} B_{\text{surf}}^2 R^6 \Omega^4 \sin^2 \alpha = \dot{E}_{\text{dipole}}. \tag{6.21}$$

This relation gives an expression for the minimum magnetic field at the poles by setting $\sin\alpha = 1$:

$$B_{\text{surf}} \geq \left(\frac{3}{8\pi^2} \frac{I}{R^6} \right)^{1/2} \cdot \left(P \cdot \dot{P} \right)^{1/2}. \tag{6.22}$$

In the following, we will denote the limit on the magnetic field from the dipole model the characteristic magnetic field B_{char} by choosing a typical radius of $R = 10\,\text{km}$ and typical moment of inertia of $I = 10^{38}\,\text{kg}\,\text{m}^2$. We define the characteristic magnetic field as

$$B_{\text{char}} = 3.20 \times 10^{19}\,\text{G} \cdot \left(\frac{P \cdot \dot{P}}{1\,\text{s}} \right)^{1/2} \cdot \left(\frac{I}{10^{38}\,\text{kg}\,\text{m}^2} \right)^{1/2} \cdot \left(\frac{R}{10\,\text{km}} \right)^{-3}. \tag{6.23}$$

For the Crab pulsar, one gets a characteristic magnetic field of $B_{\text{char}} = 3.79 \times 10^{12}\,\text{G}$ (see Table 6.1).

6.5.3 Characteristic Age

The equality of the rotational energy loss to the power of the magnetic dipole emission, Eq. (6.21), defines a differential equation of the form

$$\dot{\Omega} = -K \cdot \Omega^3, \tag{6.24}$$

with the constant $K = 2B_{\text{surf}}^2 R^6 \sin^2 \alpha/(3I)$. The solution in terms of the time t is

$$t = \frac{1}{2K\Omega^2} \left(1 - \frac{\Omega^2}{\Omega_0^2} \right), \tag{6.25}$$

where Ω_0 is the angular frequency when the pulsar is born, that is, for $t = 0$. If the pulsar is spinning rapidly at birth, one can set $\Omega_0 \ll \Omega$ and ignore the second

Table 6.1 *Data for a few selected pulsars including companion neutron stars (pulsar data from the ATNF database, see Manchester et al., 2005). The postfix (c) stands for the companion neutron star of the pulsar that is not observed as a pulsar.*

Pulsar	P (s)	$\dot{P}$	Mass ($M_\odot$)	
J0740+6620	0.00289	1.22×10^{-20}	$2.14^{+0.10}_{-0.09}$	Massive pulsar (Cromartie et al., 2019)
J0348+0432	0.0391	2.41×10^{-19}	2.01(4)	Massive pulsar (Antoniadis et al., 2013)
J1614−2230	0.00351	9.62×10^{-21}	1.908(16)	Massive pulsar (Arzoumanian et al., 2018)
J0453+1559(c)			1.174(4)	Lightest neutron star
J0453+1559	0.0458	1.86×10^{-19}	1.559(5)	(Martinez et al., 2015)
B1913+16	0.0590	8.62×10^{-18}	1.438(1)	Hulse–Taylor pulsar
B1913+16(c)			1.390(1)	(Weisberg and Huang, 2016)
J0737−3039A	0.0227	1.76×10^{-18}	1.3381(7)	Double pulsar
J0737−3039B	2.77	8.92×10^{-16}	1.2489(7)	(Kramer et al., 2006)
B1534+12	0.00379	2.42×10^{-18}	1.3330(2)	Most precisely measured
B1534+12(c)			1.3455(2)	(Fonseca et al., 2014)
B0531+21	0.0334	4.21×10^{-13}	–	Crab pulsar (Staelin and Reifenstein, 1968)
B0833−45	0.0839	1.25×10^{-13}	–	Vela pulsar (Large et al., 1968)
J1856−3754	7.06	2.98×10^{-14}	–	Closest neutron star (Tiengo and Mereghetti, 2007)

term in the brackets. Replacing the constant K by using the differential equation, Eq. (6.24), we arrive at an expression for the present age of a pulsar

$$t_{\text{char}} = \frac{P}{2\dot{P}}, \tag{6.26}$$

which is called the characteristic age. Note that the only ingredient for the expression of the characteristic age is that the change of the angular frequency is proportional to the angular frequency cubed. For the Crab pulsar, one gets a characteristic age of $T = 1,260$ years, taking the data from Lyne et al. (2015), see also Table 6.1. The Crab pulsar is the remnant of the historical supernova from 1054, so the known age of the Crab pulsar is 961 years (in 2015). Our simple estimate is not too far off.

6.5.4 The Braking Index

We noted that the change of the angular frequency with time due to dipole emission is proportional to the angular frequency to the power of three, see Eq. (6.24). One defines the so-called braking index n via the generalized expression

$$\dot{\Omega} = -K \cdot \Omega^n. \tag{6.27}$$

Observationally, the braking index can be determined by the second-order derivative of Ω via

$$n = -\frac{\Omega \cdot \ddot{\Omega}}{\dot{\Omega}^2}. \tag{6.28}$$

This formula for n coincides with the braking index, that is, the index of the power-law behavior of the time derivative of Ω given earlier. As the measurement of the braking index n requires the second time derivative of Ω, long time observations are needed. Therefore, only a few measurements are reported in the literature with values ranging between $n = 0.9$ and 2.8 (for the latest compilation, see Espinoza et al., 2017). For the Crab pulsar one has determined the braking index to be $n = 2.342(1)$. Note that all measured values of the braking index are smaller than the value for an ideal magnetic dipole of $n_{\text{dipole}} = 3$.

6.5.5 Gravitational Wave Emission

One could think about different emissions from pulsars, for example, gravitational wave emissions. The energy emitted per unit time of gravitational waves is given by Einstein's quadrupole formula

$$\dot{E}_{\text{gw}} = -\frac{1}{5} G \cdot < \dddot{Q} \cdot \dddot{Q} >, \tag{6.29}$$

where Q is the quadrupole moment of the energy distribution and $< \cdots >$ denotes a proper time average. The formula will be derived in Chapter 10. We note that the energy emission of gravitational waves involves now in total six time derivatives instead of four, the number of derivatives found in the magnetic dipole model. Let us assume that the pulsar is wobbling when rotating due to a nonzero ellipticity ϵ. The rotation frequency is Ω, the moment of inertia I. The quadrupole moment will change proportional to $\exp(2i\Omega)$.[2] By dimensional reasoning, the energy loss due to gravitational wave emission will be proportional to Ω^6:

$$\dot{E}_{\text{gw}} \approx -\frac{32}{5} G \cdot \epsilon^2 I^2 \Omega^6, \tag{6.30}$$

[2] It is generic feature of the produced gravitational waves that their frequency is twice the rotation frequency, see Chapter 10.

assuming a small ellipticity. Putting in numbers one gets

$$\dot{E}_{\rm gw} \approx 1.09 \times 10^{36}\,{\rm erg\,s^{-1}} \cdot \epsilon^2 \left(\frac{I}{10^{38}\,{\rm kg\,m^2}}\right)^2 \left(\frac{P}{1\,{\rm s}}\right)^{-6}, \qquad (6.31)$$

which for rapidly rotating pulsars can easily reach energy losses similar to electromagnetic radiation, as for example, the estimated rotational energy loss of the Crab pulsar of the order of 10^{38} erg s^{-1} (see Eq. (6.12)).

Setting the expression for the energy loss due to gravitational waves equal that due the rotational energy loss, Eq. (6.11), one arrives at a differential equation of the type

$$\dot{\Omega} = -K' \cdot \Omega^5, \qquad (6.32)$$

where K' is a constant depending on the ellipticity and the moment of inertia of the rotating neutron star. We can read off from that differential equation that the braking index for gravitational wave emission is $n_{\rm gw} = 5$. So, gravitational wave emission would shift the braking index to even higher values, not smaller ones as observed.

One is in the position to put limits on the emission of gravitational waves from pulsars by direct measurements with gravitational wave detectors. The Laser Interferometer Gravitational-Wave Observatory (LIGO) has observed 200 pulsars for their emission of gravitational waves. No signal has been found that puts stringent limits on the ellipticity of rotating neutron stars (the following numbers assume a moment of inertia of $I = 10^{38}\,{\rm kg\,m^2}$). The Crab pulsar must be a nearly perfect sphere, at most an ellipticity of $\epsilon \leq 3.6 \times 10^{-5}$ is allowed from LIGO, which has not detected any gravitational waves from the Crab pulsar. LIGO can test also if the energy emission of pulsars is solely due to gravitational radiation. If the ratio of the energy loss due to gravitational wave emission to that of rotational energy loss $\dot{E}_{\rm gw}/\dot{E}_{\rm rot}$ is found to be less than 1, one has beaten the spin-down limit of the pulsar under observation. LIGO has beaten the spin-down limit now for several pulsars (Abbott et al., 2017a). The most stringent limits have been determined for the Crab and Vela pulsars, with $\dot{E}_{\rm gw}/\dot{E}_{\rm rot} \leq 2 \times 10^{-3}$ and 10^{-2}, respectively. Hence, less than two permille of the total energy loss can be emitted in gravitation waves from the Crab pulsar. Seemingly, emission of gravitational waves is not the dominant energy loss mechanism of pulsars (see also Chapter 10).

6.6 The Pulsar Diagram ($P - \dot{P}$ Diagram)

We are now in the position to discuss the rich physics of the pulsar diagram, which is a scatter plot of the period P and the period derivative $\dot{P}$ shown in Figure 6.1. Overlaid onto the diagram are the characteristic magnetic field $B_{\rm char}$ from Eq. (6.23)

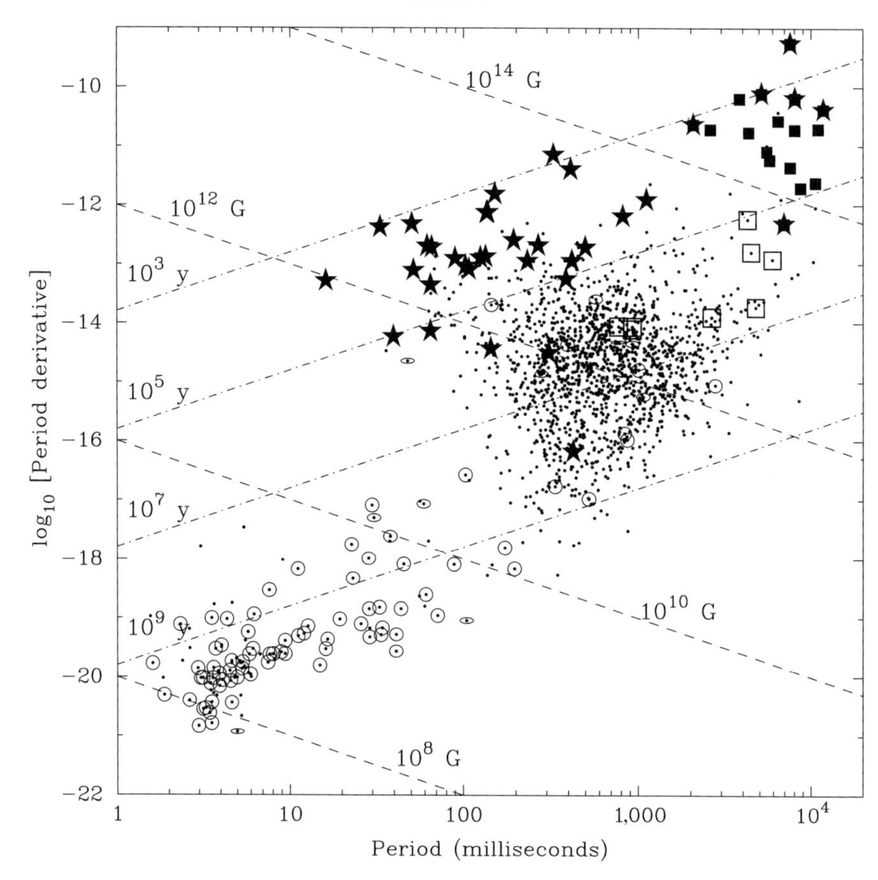

Figure 6.1 The pulsar diagram: The period and period derivative of pulsars in a scatter plot. The dashed lines show the characteristic magnetic field, the dashed-dotted lines the characteristic age of the pulsar. Stars: pulsars in supernova remnants; filled squares: magnetars; circles: binary pulsars; squares: RRATs. Reprinted from Lorimer (2011), with the permission of AIP publishing

and the characteristic age t_{char} from Eq. (6.26). One sees four distinct areas of the pulsar diagram depending on the characteristic magnetic field and the characteristic age.

- Characteristic magnetic fields between $B_{\text{char}} \approx 10^{12}$ G and 10^{14} G and characteristic ages of only $t_{\text{char}} \approx 10^3$–$10^5$ years: These are newborn pulsars that are found in supernova remnants. Prominent examples are the Crab pulsar and the Vela pulsar.
- $B_{\text{char}} > 10^{14}$ G up to about 10^{15} G and $t_{\text{char}} \approx 10^3$–$10^5$ years: These are the pulsars with the highest known characteristic magnetic fields and are therefore called magnetars. Note that their periods of several seconds are the largest ones known. It seems that the large energy loss from the extreme magnetic field has slowed down those pulsars enormously.

- $B_{\text{char}} \approx 10^{12}$ G and $t_{\text{char}} \approx 10^5$–$10^9$ years: Most of the pulsars are found in this part of the pulsar diagram. Compared to the newborn pulsars, the characteristic magnetic field reduces slightly for these older pulsars. The period increases by several oders of magnitudes to several hundred milliseconds. Obviously, pulsars spend most of their time here.
- $B_{\text{char}} \approx 10^8$ G–10^{10} G and $t_{\text{char}} \approx 10^9$ up to about 10^{10} years: Here one finds the pulsars in binary systems. They have low magnetic fields for pulsar standards, which has allowed the neutron star to accrete matter from the companion star and to spin-up. With periods of down of just a few milliseconds, they are the fastest rotation-powered neutron stars known. They are also the pulsars with the highest characteristic ages, reaching values close to the age of the universe of $t_{\text{universe}} = 13.799 \pm 0.021$ Gyrs, as determined by the PLANCK collaboration from cosmological observations (Ade et al., 2016). The prominent example of this class of pulsars is the Hulse–Taylor pulsar.

Similar to the Hertzsprung–Russell diagram, for ordinary stars, one can discuss the evolutionary path of pulsars through the $P - \dot{P}$ diagram. Pulsars are born with large magnetic fields and short periods in the upper left corner of the pulsar diagram. They slow down to periods of seconds or so within a billion years. Eventually, the pulsar mechanism shuts off for older pulsars, and they cannot be seen by radio emission. For neutron stars in binary systems, there exists an additional path through the pulsar diagram. By their strong gravitational field, they can extract matter from the companion star, forming an accretion disk around the neutron star. They spin-up as the matter in the accretion disk transports angular momentum to the neutron star. The pulsar mechanism starts again and they are seen as old, rapidly spinning recycled pulsars in the lower left corner of the pulsar diagram.

6.7 Pulsar Glitches

There is an observable obstacle in determining the long-time behavior of the spin-down of pulsars. From the precision and the extremely small change in the angular frequency, one would not expect that pulsars lose the beat from time to time, which is a so-called pulsar glitch. One has seen for many pulsars that there are sudden spin-ups of the pulse frequency (for a review and a link to an online database see Espinoza et al., 2011). The fractional change in the pulse frequency ranges from $\Delta\nu/\nu = 10^{-11}$ to 10^{-5}, where $\Delta\nu$ is the difference in the frequency before and after the glitch. The Crab pulsar shows moderate changes in the frequency due to glitches with fractional changes of less than 2×10^{-7}. The Vela pulsar shows particular strong glitches at the upper end of the given range, with almost all of them having fractional changes of more than 10^{-6}. Characteristically, the spin-down rate increases after the glitch and shows a recovery toward the spin-down rate before the glitch on typical timescales of 100 days. The change of the spin-down rate relates to a fractional change of the time derivative of the frequency in the range of $\Delta\dot{\nu}/\dot{\nu} = 10^{-4}$ to about 1. Glitches have been seen to occur repeatedly for a given pulsar within a timespan of typically a year.

So far, the origin of glitches has not been fully resolved. The glitch could be related to a change of the moment of inertia of the rotating neutron star. The crust of neutron stars consists of a solid structure, a lattice of nuclei immersed in a gas of electrons, as for white dwarfs. Stress in the crust could build up and the sudden release of the torsional energy, a star quake, could be the origin of glitches. This would be similar in origin to an earthquake that is caused by the release of stress in the crust of Earth.

Another possible explanation is that the whole solid outer crust of a neutron star gets distorted and its torsional energy is released suddenly. The connection of the outer crust to the inner part of the neutron star is realized by vortices of the neutron superfluid, which are unpinning (the glitch) and repinning (the recovery period after the glitch). If the liquid superfluid rotates slightly faster than the solid crust, stress will build up. The vortices try to keep the crust connected to the core. If the stress is too large, the vortices will unpin, the superfluid will slow down, losing angular momentum. As angular momentum is conserved, the angular momentum of the crust has to increase accordingly, thereby spinning up the crust. The pulsar spins up and the pulsar signals will increase in frequency, showing a glitch in pulsar timing. Eventually, the neutron star will slow down, the crust catches up with the superfluid, and the vortices will repin. The whole neutron star rotates in unison until the next glitch happens. More details will be discussed in the chapter on neutron stars, Chapter 7.

6.8 The Aligned Rotator

In the following, we discuss the electrodynamics of pulsars, which goes back to the work of Goldreich and Julian (1969). Consider a rotating neutron star with the angular frequency $\boldsymbol{\Omega}$ and with a magnetic field $\boldsymbol{B}$ being aligned with the rotation axis along the z-axis. The magnetic field inside the neutron star shall be the one of a uniformly magnetized sphere $\boldsymbol{B} = B_0\boldsymbol{e}_z$, where $\boldsymbol{e}_z$ is a unit vector along the z-axis (although, from here on we switch to spherical coordinates).

6.8.1 Electric Forces

We assume also that the interior of the neutron star is a perfect conductor. This is justified as we know that there is a free degenerate gas of electrons in neutron star matter. The velocity from uniform rotation is given by

$$\boldsymbol{v} = \boldsymbol{\Omega} \times \boldsymbol{r} = \Omega r \sin\theta \cdot \boldsymbol{e}_\phi, \tag{6.33}$$

where $\boldsymbol{e}_\phi$ is a unit vector along the ϕ-direction. Any charge in the neutron star will feel a Lorentz force from the magnetic field and the uniform rotation. The moving charges will generate an electric field $\boldsymbol{E}$ such that the net force acting on the charge is vanishing

$$F_L = e\,(\boldsymbol{E} + \boldsymbol{v} \times \mathrm{B}) = 0. \tag{6.34}$$

This condition results in an electric field of

$$\boldsymbol{E} = -\boldsymbol{v} \times \boldsymbol{B} = (\boldsymbol{\Omega} \times \boldsymbol{r}) \times \boldsymbol{B} = \Omega B_0 r \sin\theta (\sin\theta \boldsymbol{e}_r + \cos\theta \boldsymbol{e}_\theta). \tag{6.35}$$

As the electric field is curl-free, $\nabla \times \boldsymbol{E} = 0$, we can define an electric potential via $\boldsymbol{E} = -\nabla\phi$. Integrating the electric field along the radial direction gives the electric potential

$$\phi = \frac{1}{2}\Omega B_0 r^2 \sin^2\theta + \text{const.} \tag{6.36}$$

The magnitudes of the electric force generated by the electric field at the surface of the neutron star are

$$F_e = e \cdot E \approx e B_0 \Omega R. \tag{6.37}$$

We note that there is also a component of the electric force pointing in radial direction. The ratio of this electric force to the gravitational force is given by

$$\frac{F_e}{F_g} \approx \frac{e B_0 \Omega R}{GMm/R^2} = 3 \times 10^8 \left(\frac{B_0}{10^{12}\,\mathrm{G}}\right) \left(\frac{1\,\mathrm{s}}{P}\right) \left(\frac{R}{10\,\mathrm{km}}\right)^3 \left(\frac{1.4 M_\odot}{M}\right) \left(\frac{m_p}{m}\right), \tag{6.38}$$

where M is the mass of the neutron star and m the mass of the charged particle (the electron or the proton mass). We see that the electric force is typically several orders of magnitude larger than the gravitational force (even for protons), so charged particles can be pulled out of the neutron star in principle. These pulled-out charges will form a plasma surrounding the pulsar that is corotating with it. A vacuum solution for the region around the pulsar turns out to be unstable.

6.8.2 The Magnetosphere

The corotation stops, of course, when the charged particles reach the velocity of light at a certain distance, given by

$$R_L = \frac{c}{\Omega} = 4.71 \times 10^4\,\mathrm{km} \left(\frac{P}{1\,\mathrm{s}}\right), \tag{6.39}$$

which is the radius of the light cylinder around the pulsar. We note that for a millisecond pulsar, the radius of the light cylinder is barely 50 km, not so far from the surface of the pulsar. The region within the light cylinder is also called the magnetosphere of a pulsar.

Let us take a closer look at the magnetosphere of a pulsar. The outside solution for the magnetic field is a magnetic dipole (see Eq. (6.13)). The magnetic field lines can be calculated by considering an infinitesimal path $\mathrm{d}\boldsymbol{s} = (\mathrm{d}r, r\mathrm{d}\theta)$, where the magnetic field $\boldsymbol{B} = (B_r, B_\theta)$ does not change. Hence, the cross product vanishes: $\mathrm{d}\boldsymbol{s} \times \boldsymbol{B} = 0$, which gives the condition:

$$\frac{\mathrm{d}r}{r \cdot \mathrm{d}\theta} = \frac{B_r}{B_\theta} = \frac{2\cos\theta}{\sin\theta}. \tag{6.40}$$

The differential equation can be solved to give the solution for the magnetic field lines in the parametrized form

$$r = r(\theta) = K \cdot \sin^2\theta, \tag{6.41}$$

where K is a constant labeling the corresponding magnetic field line. The magnetic field lines will be inside the magnetosphere of the pulsar up to the point where the magnetic field lines touch the light cylinder at the equator at $r = R_L$ and $\theta = \pi/2$. The constant describing this critical magnetic field line is then given by $K_c = R_L$. The critical magnetic field lines emerge from the polar region of the pulsar at the radius of the pulsar $r = R$ with the angle θ_P relative to the rotation axis determined by

$$\sin\theta_P = \left(\frac{R}{R_L}\right)^{1/2}. \tag{6.42}$$

Magnetic field lines emerging within the angle θ_P around the poles will not close within the magnetosphere and will go beyond the light cylinder. Particles escaping from the surface of the pulsar and accelerating along the magnetic field lines will be visible from the outside. The radius of this polar cap is given by

$$R_P \approx R \cdot \sin\theta_P = R\left(\frac{R}{R_L}\right)^{1/2} = 145\,\mathrm{m}\left(\frac{R}{10\,\mathrm{km}}\right)^{3/2}\left(\frac{1\,\mathrm{s}}{P}\right)^{1/2}, \tag{6.43}$$

which is much smaller than the typical radius of a neutron star.

6.8.3 Particle Acceleration

We can estimate the energy to which particles can be accelerated from the poles by looking at the potential at the surface of the pulsar. Take the expression for the electric potential, Eq. (6.36). Within the polar cap surface, one has a potential difference from the border of the cap at $\theta = \theta_p$ to the pole of

$$\Delta\phi = \frac{1}{2}\Omega B_0 R^2 \sin^2\theta_P = \frac{1}{2}\Omega B_0 R^2 \frac{R}{R_L} = 6.58 \times 10^{12}\,\mathrm{V}\left(\frac{B_0}{10^{12}\,\mathrm{G}}\right)\left(\frac{1\,\mathrm{s}}{P}\right)^2, \tag{6.44}$$

assuming a radius of $R = 10\,\mathrm{km}$. This is the typical electric potential to be expected at the polar caps of a pulsar. A charged particle will be accelerated within such a potential difference to energies of $E_{\mathrm{acc}} = e \cdot \Delta\phi$, which gives

$$E_{\mathrm{acc}} = e \cdot \Delta\phi = 6.58\,\mathrm{TeV}\left(\frac{B_0}{10^{12}\,\mathrm{G}}\right)\left(\frac{1\,\mathrm{s}}{P}\right)^2. \tag{6.45}$$

These energies are huge. In fact, they are comparable to the energies of the Large Hadron Collider (LHC) at CERN (the European Organization for Nuclear Research), where two proton beams are accelerated up to 6.5 TeV (with a combined energy of 13 TeV). For a millisecond pulsar or a magnetar with $B_0 = 10^{15}\,\mathrm{G}$, the acceleration energy scale of a pulsar would be even orders of magnitude larger than the one of the LHC. That is why pulsars are considered to be a source of the most energetic particles in the universe seen, for example, in cosmic rays, where cosmic ray showers with an energy of up to $10^{21}\,\mathrm{eV}$ have been observed. The lightest charged particle, the electron can easily be accelerated to energies larger than twice the mass of the electron. The acceleration along the curved magnetic field lines will produce highly energetic photons, which can produce electron–positron pairs and fill the magnetopshere with charged particles.

6.8.4 Magnetic Energy Loss

Finally, let us consider the energy generated by the magnetic field lines at the light cylinder that is flowing outwards. The magnetic field at the light cylinder has the parametric dependence

$$B \sim \frac{|\boldsymbol{m}|}{R_L^3} \sim B_0\left(\frac{R}{R_L}\right)^3, \tag{6.46}$$

where $\boldsymbol{m}$ is the magnetic moment. The energy density in the magnetic field flowing outward is given by the Pointing energy flux $S = B^2/(4\pi)$. The total energy loss integrated over the light cylinder is given by

$$\dot{E}_B \approx -\frac{B^2}{4\pi} \cdot 4\pi R_L^2 \sim -\frac{B_0^2 R^6}{R_L^4} \sim -B_0^2 R^6 \Omega^4. \tag{6.47}$$

We recover the same parametric dependence of the energy loss due to dipole emission of the Lamor formula, see Eq. (6.20), in particular, the dependence on Ω to the fourth power. Hence, the dipole formula for the energy loss of a pulsar is more general and applies also for a more sophisticated model. In fact, observations of the Crab pulsar reveal that there is a substantial energy loss not only from the polar regions of the pulsar. Wisps have been seen that emerge within the equatorial plane of the rotation axis, both in the optical by the Hubble Space Telescope and in the X-ray band by the Chandra satellite (Hester et al., 2002).

6.9 Dispersion Measure

The regular radio pulses of pulsars can be used to estimate the pulsar distance to Earth. The interstellar medium can be quite hot, with temperatures of about 10^4 K, so that electrons are stripped off the atoms and a plasma of hydrogen nuclei (protons) and electrons forms. These ionized regions, HII-regions in astrophysical terminology, have typical electron densities of 10^3–10^4 cm^{-3}. If the radio pulse goes through such interstellar plasma clouds, a frequency-dependent time shift will occur that is proportional to the integrated electron density along the path of the radio pulse. The dispersion measure DM is defined as the integral along the line of sight of the electron density n_e, the column density

$$\mathrm{DM} = \int_0^D \mathrm{d}s \; n_e(s), \tag{6.48}$$

where D is the distance to the pulsar. The dispersion measure can be related to the frequency-dependent time of arrival of the electromagnetic wave going through the ionized medium.

Let us see how this works by first deriving heuristically the characteristic frequency of the electrons in a plasma. Consider a plasma of electrons with protons, that is, an interstellar ionized hydrogen cloud. If the charge distribution of the electrons is shifted by a distance x compared to the background of protons, a net force will act on the electrons. The force is equal to $F_e = e \cdot E = -m_e \cdot \ddot{x}$, where E is the electromagnetic field and m_e the mass of the electron. From the Poisson equation $\nabla E = 4\pi e^2 n_e$, with n_e being the electron density, one gets an oscillation equation for electrons in a plasma

$$\ddot{x} + \omega_p^2 \cdot x = 0, \tag{6.49}$$

with the plasma frequency defined as

$$\omega_p^2 = \frac{4\pi e^2}{m_e} \cdot n_e. \tag{6.50}$$

The dispersion relation for an electromagnetic wave changes from $\omega = k$ in the vacuum to

$$\omega^2 = \omega_p^2 + k^2 \tag{6.51}$$

in a plasma with the plasma frequency ω_p. The dispersion relation looks like one for a massive particle, a massive photon, with the mass $m = \omega_p$. Note that the frequency ω must be larger than the plasma frequency ω_p to propagate through the plasma. Frequencies below the plasma frequency are not propagated as the electrons cannot be excited to start oscillating. The group velocity of an electromagnetic wave will be frequency-dependent now, as

$$v_g(\omega) = \frac{d\omega}{dk} = \frac{k}{\omega} = \left(1 - \frac{\omega_p^2}{\omega^2}\right)^{1/2} \approx \left(1 - \frac{\omega_p^2}{2\omega^2}\right), \tag{6.52}$$

where for the last expression the approximation $\omega \gg \omega_p$ was used in a Taylor expansion. The time it takes for pulses to reach the observer will be frequency-dependent due to the presence of the plasma:

$$t(\omega) = \int_0^t dt' = \int_o^D \frac{ds}{v_g(\omega)} \approx \int_o^D ds \left(1 + \frac{\omega_p^2}{2\omega^2}\right), \tag{6.53}$$

where we used Eq. (6.52) and made again a Taylor expansion for $\omega \gg \omega_p$ in the last step. Putting in the expression for the plasma frequency, we arrive at the final expression for the frequency dependency of the time shift of the radio signal from the pulsar:

$$t(\omega) = D + \frac{2\pi e^2}{m_e\omega^2} \int_0^D ds\, n_e(s) = D + \frac{2\pi e^2}{m_e\omega^2} \mathrm{DM}. \tag{6.54}$$

For a known dispersion measure or column density of electrons, the measured time shift will result in an estimate of the distance to the pulsar D. The electron density or dispersion measure DM within our galaxy has been mapped out by radio observations so that distances to pulsars can be determined.

6.10 Neutron Star Masses

The orbital parameters of binary and triple pulsar systems, including the masses, can be very precisely determined by including corrections from general relativity. As of this writing, the most massive known neutron stars are the pulsar J0740+6620, with a mass of $2.17^{+0.11}_{-0.10}$, discovered in 2019 (Cromartie et al., 2019), the pulsar J0348+0432, with a mass of $M = 2.01(4) M_\odot$, discovered in 2013 (Antoniadis et al., 2013), and the pulsar J1614−2230 discovered in

2010 (Demorest et al., 2010). The newest release from the 11-year data of the NANOGrav collaboration reports a refined mass of J1614−2230 of $M = 1.908(16)M_\odot$ (Arzoumanian et al., 2018). For the decades before their discoveries, the Hulse–Taylor pulsar was the most massive known neutron star with a mass of $M = 1.438(1)M_\odot$, which has a companion neutron star with a mass of $M = 1.390(1)M_\odot$ (Weisberg and Huang, 2016). Presumably, this is the likely reason that the literature on neutron stars uses as a reference the pulsar mass value of the old record holder mass of $M = 1.4M_\odot$, even today after the discovery of much heavier neutron stars. The lightest neutron star known at present is the companion of the pulsar J0453+1559 with a mass of only $M = 1.174(4)M_\odot$ (Martinez et al., 2015). The most precisely measured neutron star masses are those of the double neutron star system PSR B1534+12 with $M_1 = 1.3330(2)$ and $M_2 = 1.3455(2)$ (Fonseca et al., 2014). The double pulsar J0737−3039 is the only known system of two neutron stars orbiting each other where one has detected the pulses of both neutron stars (Kramer et al., 2006). A summary of accurately measured neutron star masses has been given by Özel and Freire (2016).

How can neutron star masses be determined so precisely? One reason is that pulsars are so extremely precise clocks, sending out regular signals that allow mapping out of the orbit to a very high degree of accuracy. The other reason is that effects from general relativity, giving corrections to the Keplerian orbits, are so strong that they can be measured by pulsar timing. These corrections from general relativity depend on the warping of space time and thereby on the mass of the neutron star. The statement is therefore that the masses of neutron stars are solely determined from general relativity. No model assumptions enter the determination of neutron star masses. There are exceptions, of course, and we will discuss those exceptions also.

6.10.1 The Mass Function

Let us start with the classic way of determining star masses from binary systems. We just use Kepler's laws for elliptic orbits. Let us denote the orbital period by P_b and the semi-major axis length by a. In the center-of-mass frame, the two stars orbit each other on two ellipses. The semi-major axis of the individual ellipses a_1 and a_2 are related by the center-of-mass relation for the individual masses m_1 and m_2:

$$m_1 \cdot a_1 = m_2 \cdot a_2, \tag{6.55}$$

where also $a = a_1 + a_2$. For circular orbits, the velocity of the two stars v_1 and v_2 are constant and related to the semi-major axis by

$$v_1 = \frac{2\pi a_1}{P_b} \qquad v_2 = \frac{2\pi a_2}{P_b}. \tag{6.56}$$

Now we use Kepler's third law which states that the square of the orbital period P_b is proportional to the cube of the semi-major axis:

$$P_b^2 = \frac{4\pi^2}{G(m_1 + m_2)}. \tag{6.57}$$

Of course, we put Kepler's third law in the form that takes care of the different masses involved in the problem. Using the relation for the velocities, we get for the sum of the two masses

$$m_1 + m_2 = \frac{P_b}{2\pi G}(v_1 + v_2)^3. \tag{6.58}$$

Observationally, one can only measure the radial velocity component v_r, which is related to the velocity v by the inclination angle ι of the orbital plane relative to the plane of the sky of the observer by $v_r = \sin\iota \cdot v$. Note that $\iota = 0$ corresponds to the case where the observer looks directly onto the orbital plane, so the radial velocity vanishes. Hence, the sum of the masses is given by

$$m_1 + m_2 = \frac{P_b}{2\pi G}\frac{(v_{1r} + v_{2r})^3}{\sin^3\iota}. \tag{6.59}$$

So, if one knows the radial velocities of both stars, one knows the mass ratio of the two stars. Measuring the orbital period P_b gives from Eq. (6.59) only a limit on the total mass of the system if the inclination angle ι is not known.

In many cases, only one star of the binary system can be observed directly, as is usually the case for binary pulsars. We use now the center-of-mass relation to replace the radial velocities of the unobserved star:

$$\frac{m_1}{m_2} = \frac{a_2}{a_1} = \frac{v_2}{v_1} = \frac{v_{1r}}{v_{2r}}, \tag{6.60}$$

which holds also for noncircular orbits, see later. We arrive then at the mass function:

$$f(m_1, m_2, \sin\iota) = \frac{m_2^3}{(m_1 + m_2)^2}\sin^3\iota = \frac{P_b}{2\pi G}v_{1r}. \tag{6.61}$$

The right-hand side contains only observable quantities P_b and v_{1r}. Now two more constraints are needed to fix the individual masses m_1 and m_2 and the inclination angle ι.

For noncircular orbits, the radial velocity will change in dependence on the eccentricity e of the orbit. The amplitude of the radial velocity is then given by

$$v_{1r} = \frac{2\pi a_1}{P_b}\frac{\sin\iota}{(1 - e^2)^{1/2}} \tag{6.62}$$

and generalizes the expression for circular orbits, Eq. (6.56), by an additional common factor. This means that the ratio of the amplitude remains the same as for circular orbits and that the expression for the mass ratio, Eq. (6.61), remains the same. The eccentricity e can be extracted by making a fit to the change of the radial velocity curve as a function of the orbital phase, which determines also the orbital period P_b as well as the amplitude of the radial velocity v_{1r}. Still, two more constraints are needed to determine both masses.

For determining the masses of exoplanets, one uses the effect that the spectral lines are Doppler-shifted due to the tiny movements of the star around the center-of-mass, which determines v_{1r} and P_b. From stellar theory, one can deduce the mass of the star from its spectral features. Then, the mass of the planet can be calculated if the inclination angle ι is known. The inclination angle can be determined for eclipsing binary systems, where the exoplanet is passing between the line of sight of the star and the observer.

6.10.2 Determining Pulsar Masses

How about determining pulsar masses? The pulses of the pulsar can be used to determine the radial velocity v_{1r} and the period P_b. If the companion of the pulsar is a visible star, as for example, a white dwarf or an ordinary star, the mass ratio can be fixed by measuring, for example, v_{2r}. The mass of the star can be fixed by using stellar theory, as done for stars of exoplanets. The mass of a white dwarf can be determined using the techniques outlined in Section 5.8 by measuring the gravitational redshift from spectral lines and the surface temperature. Indeed, the mass of the massive pulsar PSR J0348+0432 has been calculated from the properties of the white dwarf companion. Observations of the Very Large Telescope (VLT) gives a radial velocity, a gravitational redshift of the Balmer lines, and the surface temperature. The measurement of the radial velocity of the pulsar with the one of the white dwarf gives the mass ratio $q = M_P/M_{\rm wd} = 11.70 \pm 0.13$. The mass of the white dwarf is estimated to be $M_{wd} = 0.172 \pm 0.003 M_\odot$, so the pulsar mass is calculated to be $M_P = 2.01 \pm 0.04$ (Antoniadis et al., 2013).

If the companion of the pulsar is unseen, one has to resort to other observables. Here the strong gravitational field around a neutron star helps, as corrections to the Keplerian orbit from general relativity depends on the mass of the neutron star. From the previous discussion, one needs two more observables to determine the mass of the neutron star from the mass function, Eq. (6.61). In fact, corrections from general relativity to the first order in an expansion in velocities to Keplerian orbits can provide five additional observables. These parameters are called the post-Keplerian (PK) parameters. As this overdetermines the system, a measurement of three or more relativistic corrections allows for a test of general relativity in the strong gravitational field around a neutron star. In the following, we list those five correction terms from general relativity. For details on the derivation of the formulae given, we refer to for example, the textbook by Straumann (2013). In the following, we give the expressions for elliptic orbits with an eccentricity e.

Einstein Time Delay

There is a characteristic delay of the pulsar signals due to the difference between the coordinate time t_e and the proper time T_e. For weak fields, the proper time is given by

$$dT_e^2 = (1 + 2\phi)dt_e^2 - (1 - 2\phi)dr^2, \tag{6.63}$$

where ϕ is the gravitational potential of the pulsar. The offset of the two different timescales has two contributions

$$\frac{dT_e}{dt_e} \approx 1 + \phi - \frac{1}{2}v^2. \tag{6.64}$$

The first term describes the classic gravitational redshift, the second term the Doppler shift. We introduce now the eccentric anomaly u, implicitly defined by the relation for the radial distance r

$$r = a \cdot (1 - e \cdot \cos u), \tag{6.65}$$

where a is the semi-major axis and e the eccentricity of the orbit. The time delay of the pulsar signals has the characteristic pattern in terms of the eccentric anomaly u as

$$t_e = T_e + \Delta_E = T_e + \gamma \cdot \sin u. \tag{6.66}$$

Here γ is the Einstein parameter, which depends on the masses m_1 and m_2 of the binary system as

$$\gamma = G^{3/2} \frac{m_2(m_1 + 2m_2)e}{(m_1 + m_2)^{4/3}} \left(\frac{P_b}{2\pi}\right)^{1/3}. \tag{6.67}$$

Measuring γ gives an additional constraint on the masses m_1 and m_2.

Shapiro Time Delay

There is an additional time delay from the gravitational field of the companion star of the pulsar. For radio pulses we use now the condition for a light-like path:

$$ds^2 = (1 + 2\phi_c)dt_s^2 - (1 - 2\phi_c)dr^2 = 0 \tag{6.68}$$

for weak gravitational fields where ϕ_c is now the gravitational potential of the companion star. When the pulses travel through the gravitational potential of the companion star, the coordinate time difference is given by

$$\mathrm{d}t_s \approx \pm(1 - 2\phi_c)\mathrm{d}r. \tag{6.69}$$

The time delay by the gravitational field ϕ_c is then

$$\Delta_s = -2 \int \mathrm{d}r\, \phi_c \tag{6.70}$$

and can be related to orbital parameters by

$$\Delta_s = -2r \cdot \log\left\{(1 - e\cos u) - s \cdot \left[\sin\omega(\cos u - e) + \sqrt{1 - e^2}\cos\omega\sin u\right]\right\}, \tag{6.71}$$

where ω stands for the periastron of the orbit. There are two parameters that can be determined from fits to the pulsar signal. The first one is the range parameter r, which is simply given by the mass m_2 alone: $r = Gm_2$. The second one is the shape parameter s, which is solely determined by the orbital inclination angle: $s = \sin\iota$. So a detection of a Shapiro delay alone is sufficient to determine the mass of a pulsar.

The Shapiro delay is a well-tested prediction of general relativity. The Cassini mission has been sent to Mars, emitting radio signals when Mars was nearly behind the Sun in superior conjunction. The radio signals had to pass through the gravitational potential of the Sun, thereby experiencing a Shapiro time delay. The prediction of general relativity could be confirmed within a level of 2×10^{-5} (Bertotti et al., 2003).

The mass of the pulsar PSR 1614−2230 has been determined by measuring the Shapiro delay (Demorest et al., 2010). At the time of the mass measurement in 2010, it was the most massive neutron star known. In fact, it was the first measured pulsar with a mass close to two solar masses, much larger compared to the old record holder, the Hulse–Taylor pulsar with a mass of $1.44 M_\odot$. The discovery of a two solar mass pulsar was simply sensational at that time. The mass measurement profited from the extremely strong Shapiro delay amplitude due to the highly inclined orbital plane: the measured inclination was determined to be $\iota = 89.17^0 \pm 0.02^0$, indicating an orbit that we see nearly perfectly edge-on.

Periastron Advance

The periastron advance of Mercury is one of the classical tests of general relativity. While the periastron advance of Mercury due to effects from general relativity alone is tiny, just 43 arcseconds in a century, the periastron advance of pulsars can be huge due to the considerably larger gravitational fields involved. The periastron advance is given by

$$\dot{\omega} = 3 \left(\frac{P_b}{2\pi} \right)^{-5/3} \frac{(G(m_1 + m_2))^{2/3}}{(1 - e^2)} \tag{6.72}$$

and depends on the total mass of the system. The measured value of the periastron advance of, for example, the double pulsar system PSR 0737−3039 is $\dot{\omega} = 16.90^0 \pm 0.01^0\,\mathrm{yr}^{-1}$ (Lyne et al., 2004).

Gravitational Wave Emission

Keplerian orbits are stable, orbits in general relativity are not. Binary systems spiraling around each other emit gravitational waves, thereby losing energy. The loss of energy leads to a decrease of the orbital period, the binary stars are getting closer in time, eventually merging with each other. The change of the orbital period is given by (see also Chapter 10):

$$\dot{P}_b = -\frac{192\pi}{5} \frac{G^{5/3} m_1 m_2}{(m_1 + m_2)^{1/3}} \left(\frac{2\pi}{P_b} \right)^{5/3} f(e), \tag{6.73}$$

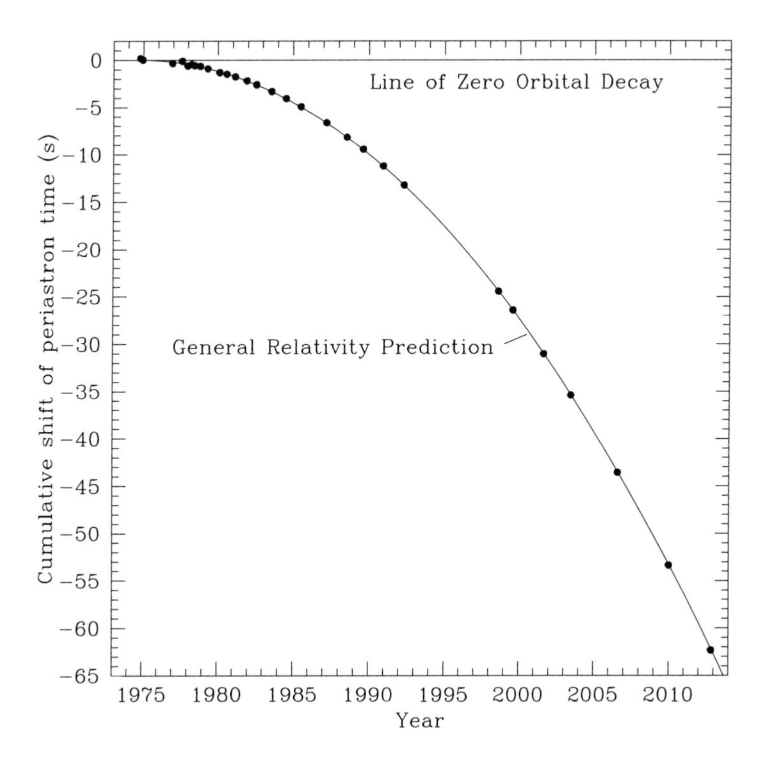

Figure 6.2 The rate of change of the orbital period P_b for the Hulse–Taylor pulsar PSR1913+16. The prediction from general relativity for gravitational wave emission is shown by the solid line and is in excellent agreement (from Weisberg and Huang, 2016). © AAS. Reproduced with permission

where $f(e)$ is function describing the dependance on the eccentricity e:

$$f(e) = \left(1 + \frac{73}{24}e^2 + \frac{37}{96}e^4\right)(1 - e^2)^{-7/2}. \tag{6.74}$$

The orbital decay for binary pulsars has been confirmed by observing the Hulse–Taylor pulsar for several decades. Figure 6.2 shows the change of the orbital period for more than three decades, from 1974 until 2016. The prediction of general relativity shown by a solid line goes right through the data points. The ratio of the observed value to the predicted one is 0.9983 ± 0.0016 (Weisberg and Huang, 2016).

6.11 The Double Pulsar

In 2003, a system of two neutron stars was discovered (Burgay et al., 2003) where both neutron stars were observed as pulsars shortly afterward (Lyne et al., 2004). The system PSR J0737−3039 is the only known example of a double pulsar. It serves as the best test of general relativity for strong gravitational fields so far, as the orbital parameters of both neutron stars can be measured by observing the two pulsars orbiting each other. In fact, for the double pulsar, all the five PK parameters mentioned could be measured (Kramer et al., 2006). In addition, the spin precession of pulsar B around the angular momentum $\Omega_{\mathrm{SO,B}}$ has been also measured (Kramer and Wex, 2009). As the radial velocity curve of both stars can be measured, the mass ratio is known automatically, see Eq. (6.60), and only one additional measurement of a PK parameter is enough to determine both masses of the pulsars separately. Hence, the measurement of six PK parameters provides a fivefold test of general relativity.

Figure 6.3 shows the diagram of the two pulsar masses. The various lines show the constraint from the measurement of the five PK parameters. For a known mass ratio R, the crossing of just one PK parameter constraint fixes the two pulsar masses uniquely. If general relativity is the correct description of gravity around pulsars, all five constraints have to meet in a common area. As one can see from the enlarged area of the crossing, all five PK parameter constraints are compatible with each other, so that general relativity passes this fourfold test in the strong gravity regime with flying colors. The most stringent test comes from the measurement of the Shapiro time-delay shape parameter s, confirming the prediction of general relativity within a precision of 0.05%.

Table 6.2 lists the numerical values of five of the six PK parameters, comparing the measurement with the prediction of general relativity. Here, the masses are fixed from the measured mass ratio R and the perihelion advance $\dot{\omega}$ first, and then the

Table 6.2 *The five experimental tests of general relativity with the double pulsar. The masses of the two pulsars are determined by the mass ratio R and the periastron advance* $\dot{\omega}$. *The four PK parameters of orbital period change* $\dot{P}$, *Einstein parameter* γ, *the Shapiro time-delay parameters for the shape s and the range r, and the spin precession of pulsar B* $\Omega_{SO,B}$, *respectively, are then predicted by general relativity (taken from Kramer and Wex, 2009).*

PK parameter	Observed	GR expectation	Ratio
$\dot{P}_b$	1.252(17)	1.24787(13)	1.003(14)
γ (ms)	0.3856(26)	0.38418(22)	1.0036(68)
s	0.99974(−39,16)	0.99987(−48,+13)	0.99987(50)
r (μs)	6.21(33)	6.153(26)	1.009(55)
$\Omega_{SO,B}$ (deg yr^{-1})	4.77(+0.66,−0.65)	5.0734(7)	0.948(13)

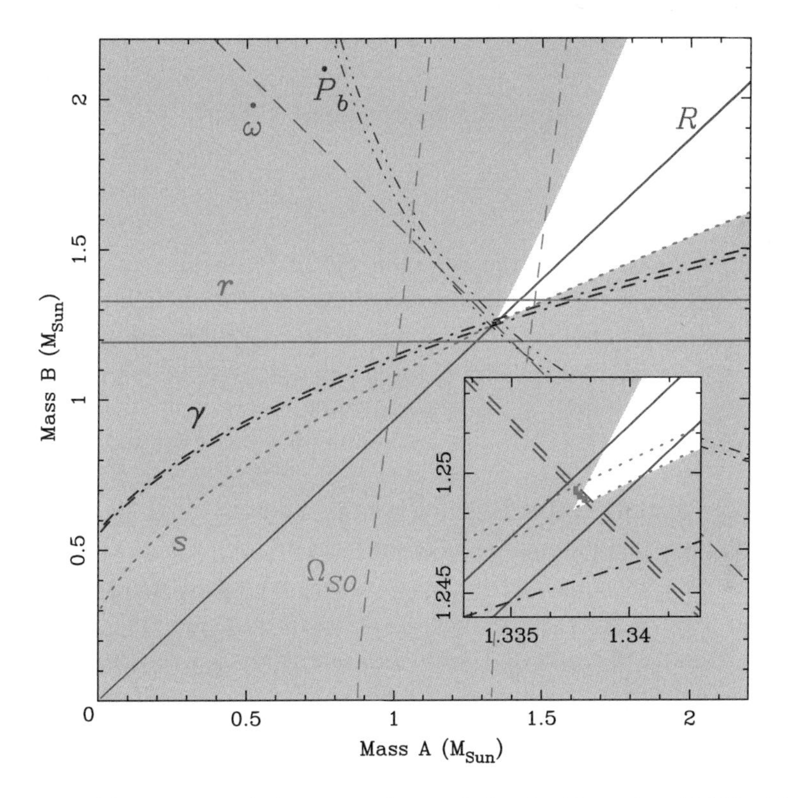

Figure 6.3 The masses of the two pulsars of the double pulsar J0737−3039 showing the mass ratio R and the five PK parameters: perihelion advance $\dot{\omega}$, change of orbital period $\dot{P}_b$, Shapiro delay range r and shape s, the Einstein parameter γ, and the spin precession of pulsar B about the orbital angular momentum Ω_{SO} (from Kramer and Wex, 2009).

PK parameters are determined from the measurement and computed from general relativity. As seen from the ratio of the measured and predicted values, the overall agreement is excellent.

Exercises

(6.1) Verify the expression for the moment of inertia of a homogenous sphere, Eq. (6.8) (hint: calculate $I = (I_x + I_y + I_z)/3$, where I_x is the moment of inertia with respect to a rotation axis along the x-axis, etc.) . For comparison, calculate the moment of inertia prefactor f for a sphere with the same mass and radius, where the density falls off with a power law toward the surface of the sphere $\rho(r) = \rho_0(1 - (r/R)^n)$.

(6.2) Calculate the total rotational energy stored in the Crab pulsar, adopting a mass of $M = 1.4M_\odot$ and a radius of $R = 10\,\mathrm{km}$. Compare your result to the total energy emitted by our Sun over its total lifetime (assume a constant luminosity of $L_\odot = 3.828 \times 10^{33}\,\mathrm{erg\,s^{-1}}$ and a lifetime of 10 Gyrs).

(6.3) Assume that the decrease in the angular frequency Ω of pulsars is a power law of the form

$$\dot{\Omega} = -K \cdot \Omega^n,$$

where K and n are constants. Solve the differential equation for arbitrary n. Assume that the initial angular frequency Ω_i is much larger than the one measured today Ω_0. Show that in the limit of $\Omega_i \gg \Omega_0$, the present characteristic timescale for slowing down is given by

$$t_{\mathrm{char}} = -\frac{\Omega_0}{(n-1)\dot{\Omega}_0}.$$

(6.4) Solve the differential equation for the magnetic field lines, Eq. (6.40), and show that the solution is of the form given in Eq. (6.41).

(6.5) Calculate the characteristic age and the characteristic magnetic field of the pulsars listed in Table 6.1. Where are the pulsars located in the diagram of Figure 6.1? Explain the correlation between the characteristic age and the surface magnetic field and derive an evolutionary scenario for pulsars.

(6.6) Estimate the maximum amount of magnetic field strength for a neutron star. Assume a constant magnetic field and calculate the total energy of the magnetic field for a typical neutron star (adopt $M = 2M_\odot$ and a radius of $R = 10\,\mathrm{km}$). Compare the energy to the gravitational binding energy. At which magnetic field strength is the energy in the magnetic field as large as the gravitational binding energy?

7

Neutron Stars

In the previous chapters we laid the foundation to discuss the properties of neutron stars. In this chapter we will explicitly construct the composition of a neutron star from the outside to the inner core, as known today.

7.1 Brief History of Neutron Stars

The history of neutron stars is particular interesting. White dwarfs were discovered before theoretical modeling in terms of a Fermi gas of electrons. For neutron stars, the story is entirely different as their existence was postulated on theoretical grounds. It took more than 30 years from the first predictions of the existence of neutron stars to their serendipitous discovery by observing pulsars that are rotation-powered neutron stars.

It was 1931 in Zurich that Landau wrote his influential article on the maximum mass of compact stars, which was published in 1932 (Landau, 1932). The neutron was discovered also in 1932 by Chadwick. Leon Rosenfeld recalls much later in 1974 in a transcribed discussion about neutron stars at a conference that he had witnessed a discussion between Niels Bohr and Lev Landau 1932 in Copenhagen shortly after the discovery of the neutron (Rosenfeld, 1974). Rosenfeld let Landau speak about neutron stars as being 'eerie stars' (in fact, he uses the corresponding German expression *'unheimliche Sterne'* as scientific discussions were also held in German at that time). However, the Niels Bohr institute keeps a record of his visitors' stays. A perusal of the archival data revealed that most likely the discussion happened in 1931 (see the nice and excellent historical treatise of Yakovlev et al., 2013). Hence, the discussion in 1931 was obviously about the work of Landau on the existence of dense stars being like a giant nucleus, which anticipated the concept of neutron stars. The term neutron stars appears for the first time in an abstract by Walter Baade and Fritz Zwicky of the meeting of the American Physical Society held in December 1933 and published in 1934 (Baade and Zwicky, 1934a, 1934b).

Baade and Zwicky reported on the recent discovery of supernovae and speculated that their origin is the collapse of a massive star to a neutron star. However, it was the work of Landau that motivated Oppenheimer and his student Volkoff to study neutron stars within general relativity. The equations to solve were laid out in the textbook of Tolman (Tolman, 1934) for fluids of spheres. Thereby, these equations are now called the Tolman–Oppenheimer–Volkoff (TOV) equations. They adopted these equations for the case of a fluid of free neutrons and solved the equations numerically. Oppenheimer with Volkoff, at Berkeley, and Tolman, at Caltech in Pasadena, were in mutual contact with each other about this project (see e.g., Thorne, 1994, for a nice historical account). Tolman was solving the equations for a sphere of fluids, making assumptions on the form of the metric functions. Their articles appeared back-to-back, making reference to each other (Oppenheimer and Volkoff, 1939; Tolman, 1939). The important numerical result of Oppenheimer and Volkoff was that the maximum mass of a neutron star composed of a free gas of neutrons was just $0.7 M_\odot$. Using Tolman's analytic solutions, they argued that the inclusion of repulsive forces between neutrons will not raise the maximum mass to values much higher value than of the order of $1 M_\odot$. The last statement is remarkable as the existence of repulsive nuclear forces at high densities was not known at that time and their statement is still correct today, about 80 years later.

In 1942, Duyvendak recollected ancient Chinese and Japanese records of events in the sky called 'guest stars'. According to these records, a guest star was observed close to the star ζ Tauris in the year AD 1054. It was so bright that it was visible during daytime for 23 days, even being visible to the naked eye at night for nearly two years (Duyvendak, 1942). Mayall and Oort put forward the astronomical implications of the new translation of Duyvendak in a follow-up paper (Mayall and Oort, 1942). They identified the observation of this guest star with the phenomenon of supernovae and associated the Crab nebula in the constellation Tauris to this historical supernova of AD 1054, which is just about one degree away from ζ Tauris. Surprisingly, it took many more years, in fact, one had to wait for the serendipitous discovery of pulsars in 1967 to associate the star in the middle of the Crab nebula with the neutron star created by the supernova of AD 1054 (Cocke et al., 1969). The neutron star in the Crab nebula, the Crab pulsar, is visible by optical telescopes, blinking in the optical with a frequency of 30 Hz. This frequency is normally to high to be observable by the human eye. However, at least two people observing the Crab nebula through telescopes reported to have seen a blinking star before 1967, which, however, was not followed up (see e.g., Bell Burnell, 2009).

Twenty years after the work of Oppenheimer and Volkoff, the theoretical investigation of neutron stars took up momentum again. A unified equation of state (EOS), connecting dense matter for white dwarfs to neutron star core densities, was developed by Harrison and Wheeler starting in 1958. They used an empirical mass formula for nuclei, the liquid drop model, and calculated the EOS for a mixture of nuclei together with electrons and a free gas of neutrons by minimizing the total energy in β-equilibrium. Neutron stars consisted then of a crust of nuclei and a liquid core of a gas of neutrons, protons, and electrons. They wrote up their results with additional material several years later. The amount of material collected at the end was so much that the results were not published in a regular article but in a small booklet together with Wakano and Thorne (Harrison et al., 1965). The early work of Harrison and Wheeler motivated Cameron to study the properties of neutron stars, taking into account interactions between neutrons (Cameron, 1959). He used the model of Skyrme for describing the nuclear interactions in terms of an effective potential in the nucleon density adapted for the case of neutron matter (Skyrme, 1959). He found a maximum mass of $2.0M_{\odot}$ at a radius of 8.1 km, which is a considerable increase in the maximum mass compared to the findings of Oppenheimer and Volkoff of $0.7M_{\odot}$ for the noninteracting case. Cameron also points out in his article that hyperons will eventually appear at high densities. Salpeter discusses the EOS for neutron stars using a different model for the nuclear interactions, where he also includes hyperons. Assuming a universal potential for nucleons and hyperons, he finds that hyperons coexist above 10^{15} g cm^{-3} with neutrons and protons (Salpeter, 1960). At the same time, Ambartsumyan and Saakyan investigate the appearance of hyperons in dense neutron star matter for a free Fermi gas (Ambartsumyan and Saakyan, 1960). They find that the Σ^- hyperon appears at a density of $0.64\,\text{fm}^{-3}$, the Λ hyperon at a density of $1.27\,\text{fm}^{-3}$, which are about four times and eight times the saturation density of $n_0 = 0.15\,\text{fm}^{-3}$. In a subsequent paper (Ambartsumyan and Saakyan, 1962), they calculate the mass–radius relation of neutron stars including hyperons for the free Fermi gas. They even compute the EOS for nucleons and hyperons with an interaction term being proportional to the number density squared. They find that hyperons appear abundantly in the core of the massive neutron star configurations. The maximum mass of neutron stars in the free case is $0.63M_{\odot}$ at a radius of 11 km, and for the interacting case $1.0M_{\odot}$ at a radius of only 5.4 km.

In the following, we discuss the present modern picture of the structure of neutron stars working from the outside, low densities, to the core, high densities. Surprisingly, we will find that many basic aspects can be traced back to the early papers on neutron stars. The outline is clear: we need to discuss the crust of a neutron star, the neutron matter EOS with interactions, and the possible transition to hyperons at high densities.

7.2 Neutron Star Crust

In the following, we will explore the structure of the outer layers of a neutron star. Besides the top layer, the atmosphere, the outer layers consist of a solid structure due to the formation of a lattice of nuclei. We will cover several aspects of the neutron star crust that can be taken over from the study of white dwarf material. In fact, the outer crust has similarities to the core of white dwarfs.

7.2.1 The Atmosphere

The outermost layer of a neutron star is the atmosphere consisting of atoms. We have estimated in the section on white dwarfs in Eq. (5.68) that the thickness of the atmosphere of a neutron star is tiny, ranging from 0.3 mm for a hydrogen atmosphere to 10 cm for an iron atmosphere. The atmosphere can be safely ignored then for discussing the radius of neutron stars.

However, the atmosphere determines the spectral features of a neutron star, as the mean-free path of photons is much smaller than the size of even such a tiny atmosphere. Determining the radius of a neutron star by spectroscopic methods therefore has to take into account the properties of the atmosphere. The composition of the atmosphere is not fixed and can change from neutron star to neutron star, depending on the production mechanism and the surroundings of the neutron star. For a binary neutron star system with an ordinary star, the atmosphere will consist mainly of hydrogen, as this is the material accreted from the companion star's outer layer. For an isolated neutron star moving through the interstellar medium, the accumulated atmosphere will be also mainly hydrogen. A newly born neutron star will have more likely a composition related to the fallback material from the supernova, which is dominated by heavier elements as, for example, silicon.

7.2.2 Outer Crust

The approach outlined here for describing the outer crust of a neutron star is called the BPS model as it was first introduced by Baym, Pethick, and Sutherland (Baym et al., 1971).

In the following, we will assume that the outer crust is in β-equilibrium. This is different to the case for matter in white dwarfs, where white dwarfs with different compositions, a helium or a carbon-oxygen white dwarf, can coexist. Neutron stars, however, can harbor much heavier elements in the crust. For one thing, heavy elements are produced in a core-collapse supernova. Neutrons are abundantly present and can be easily captured by nuclei, as there is no Coulomb barrier present for neutrons. And moreover, the densities reached in a neutron star are several orders of magnitude larger compared to the densities in a white dwarf. We note, however, that the element distribution in the outer crust might not be the one of β-equilibrium. The composition will depend on the formation history of the neutron star. To be absolutely sure that β-equilibrium has been established, free neutrons have to be present, as outlined earlier, which is not the case for the outer crust. Having said that, we proceed and establish the composition of the outer crust as if it is in β-equilibrium.

With increasing density, the Fermi energy of electrons will eventually be comparable to the atomic binding energy. For the most stable element, ^{56}Fe, we estimated the critical energy density to be 1.42×10^4 g cm^{-3}, see the estimates after Eq. (5.56). Above this critical density, the electrons form a free Fermi gas and the nuclei will minimize their energy by forming a lattice. The outer crust is therefore very similar in structure to white dwarf material. The contribution of the lattice to the EOS is small, as estimated for white dwarfs in Eq. (5.49). A lattice contribution of 10% or more will require nuclear charges of more than $Z \approx 50$. As we will see, the nuclear charges present in the crust are usually lower. The pressure is then dominated by the degeneracy pressure of the electrons, the energy density is dominated by the mass density of the nuclei. Hence, the EOS of the outer crust of neutron stars is the same as for matter in white dwarfs.

Electrons will get relativistic at a mass density of 2.1×10^6 g cm^{-3} for ^{56}Fe, see Eq. (5.57) and following text. The composition will change when it is favorable to transform protons and electrons to neutrons. This transformation will occur when the Fermi energy of electrons gets comparable to the energy difference of neutrons and protons. We did a rough estimate in Eq. (5.58) for the corresponding mass density and found a value of $\rho_n = 3.37 \times 10^7$ g cm^{-3} for ^{56}Fe. In reality, one has to scan over all nuclei and check which nucleus is the most energetically favored one at each given density. In order to do so, one has to minimize the total energy of the system in dependence on the mass number A and the charge Z of the nucleus. For simplicity, one assumes a Wigner–Seitz cell, describing the lattice that approximates the unit cell by a sphere with radius R. For a body-centered cubic (bcc) lattice, there will be two nuclei present in one unit cell (one at the center and $8 \times 1/8 = 1$ at the corners). The number density of nuclei is then related to the lattice size a and the radius of the Wigner–Seitz cell r_0 by

$$n_N = \frac{2}{a^3} = \frac{3}{4\pi r_0^3} \tag{7.1}$$

so that the radius r_0 is about a factor two smaller than the lattice size a. The charges of the nuclei are balanced by the charge of the surrounding electrons within the radius of the Wigner–Seitz cell, that is, $n_e = Z n_N$. The total energy density of the Wigner–Seitz cell has then contributions from the electron energy density ε_e, the lattice energy W_L, and the mass of the nucleus W_N:

$$\varepsilon(A, Z, n_N) = \varepsilon_e + n_N \cdot (W_L + W_N). \tag{7.2}$$

The term from the electrons ε_e is given by the standard expression of a free Fermi gas, see Eq. (3.110).

The contribution from the lattice, W_L, we encountered before when discussing white dwarf material, see Eq. (5.45). It has contributions from the Coulomb energy of the electrons and of the ions:

$$E_c = E_{ee} + E_{\text{ion}} = -f \cdot \frac{Z^2 e^2}{r_0}. \tag{7.3}$$

The prefactor f was found to be $9/10$ when summing up the two Coulomb energy contributions. The exact values of the electrostatic potential can be computed numerically by taking into account that the electrons form a Wigner lattice: for a bcc lattice $f = 0.895929$, for a face-centered cubic (fcc) lattice $f = 0.895874$, and for a hexagonal close packed (hcp) lattice $f = 0.895838$ (Foldy, 1978). The numerical values of f are all close to the simple estimate of $f = 9/10$. The bcc lattice has the highest electrostatic potential and is therefore the most stable configuration. However, the energy differences to the fcc and the hcp lattices are so tiny that the lattice structure in the outer crust of a neutron star could well be different from a bcc lattice. The lattice structure is important for the transport properties of the outer crust but has little influence on the lattice energy. In the following, we take the case of the bcc lattice for the lattice energy. Recall from our discussion of the EOS for white dwarfs that the contribution of the lattice to the EOS is small, see Eq. (5.49). However, the lattice contribution turns out to be important for determining the composition of the outer crust.

We turn our attention now to the most important input for the total energy of the outer crust of neutron stars, the contribution from the nuclear mass W_N. The mass of the nucleus for a given mass number A and charge Z depends on the binding energy $E_b(A, Z)$, which is the essential input:

$$W_N = Z \cdot m_p + (A - Z) \cdot m_n + E_b(A, Z). \tag{7.4}$$

Here m_p and m_p stand for the proton and the neutron mass, respectively. The binding energy of more than 3,000 nuclei is well known and compiled in the atomic mass evaluation, the latest one being from 2016 (AME2016, see Huang et al., 2017; Wang et al., 2017). Typical binding energies are around -8 MeV/A, with nuclei around $A \sim 60$ having the highest binding energies. The nucleus ^{56}Fe is the most stable one, meaning that it has the lowest total energy per nucleon of all known nuclei. Its binding energy is -8.790 MeV/A. The highest binding energy per nucleon, however, has the nucleus ^{62}Ni with a value of -8.794 MeV/A. Still, it is not the most stable one, as it has a higher neutron-to-proton ratio than ^{56}Fe, which lifts its total energy up. Recall that the proton has a mass of $m_p = 938.27\,\text{MeV}$, the neutron has a higher mass of $m_n = 939.57\,\text{MeV}$.

However, for the extreme situation we encounter in the crust of the neutron star, this experimental nuclear mass table is not enough. The missing values for the unknown nuclei have to be taken from nuclear model calculations. One might suspect that this will lead to a substantial spread in the prediction of the nuclear composition in the neutron star crust. As we will show now, there are arguments that the sequence of nuclei follows a characteristic pattern dictated by the existence of 'magic numbers', numbers of closed shells of neutrons and protons. Nuclei with closed shells are particularly stable, having higher binding energies compared to nuclei in the neighborhood of the nucleus in the nuclear chart. The pattern of magic numbers was explained independently by Maria Goeppert-Mayer and Hans Daniel Jensen, who got the Nobel prize in physics in 1963.[1]

In the nuclear shell model, the levels are labeled according to the main quantum number $n = 1, 2, 3, \ldots,$ the orbital angular momentum $l = 0, 1, 2, \ldots,$ denoted by $l = s, p, d, \ldots$ and so on in alphabetical order, and the total angular momentum $j = l \pm \frac{1}{2}$ in the form: $nl_j = 1\text{s}_{1/2}, 1\text{p}_{3/2}, 1\text{p}_{1/2}, \ldots$. If there is no interaction coupling to spin and orbital angular momentum, the energy levels are degenerate for different j and it suffices to label them according to n and l alone. Each level can be filled with two nucleons (spin up and spin down) times the degeneracy factor for the orbital angular momentum, which gives in total $2(2l+1) = 4l+2$ states. Table 7.1 shows in the upper half the level ordering according to the shell model without a spin–orbit interaction. The shell closures turn out to be 2, 8, 20, 40, 70, 112, ... The observed numbers of nucleons for particular stable nuclei are, however, 2, 8, 20, 28, 50, 82, 126, ..., in obvious contradiction for the high numbers. The solution is to introduce a strong spin–orbit splitting for the $n = 1$ states. The 1f-state splits into the states $1\text{f}_{7/2}$ with 8 states and $1\text{f}_{5/2}$ with 6 states. The former states add to the shell closure at 20, so that the additional 8 states result in the magic number 28. The magic number 50 is built from 10 additional states to the shell closure at 40, the magic number 82 from adding 12 states to the shell closure at 70, and the magic number 126 from adding 14 states to the shell closure at 112, each of them by the splitting of the higher lying $n = 1$ states. This is, indeed, what can be inferred from the lower part of Table 7.1, which takes into account a strong spin–orbit splitting of the energy levels.

We expect from our discussion of the stability of nuclei with magic numbers of nucleons that those magic numbers will also appear in the sequence of nuclei

[1] The term 'magic number' was coined by Eugene Wigner, who shared the Nobel prize with Goeppert-Mayer and Jensen in 1963.

Table 7.1 *Levels in the nuclear shell model with the magic numbers for the case without (upper part) and with spin–orbit splitting (lower part).*

Shell:	1s	1p	1d 2s		1f 2p	1g 2d 3s	1h 2f 3p	1i 2g 3d 4s
Degeneracy:	2	6	10+2		14+6	18+10+2	22+14+6	26+18+10+2
Magic number:	2	8	20		40	70	112	168
Shell:	1s	1p	1d 2s	$1f_{7/2}$	$1f_{5/2}$ 2p $1g_{9/2}$	$1g_{7/2}$ 2d 3s $1h_{11/2}$	$1h_{9/2}$ 2f 3p $1i_{13/2}$	$1i_{11/2}$ 2g 3d 4s $1k_{15/2}$
Degeneracy:	2	6	10+2	8	6+6+10	8+10+2+12	10+14+6+14	12+18+10+2+16
Magic number:	2	8	20	28	50	82	126	184

present in the outer crust of a neutron star. The recipe to calculate the sequence of nuclei as a function of density is as follows:

(1) Set up a table of binding energies for nuclei for a given mass number A and charge Z from the presently available atomic mass evaluation.
(2) Put in the unknown values beyond the known range of A and Z of nuclei by using a nuclear model.
(3) Minimize the thermodynamic potential with respect to A and Z.

However, the sequence of nuclei will jump discontinuously. If the sequence of nuclei switches to another more stable nucleus, there will be a small jump in the energy density due to the difference in the binding energy, indicating a first-order phase transition. In the following, we assume that this phase transition can be treated by using a Maxwell construction, which implies that the pressure at the phase boundary does not jump. We note that in hydrostatic equilibrium, the pressure will be a continuous function of the radius (or of the energy density), so any regime in the EOS with constant pressure will shrink to zero extension in the radius. Hence, in this picture there is a direct jump from a lattice of one nucleus to the next nucleus, which is more stable at higher energy density. It is easier to consider the sequence of nuclei with increasing pressure instead of increasing energy density. The relevant thermodynamic potential is the Gibbs energy G, which is a function of temperature, pressure, and conserved particle number N

$$G = G(p, T, N). \tag{7.5}$$

The Gibbs energy is particularly simple if one considers thermodynamic systems at constant pressure and constant temperature, which we intend to do. To get from the internal energy U, which is a function of entropy S, volume V, and conserved number N to the Gibbs energy G, one needs to do Legendre transformations from the entropy to the temperature and from the volume to the pressure so that the Gibbs energy can be defined via the thermodynamic relation

$$G = U + P \cdot V - T \cdot S. \tag{7.6}$$

Taking the total differential of this expression, one can convince oneself that indeed

$$\mathrm{d}G = V \cdot \mathrm{d}P - S \cdot \mathrm{d}T + \mu \cdot dN, \tag{7.7}$$

expressing that G is a thermodynamic potential depending on p, T, and N. With the use of the Euler relation

$$U = -p \cdot V + T \cdot S + \mu \cdot N, \tag{7.8}$$

the definition of the Gibbs energy G, Eq. (7.6), can be rewritten in the form

$$G = \mu \cdot N. \tag{7.9}$$

This means that minimizing the thermodynamic potential G for a fixed number N corresponds to minimizing the chemical potential μ. In our case, the temperature can be considered to be zero, so that the Euler relation gives the following expression for the baryon chemical potential

$$\mu_b = \frac{\varepsilon_{\text{total}} + P_{\text{total}}}{n_b}. \tag{7.10}$$

The total pressure P_{total} can be derived from the total energy $\varepsilon_{\text{total}}$, Eq. (7.2), by the relation

$$P_{\text{total}} = n_b^2 \left. \frac{\partial(\varepsilon_{\text{total}}/n_b)}{\partial n_b} \right|_{A,Z}, \tag{7.11}$$

which we already encountered before when discussing thermodynamic potentials, see Eq. (3.69). Now, here the baryon number density is $n_b = A \cdot n_N$ and the relation for the total pressure results in

$$P_{\text{total}} = P_e + \frac{1}{3} W_L \cdot n_N, \tag{7.12}$$

where P_e is the pressure of the electron Fermi gas. One realizes that the total pressure is solely given by the electron pressure and a contribution from the lattice. There is no contribution from the nuclei, which are held fixed in their lattice structure.

The EOS can now be computed by minimizing the Gibbs energy or equivalently the baryon chemical potential with respect to the mass number A and charge Z of the nucleus. All we need are the binding energies of the so far experimentally undetected nuclei by resorting to nuclear models.

One can define three different types of nuclear models to describe the binding energy of nuclei.

Parametrized mass models: Specifically designed to describe nuclear masses. Historically, these models go back to the liquid drop models, where the nucleus is considered an incompressible fluid. The Bethe–Weizsäcker mass formula (Weizsäcker, 1935; Bethe and Bacher, 1936) parametrizes the binding energy in the form

$$E(Z, A) = a_{\text{vol}} \cdot A + a_{\text{surf}} \cdot A^{2/3} + a_{\text{Coul}} \cdot Z^2 A^{-1/3} + a_{\text{sym}} \cdot \frac{(N-Z)^2}{A} + \delta(A). \tag{7.13}$$

Here, $N = A - Z$ stands for the number of neutrons. Noting that empirically the charge radii of nuclei follow the simple law

$$R = r_0 \cdot A^{1/3} \qquad r_0 = 1.0 \ldots 1.2\,\text{fm}, \tag{7.14}$$

one can easily see that the first term corresponds to a volume term proportional to R^3, the second to a surface term ($\propto R^2$), and the third one to a Coulomb term ($\propto Z^2/R$). The fourth term is the symmetry energy term, which is proportional to the difference of the neutron and proton number squared. It describes the effect from the isospin-dependence of the nuclear force. The symmetry energy counterbalances the effect from the Coulomb repulsion of protons such that light to medium mass nuclei with equal numbers of protons and neutrons are the most stable configurations. Eventually, the Coulomb repulsion wins for heavy and highly charged nuclei so that nuclei with an excess in the number of neutrons compared to the number of protons are more stable. The heaviest stable nucleus is then ^{208}Pb with $N = 126$ and $Z = 82$. The prefactor a_{sym} stands for the symmetry energy.

The last term, $\delta(A)$, denotes effects from the difference of even numbers compared to odd numbers of protons and neutrons. Specifically, it describes the pairing energy, the additional bond energy when neutrons or protons are paired up

$$\delta(A) = \begin{cases} a_{\text{pair}} \cdot A^{-1/2} & \text{even N and even Z} \\ 0 & \text{odd N or Z} \\ -a_{\text{pair}} \cdot A^{-1/2} & \text{odd N and odd Z} \end{cases}, \tag{7.15}$$

where a_{pair} is the pairing energy coefficient.

Fits to the measured nuclear masses gives the coefficients shown in Table 7.2 (see e.g., Chowdhury and Basu, 2006).

We refrain from giving more than two significant digits, as the Bethe–Weizsäcker mass formula cannot describe shell effects nor effects from the deformation of nuclei. Therefore, there exists several extensions that incorporate these effects on a microscopic basis, as the finite-range droplet model (FRDM) (Möller et al., 2012, 2016) or the Duflo–Zuker model (Duflo and Zuker, 1995). In fact, these parametrized mass models give so far the best fits to nuclear masses available.

Skyrme models: These models make a parametric ansatz for the nucleon mean effective potential within the Schrödinger equation and as introduced by Skyrme

Table 7.2 *The coefficients of the Bethe–Weizsäcker mass formula of Eq. (7.13).*

a_{vol}	a_{surf}	a_{Coul}	a_{sym}	a_{pair}
−16 MeV	+18 MeV	+0.70 MeV	+23 MeV	−12 MeV

(Skyrme, 1959). The potential terms have zero range, are local point-like interactions, and can be given in terms of the proton and neutron density and gradients thereof. In addition to calculating nuclear masses, these models are able to describe other properties of nuclei as charge radii, surface thickness, deformations, and so on. They can also be applied for bulk nuclear matter by setting the mass number $A \to \infty$ and, of interest to us, to bulk neutron matter. For example, the parameter set BSk8 used later on has been fitted to the atomic mass evaluation data table, while SLy4 to properties of selected spherical nuclei (masses and radii) and the EOS of pure neutron matter.

Relativistic mean-field models: These models are motivated from the meson-exchange picture of the nuclear forces as proposed by Yukawa and go back to the original work of Walecka (Walecka, 1974).[2] The nucleon potential in the Dirac equation is mediated by meson fields, which themselves are calculated by solving the Klein–Gordon equation, with the nucleon densities being the source terms of the meson fields. The set of equations are solved iteratively, that is, self-consistently, by setting the meson fields on their mean expectation values and treat them as classical fields (that is where the naming of the model comes from). The nucleon interaction has a range given by the meson masses and is set up by using a Lagrangian of the nucleon and meson fields with different coupling terms, including interaction terms between the meson fields. The models can be applied also to calculate nuclear matter and neutron matter. For example, the parameter sets NL3 and TMA used after are fitted to properties of selected spherical nuclei (masses, radii, and surface thicknesses).

Table 7.3 shows the sequence of nuclei for the crust of neutron stars in β-equilibrium. The sequence starts at low densities with the most stable nucleus we know of, ^{56}Fe. For increasing densities, the baryon density and correspondingly the charge-balancing electron density increases, which can also be seen by the increasing chemical potentials shown in the two first columns. Eventually, it will be energetically more favorable to decrease the amount of electrons in the system by converting electrons and protons to neutrons. Therefore, nuclei will be getting more neutron-rich with increasing density. Starting at a mass density of $\rho \approx 8 \times 10^6\,\mathrm{g\,cm^{-3}}$, other nuclei, here the nucleus ^{62}Ni, will be energetically favored. For even higher densities, heavier nickel isotopes will be present in the outer crust. Note that nickel has the magic number $Z = 28$. At a density of about $3 \times 10^9\,\mathrm{g\,cm^{-3}}$, a sequence of nuclei starts to appear with the magic number $N = 50$. The nucleus ^{80}Zn is the last nucleus, where the experimental binding energy is known. Beyond the density of about $6 \times 10^{10}\,\mathrm{g\,cm^{-3}}$, the nuclei are

[2] Interestingly, Walecka states in his article that his motivation to develop his model of nuclear interaction is to describe the properties of neutron stars.

Table 7.3 *The sequence of nuclei in the outer crust of neutron stars in β-equilibrium. Upper part: using experimental nuclear binding energies, lower part: using nuclear binding energies calculated with the relativistic mean-field model TMA. Reprinted table from Rüster et al. (2006). Copyright (2006) by the American Physical Society.*

μ_b (MeV)	μ_e (MeV)	ρ (g cm^{-3})	P (dyne cm^{-2})	n_b (cm^{-3})	Element	Z	N
930.60	0.95	8.02×10^{6}	5.22×10^{23}	4.83×10^{30}	^{56}Fe	26	30
931.32	2.61	2.71×10^{8}	6.98×10^{25}	1.63×10^{32}	^{62}Ni	28	34
932.04	4.34	1.33×10^{9}	5.72×10^{26}	8.03×10^{32}	^{64}Ni	28	36
932.09	4.46	1.50×10^{9}	6.44×10^{26}	9.04×10^{32}	^{66}Ni	28	38
932.56	5.64	3.09×10^{9}	1.65×10^{27}	1.86×10^{33}	^{86}Kr	36	50
933.62	8.38	1.06×10^{10}	8.19×10^{27}	6.37×10^{33}	^{84}Se	34	50
934.75	11.43	2.79×10^{10}	2.85×10^{28}	1.68×10^{34}	^{82}Ge	32	50
935.93	14.71	6.21×10^{10}	7.86×10^{28}	3.73×10^{34}	^{80}Zn	30	50
937.28	18.64	1.32×10^{11}	2.03×10^{29}	7.92×10^{34}	^{78}Ni	28	50
937.63	19.80	1.68×10^{11}	2.55×10^{29}	1.01×10^{35}	^{124}Mo	42	82
938.13	21.38	2.18×10^{11}	3.48×10^{29}	1.31×10^{35}	^{122}Zr	40	82
938.67	23.19	2.89×10^{11}	4.82×10^{29}	1.73×10^{35}	^{120}Sr	38	82
939.18	24.94	3.73×10^{11}	6.47×10^{29}	2.23×10^{35}	^{118}Kr	36	82
939.57	26.29	4.55×10^{11}	8.00×10^{29}	2.72×10^{35}	^{116}Se	34	82

becoming so extremely neutron-rich that their experimental masses are no longer known. From here on, one has to resort to nuclear mass models. The table shows below the horizontal line the sequence of nuclei as predicted by the relativistic mean-field parameter set TMA. It predicts the presence of the doubly-magic nucleus ^{78}Ni and that the sequence of nuclei continues along the neutron magic number $N = 82$. The nucleus ^{116}Se is predicted to be the last nucleus in the outer crust that is present up to a density of 4.55×10^{11} g cm^{-3}. It is the last neutron-rich nucleus predicted to be stable, even more neutron-rich nuclei are not stable in the mass tables of the parameters set TMA. We can take a look at the baryon chemical potential to see what is happening at that density: the baryon chemical potential has reached the neutron mass, that is, $\mu_b = m_n = 939.57$ MeV. Hence, free neutrons can be added to the system, which are not bound to the nucleus anymore. The density at that point is called the neutron-drip density, as pictorially unbound neutrons are dripping out of the nucleus. Note, that the neutron-drip density is not precisely known, it depends on the nuclear model used. Predictions range from 4×10^{11} g cm^{-3} to 5×10^{11} g cm^{-3} (Rüster et al., 2006). The neutron-drip density marks the end of the outer crust and the beginning of the inner crust.

The sequence of nuclei in the outer crust for different nuclear models is shown in Figure 7.1 in the nuclear chart diagram as a scatter plot for the charge number Z and the neutron number N. Crosses denote combinations of Z and N, where the nuclear binding energy has been experimentally determined. The sequence of nuclei is the same for those experimentally known nuclei, of course, for all models. Then there are slight variations for nuclei where the binding energy has to be calculated by using nuclear models. However, one can see that the sequence of nuclei mostly follows the neutron magic numbers $N = 50$ and $N = 82$. From our discussion of the feature of magic numbers, we know that nuclei with a magic number of nucleons, here neutrons, are particularly stable in that they have an increased binding energy compared to nuclei in the neighborhood within the nuclear chart. One should caution, of course, that the presence of magic numbers of neutrons is due to the pronounced shell effects predicted by the nuclear models, even for neutron-rich nuclei. It can well be that that the shell effects are diminished for extremely neutron-rich nuclei, that is, that shell effects are quenched. Here, only future experiments will tell what the properties of neutron-rich nuclei toward the neutron-drip line are, such as the Facility for Rare Isotope Beams (FRIB) at Michigan State University and the Facility for Antiproton and Ion Research (FAIR) at GSI, Darmstadt. That indeed mass spectrometry can elucidate the sequence of nuclei has been demonstrated by the ISOLTRAP collaboration by measuring the binding energy of the nucleus ^{82}Zn. In some nuclear models, this nucleus was part of the sequence of nuclei in the outer crust of neutron stars. After the measurement, it turned out that ^{82}Zn is no longer present in those sequences of nuclei (Wolf et al., 2013). We note

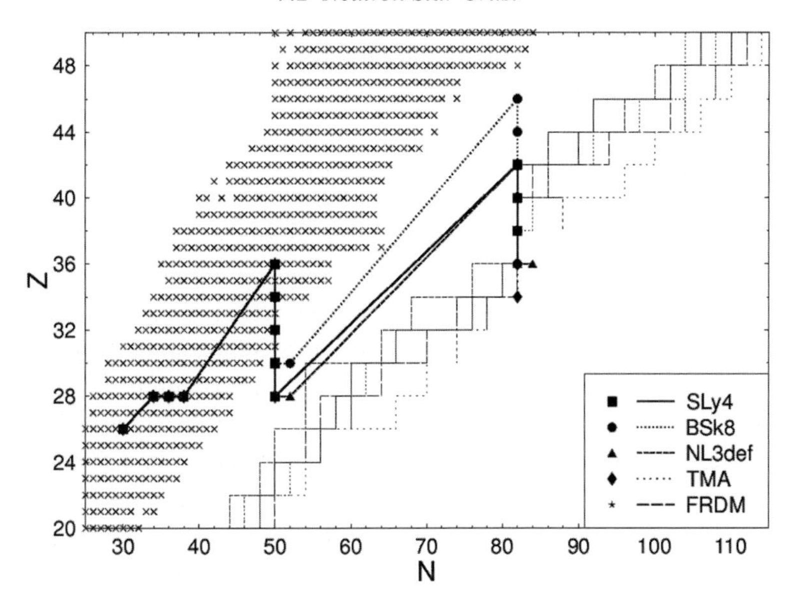

Figure 7.1 The sequence of nuclei in the outer crust of a neutron star in the nuclear chart for different nuclear models ($\times$, known nuclei; lines, neutron-drip line). The sequence of nuclei starts at ^{56}Fe, follows along the magic number of protons ($Z = 28$) and then along the magic numbers of neutrons, $N = 50$ and $N = 82$, rather independent of the nuclear model used. Reprinted figure with permission from Rüster et al. (2006). Copyright (2006) by the American Physical Society

that the nucleus ^{82}Zn has $N = 52$ neutrons and is close to the magic sequence at $N = 50$ seen earlier.

Let us turn our attention to the EOS, which determines the global properties of neutron stars. From the discussion of matter in white dwarfs, we have seen that the pressure of the outer crust is dominated by the electron degeneracy pressure, while the energy density is dominated by the mass of the nucleus. This feature does not change for the case of β-equilibrium we applied for the outer crust of neutron stars. Indeed, Figure 7.2, showing the pressure versus the energy density, does not indicate a significant variation of the EOS for the different nuclear models used. We remind the reader that the nuclear binding energy is on the scale of 8 MeV, which is a small correction to the nucleon masses of about 1 GeV. Therefore, differing binding energies will not significantly affect the total mass or energy density. The charge of the nucleus is dictated by β-equilibrium. Above the fiducial density of about 6×10^{11} g cm^{-3}, a slight variation in the nuclear charge occurs due to the different predictions of the nuclear models. As Figure 7.2 shows, even at those densities, the differences in the EOS are small. Typically, the charges of the nuclei in the outer crust are between $Z = 26$ and 46, see Figure 7.1.

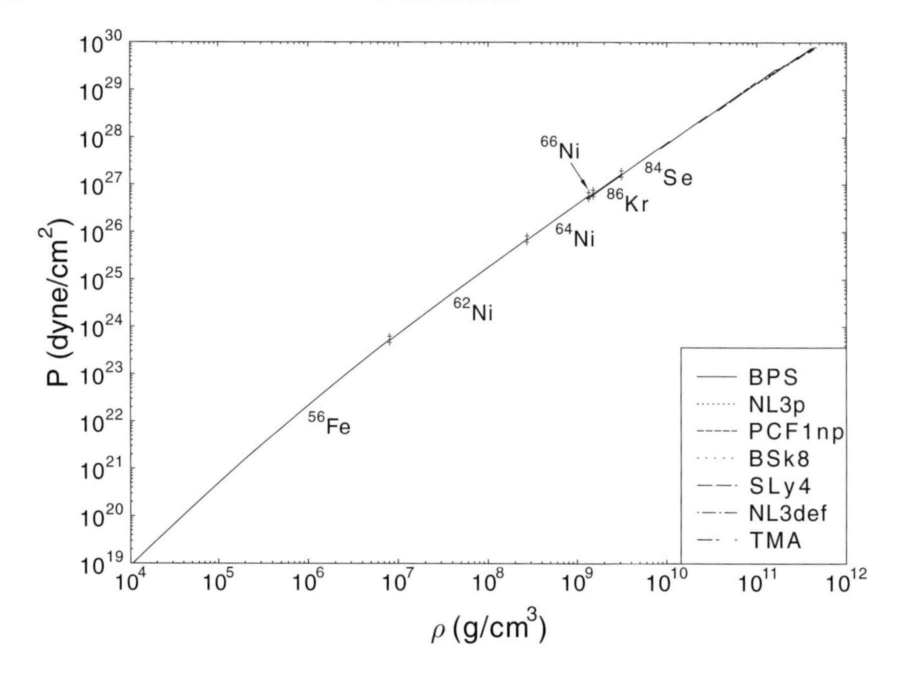

Figure 7.2 The EOS of the neutron star's outer crust for different nuclear models. The double daggers indicate the transition from one nucleus to another. The EOS is rather insensitive to the nuclear model used. Reprinted figure with permission from (Rüster et al., 2006). Copyright (2006) by the American Physical Society

As the EOS of the outer crust of neutron stars is basically quite well determined, we can make generic statements about the thickness and the mass of the outer crust. The trick is to solve the TOV equation with modified initial conditions. Instead of taking zero mass at zero radius, the integration of the TOV equation starts with a given mass and a corresponding radius for the matter inside the outer crust. The EOS of the outer crust is then added as a shell on top of this neutron star core. Figure 7.3 shows the mass and the thickness of the outer crust for different core masses M_0 as a function of the core radius R_0. For typical radii of neutron stars in the range of 10–15 km, the characteristic mass of the outer crust is only around $10^{-4} M_\odot$, so it does not give a sizable contribution to the overall mass of a neutron star. The thickness of the outer crust, however, is in the range of several hundred meters to 1 km, which is not that small compared to the core radius of 10–15 km. We notice that the thickness of the outer crust decreases with increasing neutron star mass and decreasing core radius R_0. For a typical neutron star mass of $1.4 M_\odot$ and a core radius of 15 km, the thickness of the outer crust is about 1 km, for a mass of $2.0 M_\odot$ and a core radius of 10 km, the outer crust measures only about 100 m.

We can understand the thickness of the crust by resorting to the procedure we have used when discussing the thickness of layers of white dwarfs, see Eq. (5.62),

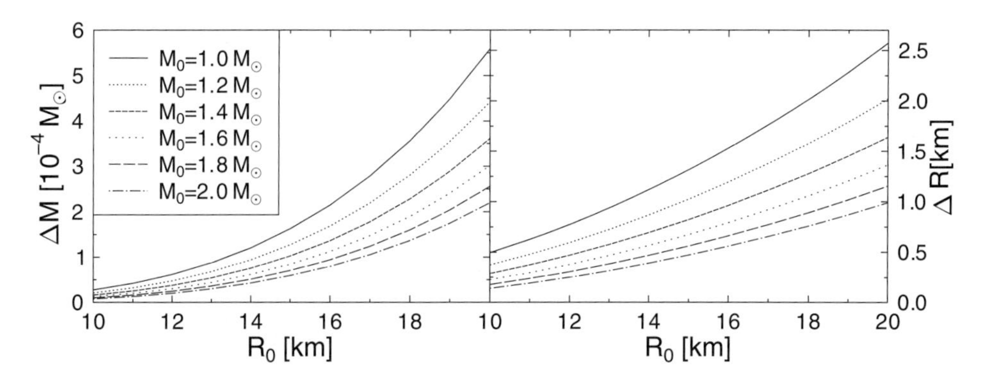

Figure 7.3 The mass (left plot) and the thickness (right plot) of the outer crust of neutron stars for different values of the total mass of the neutron star. Taken from Hempel and Schaffner-Bielich (2008)).

which we derived from the equation of hydrostatic equilibrium. In general relativity, one has to replace the mass density by the energy density for neutron stars and one has to consider that there are additional terms from general relativity that cannot be ignored beforehand for neutron stars. We start by having a look at the TOV equation, the equation of hydrostatic equilibrium with three correction terms from general relativity, see Eq. (2.149). We assume that the thickness of the crust ΔR is small compared to the radius of the neutron star R and that the total mass of the neutron star M is dominantly given by the mass interior to the crust. Moreover, we assume that the pressure is much smaller than the energy density in the crust of the neutron star. We note that the first two correction terms from general relativity are related to the pressure, one relative to the energy density, the other relative to the mass density, which we can ignore. We keep the third correction term, the Schwarzschild factor, though, and rewrite it in terms of the compactness $C = GM/R$ as

$$\left(1 - \frac{2GM}{R}\right)^{-1} = (1 - 2C)^{-1} . \tag{7.16}$$

Let us approximate the pressure gradient on the left-hand side of the TOV equation by the ratio of the pressure at the inner boundary P_b of the crust over the thickness of the crust ΔR as the pressure vanishes at the surface. We finally arrive at the approximative formula for hydrostatic equilibrium with effects from general relativity

$$-\frac{\Delta P}{\Delta R} = \frac{P_b}{\Delta R} \approx \frac{GM\varepsilon_b}{R^2(1 - 2GM/R)} = \frac{C}{1 - 2C}\frac{\varepsilon_b}{R}, \tag{7.17}$$

where ε_b is the energy density at the inner boundary of the crust. The relative thickness of the crust is then estimated to be

$$\frac{\Delta R}{R} \approx \frac{1 - 2C}{C} \frac{P_b}{\varepsilon_b}. \tag{7.18}$$

As for the case of white dwarfs, the relative thickness of the crust is approximately determined by the ratio of the pressure to the energy density and the compactness C. We remark, that there is the additional factor from the Schwarzschild factor $(1 - 2C)$, which will shrink the crust by a factor of around two for a compactness of $C = 0.2$–0.3, the typical range for compact stars. As we see, the corrections from general relativity make the gravitational pull substantially stronger compared to the Newtonian case, leading to a crust that is squeezed to about half its size in compact stars compared to stars treated in Newtonian theory.

Next, we need to determine the pressure-to-energy density ratio. For the inner boundary of the outer crust, we can assume that the pressure is approximately given by the one of relativistic electrons, see Eq. (3.85), the energy density by the mass density from the nuclei. Hence,

$$\begin{aligned} \frac{P^{(rel)}}{\rho_N} &= \frac{1}{4}(3\pi^2)^{1/3} \frac{n_e^{4/3}}{m_u \cdot n_N} \\ &= \frac{1}{4}(3\pi^2)^{1/3} \frac{Z}{A} \frac{n_e^{1/3}}{m_u} \\ &= \frac{1}{4} \frac{Z}{A} \frac{\mu_e}{m_u}, \end{aligned} \tag{7.19}$$

where n_e is the electron number density, n_N the number density of nucleons, and μ_e the electron chemical potential. In the second equation, we used the charge neutrality condition $n_e = (Z/A) n_N$. In the last equation, we used the relation between the number density and the chemical potential in the relativistic limit

$$\mu_e \approx k_{F,e} = (3\pi^2 n_e)^{1/3}. \tag{7.20}$$

So, Eq. (7.19) shows that the ratio of the pressure to the energy density is controlled by the ratio of the Fermi energy of the electrons to the nucleon mass. From Table 7.3 we can take now the values of the last entry that is at the neutron-drip density, that is, at the inner boundary of the outer crust. With $Z/A = 34/82 \approx 0.3$ and $\mu_e = 26.29$ MeV, we find for the outer crust that

$$\left.\frac{P_b}{\varepsilon_b}\right|_{\text{(outer crust)}} \approx \frac{Z\mu_e}{4Am_u} \approx 3 \times 10^{-3}, \tag{7.21}$$

which justifies a posteriori our assumption that the pressure is small compared to the energy density. Finally, the thickness of the outer crust is estimated to be

$$\Delta R_{\text{(outer crust)}} \approx \frac{1-2C}{C}\frac{Z\mu_e}{4Am_u}\cdot R, \tag{7.22}$$

which gives about 40 m for a $2M_\odot$ neutron star with a radius of 10 km ($C = 0.3$) and 200 m for a $1.4M_\odot$ neutron star with a radius of 15 km ($C = 0.15$). Comparing with the numerical result of Figure 7.3, our estimates are smaller but in the right ballpark. The mass contained in the outer crust can be estimated by resorting to the equation for mass conservation:

$$\begin{aligned}\Delta M &= 4\pi\int_0^{\varepsilon_b} \mathrm{d}r\, r^2\varepsilon(r) \approx 4\pi R^2\rho_b\cdot\Delta R\\ &\approx 2.5\times 10^{-4}M_\odot\left(\frac{R}{10\,\mathrm{km}}\right)^2\left(\frac{\rho_b}{4\times 10^{11}\mathrm{g\,cm^{-3}}}\right)\left(\frac{\Delta R}{1\,\mathrm{km}}\right),\end{aligned} \tag{7.23}$$

which is in reasonable agreement with the numerical results of Figure 7.3.

7.2.3 Inner Crust

The inner crust starts at the neutron drip density where neutrons drip out of the nuclei and form a separate liquid. According to the last section, the neutron drip density is the point where the baryon chemical potential equals the vacuum neutron mass. The neutron drip density is model dependent and varies between $4\times 10^{11}\,\mathrm{g\,cm^{-3}}$ and $5\times 10^{11}\,\mathrm{g\,cm^{-3}}$. The variation in the neutron drip density, although the critical condition is precisely set with the neutron mass, originates from the fact that the relation between the baryon density and the baryon chemical potential is not unique. Consider a neutron at a given baryon density. The baryon chemical potential is the energy needed to add or subtract one baryon (a neutron) from the medium, which depends on the potential energy, the baryons feel in the medium. Hence, differing interactions with differing nucleon potentials can give different chemical potentials, even when evaluated at the same baryon density. Even small changes in the nucleon potential can give a sizable difference in the resulting baryon density. One also sees from Table 7.3 that changes in the baryon chemical potential of about 1 MeV can change the baryon density by a factor of two or more. For the outer crust, the potential energy of the baryon is just the binding energy of the nucleus. So, the binding energy of the last nucleus in the sequence has to be known to a high precision to determine the neutron-drip density. The binding energy of the last nucleus in the sequence as well as the location of the neutron-drip line are not known at present. Therefore, the value of the neutron-drip density can only be estimated from nuclear mass models.

The structure of the inner crust is similar to the outer crust, insofar as there is still a lattice of nuclei with surrounding electrons to balance the charge of the protons in the nucleus. However, while the protons are still being localized within the nucleus, the neutrons are present also outside the nucleus, forming a separate liquid. The nuclei are now immersed in a sea of electrons and neutrons. The proper theoretical description of the inner crust turns out to be challenging. Imagine a system like in solid state physics at a density a billion times higher set in a quantum mechanical liquid whose properties can only be modeled in an effective model. The properties of the outer crust are therefore not known in detail.

To add more to the complexity of the inner crust, the neutron fluid will be superfluid. According the Cooper theorem, the basis of the BCS theory for superconductivity is that degenerate fermions will form bound pairs if there is an attractive potential between them. Hence, there just has to be one attractive channel of the neutron–neutron interaction to make neutron matter a superfluid, even if the sum of all interaction channels is repulsive. From studies of nucleon–nucleon scattering, one knows that the partial wave $^1S_0$ channel is attractive. Here, the standard notation is $^{2s+1}L_J$, where s stands for the total spin, L for the angular momentum, and J for the total angular momentum of the nucleon–nucleon pair. The angular momentum is denoted by letters mapping $L = 0, 1, 2, 3, \ldots$ to $L = S, P, D, F, \ldots$, which originates from the notation in atomic spectroscopic (the notation continues in alphabetic order after $L = F$). So, the $^1S_0$ channels stands for zero spin, zero angular momentum, and zero total angular momentum, which is the lowest possible one. Models of the nuclear interaction extrapolated to the neutron fluid present in the inner crust predict an attractive potential between two neutrons, which is about 1 MeV. So, a neutron superfluid forms where neutrons pair up in the $^1S_0$ channel. At higher density, other channels can turn attractive, changing the pairing properties of the neutron superfluid. But there is more to it. As the neutron star can be rapidly rotating, as it is the case for millisecond pulsars, the neutron superfluid is corotating. It is known from laboratory experiments that rotating superfluids form vortices, which are quantized carrying the quantum number of angular momentum. Stunningly, a rotating superfluid is irrotational everywhere, except for the core of the vortices. The quantization condition for the angular momentum is such that it is given by a multiple of Planck's constant h:

$$m \cdot \oint v_s \cdot \mathrm{d}l = s \cdot h, \tag{7.24}$$

where s is an integer number, v_s is the superfluid velocity, and m the mass of the superfluid particles. The integration of the angular momentum is around the vortex and does not depend on the chosen contour. For uniform rotation, the overall rotational velocity of the fluid is given by $\mathbf{v} = \mathbf{\Omega} \times \mathbf{r}$, so that

$$\nabla \times \mathbf{v} = \nabla \times (\mathbf{\Omega} \times \mathrm{r}) = 2\mathbf{\Omega} \tag{7.25}$$

holds. Associating the overall velocity of the fluid with the superfluid velocity $v = v_s$ for $S = 1$ of the vortices, one can combine Eqs. (7.24) and (7.25). With the help of Stokes' theorem, one finds for the number of vortices N_V per unit area A:

$$\frac{N_V}{A} = \frac{2\Omega m}{h}. \tag{7.26}$$

The same type of vortices will form in the neutron superfluid, generating a regular two-dimensional array perpendicular to the rotation axis in rotating neutron stars. In terms of the rotation period P of the pulsar, the number per unit area is given by

$$\frac{N_V}{A} = \frac{8\pi m_n}{h \cdot P} = 6.35 \times 10^7 \mathrm{m}^{-2} \left(\frac{1\,\mathrm{s}}{P}\right), \tag{7.27}$$

where we assume that the mass of the superfluid particle is given by the mass of the neutron pair, that is, $m = 2m_n$. We can safely ignore binding energy effects, as the binding energy of the pair of about 1 MeV is much smaller than the neutron mass. You are asked as a simple exercise to convince yourself that the mean spacing between the vortices is much smaller than the size of a neutron star and much larger than the microscopic lattice scale of the outer crust.

The presence of this vortex structure in the inner crust of a neutron star has important consequences for the transport properties of neutron stars. As the vortices carry angular momentum, they have been connected with the phenomenon of pulsar glitches, where a pulsar suddenly rotates faster and returns then slowly to his original rotation rate. If there is an increasing strength being built up between the solid structure of the crust and the liquid core, vortices may unpin transporting angular momentum to the crust, which speeds up (see our discussion of pulsar glitches in Section 6.7). Another phenomenon concerns the cooling of neutron stars. Young neutron stars cool by emitting neutrinos. For this to happen, pairs of bound neutrons have to be broken up. The probability for breaking neutron pairs at low but nonzero temperature is suppressed exponentially by a Boltzmann factor, which also suppresses the neutrino emission rate and thereby the cooling rate of neutron stars. Then, the recombination of two neutrons to a pair emits neutrino–antineutrino pairs and constitutes an important cooling process. Hence, the neutron pairing energy is a decisive input into cooling simulations of neutron stars. More than that, recent observations of cooling neutron stars were able to set limits on the pairing properties of neutrons, in particular on the critical temperature of the superfluid neutron matter at which the superfluid phase disappears. Reviews on the physics of superfluidity in neutron stars can be found in Sedrakian and Clark (2018) and on cooling in Yakovlev and Pethick (2004).

In view of the complex features of the inner crust, it seems hopeless to pin down its EOS. However, the features already discussed concern the transport properties that do not effect significantly the bulk properties. So, one can make some generic statements about the inner crust. It turns out that the EOS of the pure neutron fluid can be calculated quite precisely by using a model independent approach, which we will discuss in the next section in detail. As the EOS of the lattice and the electrons can be also pinned down using the methods for the outer crust, the EOS of the inner crust is quite well determined.

The composition and the structure of the inner crust was investigated by Negele and Vautherin in a pioneering work (Negele and Vautherin, 1973). They performed a self-consistent Hartree–Fock calculation of Wigner–Seitz cells in spherical symmetry. For the nucleon interactions, they used an effective two-body nucleon–nucleon potential based on microscopic calculations for asymmetric nuclear matter, including pure neutron matter.[3] The absolute ground state configuration was found by minimizing the total energy with respect to the number of protons and neutrons and the radius of the cell. The resulting density profile of the nucleon densities shows a localized region in the center for the protons. The neutron distribution also shows a plateau at the center, where the protons are residing, but continues with a flat nonvanishing distribution toward the boundary of the cell. With increasing density, the density of the neutrons outside the central region is increasing, while the density in the central region remains nearly unchanged. For higher densities, the cell size also decreases, so that the neutron density distribution approaches an overall constant profile with just a small increase in the central region of the cell. Strong shell effects were found for the protons. For lower densities, the number of protons in the cell stayed at $Z = 40$, while at higher densities, the proton number switches to $Z = 50$. We note from Table 7.3 that the charge of the nuclei in the outer crust is around $Z = 26-46$, so the charges in the inner crust are not that different. We also note from Table 7.1 that $Z = 40$ corresponds to a magic number for the shell model without a strong spin–orbit splitting and to a closed subshell for the case of strong spin–orbit splitting. The value $Z = 50$ we immediately recognize as a magic number for the shell model of nuclei, now for protons not for neutrons as for the inner crust. The underlying reason for proton numbers staying around 40–50 is a balance between the surface energy and the Coulomb energy of the Wigner–Seitz cell. We will discuss this in more detail in the next section when discussing more complicated geometrical structures in the inner crust. The total neutron number within the cell, however, reaches values of about $N = 1,800$.

[3] Still, the EOS of pure neutron matter used at the time of Siemens and Pandharipande (1971) is close to the one of Friedman and Pandharipande (1981), which is compatible with our present knowledge using modern approaches and interactions, see Pethick et al. (1995).

The inner crust stops at around half of the saturation density of $n_0 = 0.15\,\text{fm}^{-3}$, which is in terms of the baryon density $0.5n_0 \approx 0.075\,\text{fm}^{-3}$ or in terms of the energy density $0.5\varepsilon_0 \approx 0.5m_n \cdot n_0 \approx 70\,\text{MeV fm}^{-3}$. The lattice size shrinks from about 100 fm at the beginning of the inner crust to about 30 fm at the end of the inner crust, which is just a few times larger than the sizes of heavy nuclei. The charge radius of ^{208}Pb, for example, is $R \approx 5.5\,\text{fm}$.

For higher densities, one reaches the liquid outer core consisting of a liquid of neutrons with a small admixture of protons and electrons. The lattice disappears in the outer core. The precise value of the transition from the inner crust to the core is not known precisely, as it depends, again, on the nuclear model used.

We can estimate the thickness of the total crust, the inner plus the outer crust, using similar methods as before for estimating the thickness of the outer crust. For the pressure at the inner boundary of the inner crust, we take now the pressure of the nonrelativistic neutron gas instead of the pressure of the relativistic electrons we used for the outer crust. We assume that the expression for the pressure of the neutron gas is given by a free gas of neutrons, see Eq. (3.92), times a correction factor ξ. As we will see later when discussing pure neutron matter, this is not a particular bad choice. The ratio of the pressure to the energy density at the inner boundary of the inner crust then reads

$$\left.\frac{P_b}{\varepsilon_b}\right|_{\text{(crust)}} \approx \frac{\xi \cdot P_n^{\text{nonrel}}}{\rho_n} = \frac{\xi}{5}(3\pi^2)^{2/3}\frac{n_n^{2/3}}{m_n^2} = \frac{\xi}{5}\frac{k_{F,n}^2}{m_n^2}, \tag{7.28}$$

where n_n is the neutron number density and $k_{F,n}$ the Fermi momentum of the neutrons. Associating $k_F^2/(2m)$ with the nonrelativistic Fermi energy, we discover again that the ratio of the pressure to the energy density is controlled by the ratio of the neutron Fermi energy over the neutron mass. For saturation density $n_0 = 0.15\,\text{fm}^{-3}$, one finds a Fermi momentum of the neutrons of $k_{F,n} = 324\,\text{MeV}$. As the crust ends at about $n_0/2$, the neutron Fermi momentum to be used for our estimate is $k_{F,n} = 257\,\text{MeV}$, which coincides with the Fermi momentum of nuclear matter at saturation density. The ratio of the pressure to the energy density at the inner end of the crust is then approximately 6×10^{-3}, where we used a value of $\xi = 0.4$. The thickness of the total crust is then approximately, see also Eq. (7.18):

$$\Delta R_{\text{(crust)}} \approx \frac{1-2C}{C}\frac{\xi k_{F,n}^2}{5m_n^2} \cdot R, \tag{7.29}$$

which gives about 80 m for a $2M_\odot$ neutron star with a radius of 10 km ($C = 0.3$) and 400 m for a $1.4M_\odot$ neutron star with a radius of 15 km ($C = 0.15$). Again, the numbers are on the low side, numerical calculations give thicknesses that are larger, more in the range of a few hundred meters to a kilometer for the total crust (Grill et al., 2014). We can also estimate the mass contained in the crust via

$$\Delta M \approx 4\pi R^2 \rho_b \cdot \Delta R = 7.9 \times 10^{-2} M_\odot \left(\frac{R}{10\,\mathrm{km}}\right)^2 \left(\frac{\rho_b}{0.5\rho_0}\right) \left(\frac{\Delta R}{1\,\mathrm{km}}\right), \quad (7.30)$$

where $\rho_0 = m \cdot n_0$ is the mass density at saturation density. The crust gives a small contribution to the total mass of the neutron star, which is in the range of a few percent of the total mass.

The EOS for the inner crust of Negele and Vautherin is still in use today. An improvement of their model had to wait nearly three decades, when Hartree–Fock calculations of the inner crust were performed on supercomputers in full three-dimensional geometry. They revealed a fascinating new phenomenon of the inner crust: the existence of pasta phases to which we turn our attention to now.

7.2.4 Pasta in the Crust

The theoretical discovery of the pasta phases goes back to the work of Ravenhall et al. (1983). As they describe the new phases as being spaghetti-like and lasagna-like in their original paper, the name pasta phases stuck in follow-up papers.[4] So, what are those pasta phases in the crust of neutron stars?

As we discussed earlier, models for the inner crust usually resort to the Wigner–Seitz cell approximation where cells are treated in spherical symmetry. Now, Ravenhall et al. (1983) extended the discussion of the properties of the inner crust beyond spherical symmetry and discovered that more stable forms exist than bubbles of nuclei immersed in a background of an electron and neutron fluid. If spheres are considered to be three-dimensional in nature, the new phases are two- and one-dimensional structures. Imagine the bubbles of nuclei getting closer and closer together with increasing density. Eventually, bubbles could merge along one direction so that a chain of spheres forms a rod of nuclear matter. As this will happen for several chains along the same direction, the rods form a spaghetti-like structure. The rods are surrounded by the neutron liquid.

Now imagine that the density increases even further, so that a line of neighboring rods forms a plate. Again, several plates neatly stacked on top of each other could build up, forming a lasagna-like structure of nuclear matter plates alternating with pure neutron matter plates.

For even higher density, the role of the two phases will interchange. The lasagne phase is its own inverted phase. The inverted spaghetti phase, the anti-spaghetti

[4] The authors even write about the cooking of spaghetti, which, of course, is meant in strictly scientific terms, by looking at the spaghetti-like phase at nonvanishing temperature

phase, would be built out of rods of neutron matter surrounded by nuclear matter. The inverted bubble phase would consist of holes of neutron matter within nuclear matter.

The pasta phases will occur, if they exist, at the high-density inner boundary of the inner crust. Note that the regular Coulomb lattice still exists in all phases discussed. One can formulate a Wigner–Seitz cell for all pasta phases that is repeated in all directions. In the spaghetti phase, the nuclei or nuclear matter phase extends along a rod from the bottom to the top of the cell. In the lasagna phase, the nuclear matter phase consists of plates within the cell. In the hole phase, the cell can be imagined as having the holes at the corners of the cell. Note that the lattice has to vanish to form the structureless liquid of the outer core, which can only occur via a first-order phase transition.

The electrons will not be able to adapt to the pasta structures, as the Compton length of electrons of $\lambda = 1/(2\pi m_e) \approx 2 \times 10^{-12}$ m is much larger than the typical size of the cells, which is found to be about 10–20 fm. Hence, the electron liquid will be a uniform background, ensuring charge neutrality.

The condition for the stability of the pasta phases can be derived by looking at the energy of a single cell. The energy per volume consists of contributions from a bulk, a surface, a Coulomb, and an electron term:

$$\frac{E_{\text{total}}}{V} = \varepsilon_{\text{total}} = \varepsilon_{\text{bulk}} + \varepsilon_{\text{surf}} + \varepsilon_{\text{Coul}} + \varepsilon_e. \tag{7.31}$$

The total energy is minimized with respect to the radius of the nuclear structures R_N. The bulk and electron contributions do not depend on this radius, so only the surface and the Coulomb terms contribute to the stability condition. Considering spheres for simplicity, the surface and Coulomb energy are proportional to R_N^2 and R_N^{-1}. The corresponding energies per unit volume read then

$$\varepsilon_{\text{surf}} = \frac{C_{\text{surf}}}{R_N} \quad \text{and} \quad \varepsilon_{\text{Coul}} = C_{\text{Coul}} \cdot R_N^2, \tag{7.32}$$

as the volume scales as R_N^3. Minimization of the sum of the surface and Coulomb energy per unit volume results in the condition that

$$\varepsilon_{\text{surf}} = 2\varepsilon_{\text{Coul}}. \tag{7.33}$$

Interestingly, extending these considerations to one and two dimensions, one finds exactly the same condition for stability. The underlying reason is that the change in the dimensionality of the structure cancels out when looking at ratios with respect to the volume of the structure. The overall dependence on R_N of Eq. (7.32) remains unchanged for rods and plates. For two-dimensional rods, the dimensionality changes by one unit compared to the three-dimensional bubbles. The surface of rods becomes proportional to R_N while the rods occupy space proportional to R_N^2. The ratio remains to be proportional to R_N^{-1}. Similar considerations hold for the Coulomb term. The argument can be repeated for the one-dimensional plate structure. Eq. (7.33) shows that the stability of the pasta phase is controlled by an interplay between the nuclear force controlling the surface energy of nuclei and the Coulomb interaction. Determining the surface energy of nuclei is already difficult, but now for the pasta phases the surface energy needs to be known for neutron-rich systems. So, the density-dependence of the asymmetry energy enters also as an additional essential input for the description of the pasta phases. On the other hand, the contribution from Coulomb interactions seems to be well known. However, the charged structures of nonspherical nuclei are rather intricate, making the determination of the Coulomb term also a nontrivial task.

Full three-dimensional numerical simulations of the pasta phases became available only many years after the original discovery. These simulations were performed on supercomputers within three-dimensional Skyrme–Hartree–Fock methods or using techniques from molecular dynamics. Many new pasta phases were discovered with sometimes quite unusual shapes as mixtures of the pasta phases, dumbbell shaped and diamond-like structures (Okamoto et al., 2012), cross-rod phase (Pais and Stone, 2012), and waffle phase, a phase with perforated plates (Schneider et al., 2014). Some structures were found in the simulations that resembled structures seen in completely different systems. A so-called parking-garage structure was seen by Berry et al. (2016), consisting of flat sheets with helical ramps. This type of structure is familiar in biological cell structures. A gyroid phase was described in Schuetrumpf et al. (2015), where the two phases were intertwined in a complex, but regular, pattern. Gyroid phases have been seen before in nanostructured soft-matter systems. Whether or not pasta phases exist in the inner crust of neutron stars is still not fully settled and remains a field of active research.

7.3 Neutron Matter: The Outer Core

For discussing the core of a neutron star, we need to introduce the concept of bulk matter. In the outer core of a neutron star, the density is so high that the lattice structure of the crust disappears and the localized structures of neutrons and protons is absent, that is, nuclear structures or nuclei are dissolved. So, the neutron star material in the outer core has no size scale, it can be considered as infinite matter or bulk matter.

The matter in the outer core consists mainly of neutrons with a small admixture of protons and electrons. Hence, the outer core harbors the material that gave neutron stars its name: neutron matter. Before we discuss the detailed properties of neutron matter in bulk, we are going to introduce the concept of nuclear matter.

7.3.1 Nuclear Matter

Models of the nuclear interaction are derived from experimental data of nucleon–nucleon scattering and from measurements of nuclei. Nuclear matter, however, is a hypothetical form of matter that cannot be assessed by nuclear experiments. It is a theoretical construction that uses nuclear models adjusted to nuclear data extended to infinite extensions or to infinite atomic mass number A.[5]

Let us take a look at the Bethe–Weizsäcker mass formula given in Eq. (7.13), which gives the binding energy of a nucleus with a given atomic mass number A and charge Z. The binding energy per particle reads

$$\frac{E_b}{A} = a_{\mathrm{vol}} + a_{\mathrm{surf}} \cdot A^{-1/3} + a_{\mathrm{Coul}} \cdot Z^2 A^{-2/3} + a_{\mathrm{sym}} \cdot \frac{(N-Z)^2}{A^2} + \frac{\delta(A)}{A}. \quad (7.34)$$

For the infinite matter limit, that is, for $A \to \infty$, we see that the surface term and the Coulomb term will vanish. Also, the pairing energy will be negligibly small. We set also the number of protons and neutrons to be equal, $N = Z$, considering so-called symmetric nuclear matter. What remains in the limit $A \to \infty$ for the binding energy per particle of nuclear matter is simply the volume term of the Bethe–Weizsacker mass formula:

$$\left.\frac{E_b}{A}\right|_{\mathrm{n.m.}} = a_{\mathrm{vol}} = -16\,\mathrm{MeV}. \quad (7.35)$$

Hence, the binding energy of nuclear matter is -16 MeV. We notice that the binding energy of nuclear matter is much higher than the typical binding energy of nuclei, which is about -8 MeV. As the pairing energy is a small correction to the binding energy of nuclei, the surface and the Coulomb terms are responsible for the binding energy difference of -8 MeV to the bulk value of -16 MeV. In an ideal world without finite size correction, nuclei would be bound by -16 MeV. Corrections from the repulsive Coulomb force and the surface contribution shift the binding energy for nuclei to about half that value. The sizable correction from finite size effects shows that it is in principle a nontrivial task to extrapolate from the properties of nuclei with up to $A \sim 200$ to bulk matter with $A \to \infty$.

[5] For neutron stars, we recall that the baryon number or atomic mass number is about 10^{57}. So, while nuclear matter is a theoretical construction, neutron matter is realized in nature to a very good approximation in the outer core of neutron stars.

The number density of nuclei has a characteristic value, as one can see from the empirical law of the charge radius of nuclei, Eq. (7.14). From this relation, it follows that there is a constant number density of nuclei of

$$n_0 = \frac{A}{V} = \frac{3A}{4\pi R^3} = \frac{3}{4\pi r_0^3} = \text{const.} \tag{7.36}$$

Note that this number density is independent of the atomic mass number A. Of course, the value of the number density n_0 will be affected by surface and Coulomb effects when studying nuclei. But, in fact, scattering data on the heaviest stable nucleus ^{208}Pb reveal that the number density reaches a constant value below the surface of the nucleus. The surface thickness is typically about 1 fm, the radius of ^{208}Pb is about 5.5 fm, so that most of the nucleon density distribution is indeed at a number density of $n = n_0$.

The constant number density n_0 can be explained by a balancing effect: the nucleon–nucleon interaction is attractive at long distances, which holds the nucleons together in the nucleus. At short distances, the nucleon–nucleon interaction turns repulsive, the nucleons cannot be squeezed together indefinitely. There is a magic distance where the attractive and repulsive forces balance each other. This magic distance corresponds to the characteristic number density introduced earlier. As the number density saturates at the characteristic number density, the number density n_0 is also called the saturation density of nuclear matter.

A more precise value for the saturation density n_0 can be derived by fitting nuclear models to radii and surface thicknesses of nuclei. Fits from mean-field models find values of the saturation density of

$$n_0 = 0.15 \ldots 0.16\,\text{fm}^{-3} \tag{7.37}$$

or in terms of the mass density

$$\rho_0 = m_N \cdot n_0 = 140 \ldots 150 \text{ MeV fm}^{-3} = 2.5 \ldots 2.7 \times 10^{14} \text{ g cm}^{-3}, \tag{7.38}$$

where $m_N = 939\,\text{MeV}$ is the averaged nucleon mass. The corresponding Fermi momentum at saturation density is

$$k_{F,0} = \left(\frac{6\pi^2 n_0}{\gamma}\right)^{1/3} = 260 \text{ MeV}. \tag{7.39}$$

The degeneracy factor for neutrons plus protons is built from the spin degree of freedom and the degree of freedom of protons and neutrons. Protons and neutrons can be described as two different states of isospin, where the proton has isospin +1/2 and the neutron $-1/2$, respectively. The total degeneracy factor for nucleons is then given by $\gamma = (2I + 1)(2s + 1) = 4$ for a spin of $s = 1/2$ and an isospin of $I = 1/2$. Fits based on the Skyrme model find higher values of n_0, those based on relativistic mean-field models, lower values of n_0 (Stone and Reinhard, 2007). There is a bewildering variety of parameterizations based on the Skyrme model (Dutra et al., 2012) and based on the relativistic mean-field model (Dutra et al., 2014), with more than 200 parameter sets each (however, most of those parameter sets have been fitted to nuclear matter properties).

The EOS of nuclear matter can be expressed in the general density dependent form $\varepsilon = \varepsilon(n_B)$, where n_B is the baryon number density. The binding energy per nucleon can be written as

$$\frac{E_b}{A}(n_B) = \frac{\varepsilon(n_B)}{n_B} - m_N. \tag{7.40}$$

We know that the binding energy has to be $-16\,\text{MeV}$ at the saturation density $n_B = n_0$:

$$\frac{E_b}{A}(n_0) = -16\text{ MeV}. \tag{7.41}$$

As nuclear matter is stable at that point, it represents a minimum in the energy as a function of density. Hence, the first derivative of the energy with respect to the density has to vanish, which results in the condition that the pressure has to vanish

$$P(n_0) = 0 \tag{7.42}$$

because of the thermodynamic relation

$$P = n_B^2 \frac{\mathrm{d}(E/A)}{\mathrm{d}n_B}, \tag{7.43}$$

see Eq. (3.69). Next, we can expand further around the saturation density. The second derivative of the energy with respect to the density gives the curvature of the EOS around the saturation point. The corresponding coefficient is the incompressibility of nuclear matter. Historically, the second derivative was introduced in terms of the Fermi momentum k_F instead of the number density, so the incompressibility coefficient, or incompressibility in short,[6] of nuclear matter at saturation density is defined as

$$K_\infty = k_{F,0}^2 \cdot \left.\frac{\mathrm{d}^2(E/A)}{\mathrm{d}k_F^2}\right|_{k_{F,0}} = 9n_0^2 \cdot \left.\frac{\mathrm{d}^2(E/A)}{\mathrm{d}n_B^2}\right|_{n_0}, \tag{7.44}$$

with its ubiquitous prefactor 9 in terms of the second derivative with respect to the baryon number density n_B. The incompressibility can be related to the excitation energy of nuclei, in particular to giant monopole and giant dipole resonances with

[6] Sometimes, K_∞ is denoted as the compressibility. However, a larger value of K_∞ corresponds to nuclear matter, which is less compressible, so it is a measure of the incompressibility of nuclear matter at saturation density.

excitation energies in the range of several megaelectronvolts. The giant resonances can be interpreted as compression modes of the nucleus. Using fits to the excitation spectrum and correcting for surface effects within an expansion scheme called the leptodermous expansion, one has constrained the incompressibility to $K_\infty = 240 \pm 20\,\text{MeV}$ (Shlomo et al., 2006). In fits to the excitation spectra by using nuclear models, one finds that Skyrme models usually give values at the lower end of this range, while fits using relativistic mean-field models give larger values for K_∞. Note, that the incompressibility is not a directly measurable quantity, as are all quantities describing nuclear matter properties.

For a known incompressibility, one is tempted to write down a simple EOS of the form

$$\frac{E_b}{A} = \frac{K_\infty}{18}\,(n - n_0)^2\,, \tag{7.45}$$

which would relate the incompressibility to the high density behavior of nuclear matter, which is what we need to describe neutron stars. However, relating K_∞ to the high-density EOS is a pitfall of our ansatz, Eq. (7.45). Just think about higher order terms in the density as

$$\frac{E_b}{A} = \frac{K_\infty}{18}\,(n - n_0)^2 + K_\gamma\,(n - n_0)^\gamma\,, \tag{7.46}$$

with $\gamma > 2$, which shows that the incompressibility K_∞ is in general not controlling the high-density EOS of nuclear matter. In fact, it is well known from relativistic mean-field models that the high-density EOS does not depend on the incompressibility K_∞ (Boguta and Stöcker, 1983). We show now that the high-density EOS is related to the effective mass at saturation density within Dirac phenomenology by using the Hugenholtz–van Hove theorem (Hugenholtz and van Hove, 1958).

The Hugenholtz–van Hove theorem states that the Fermi energy of a Fermi gas with interactions at zero temperature is equal to the average energy plus the pressure per particle

$$E_F = \frac{d\varepsilon}{dn_B} = \frac{\varepsilon + P}{n_B}. \tag{7.47}$$

Using the thermodynamic relation $\varepsilon = -P + \mu \cdot n$, we realize that the Hugenholtz–van Hove theorem states that for a Fermi gas, the chemical potential is equal to the Fermi energy even when interactions are included. The pressure vanishes at saturation density, $P = 0$, so that the Hugenholtz–van Hove theorem relates the binding energy to the Fermi energy at saturation density

$$E_F(n_0) = \left.\frac{\varepsilon}{n_B}\right|_{n_0} = \frac{E_b}{A}(n_0) + m_N = 923\ \text{MeV}\,, \tag{7.48}$$

see Eq. (7.40) and the following text.

In Dirac phenomenology, the relativistic Fermi energy is given by $E_F = \sqrt{k_F^2 + m^2}$, which will be modified when including interactions. Consider that there is a shift in the mass of the particle by some scalar potential $S(n)$ to an effective mass m^*:

$$m^* = m - S(n). \tag{7.49}$$

The four-momentum of a Dirac particle can be shifted by a vector potential $V^\mu(n)$ to an effective four-momentum

$$k^{\mu *} = k^\mu - V^\mu(n). \tag{7.50}$$

Let us consider just a shift in the energy of the Dirac particle. There is no preferred direction in nuclear matter, so we can set the spatial components of the vector field to be zero. The energy–momentum relation with interactions is modified to

$$E_F = \sqrt{k_F^2 + (m - S(n))^2} + V^0(n) = \sqrt{k_F^2 + m^{*2}} + V^0(n). \tag{7.51}$$

Saturation of nuclear matter is achieved by the interplay of the scalar and the vector potentials. At low densities, the attractive scalar potential dominates, so nuclear matter is bound. At high densities, the repulsive vector potential becomes larger than the scalar potential, so nuclear matter gets unbound. At saturation density n_0, we know that $E_{F,0} = 923$ MeV. If we fix the effective mass m_0^* at saturation density, the relation (7.51) determines the repulsive vector potential $V^0(n_0)$. The lower the effective mass m_0^*, the larger is the scalar potential and the larger is the vector potential. Hence, m_0^* determines the high-density behavior of the nuclear EOS. The scalar and vector potentials at saturation density are quite large, as already noted in the pioneering work of Duerr (1956), and are in the range of several hundred megaelectronvolts (Furnstahl and Serot, 2000). The combined potential felt by the nucleons within a nucleus at zero momentum is just

$$U_N(n_0) = S(n_0) + V^0(n_0) \approx -50 \text{ MeV}, \tag{7.52}$$

as the two potentials largely cancel each other. The value of U_N at saturation density can be read off from the Hugenholtz–van Hove theorem. In the nonrelativistic limit, the binding energy has a contribution from the kinetic energy and the potential energy

$$\frac{E_b}{A} = E_{\text{kin}}(n_0) + U_N(n_0) = \frac{k_{F,0}^2}{2m} + U_N(n_0), \tag{7.53}$$

which gives

$$U_N(n_0) = \frac{E_b}{A} - \frac{k_{F,0}^2}{2m} = (-16 - 35) \text{ MeV} \approx -50 \text{ MeV} \tag{7.54}$$

at saturation density with $k_{F,0} = 260$ MeV.

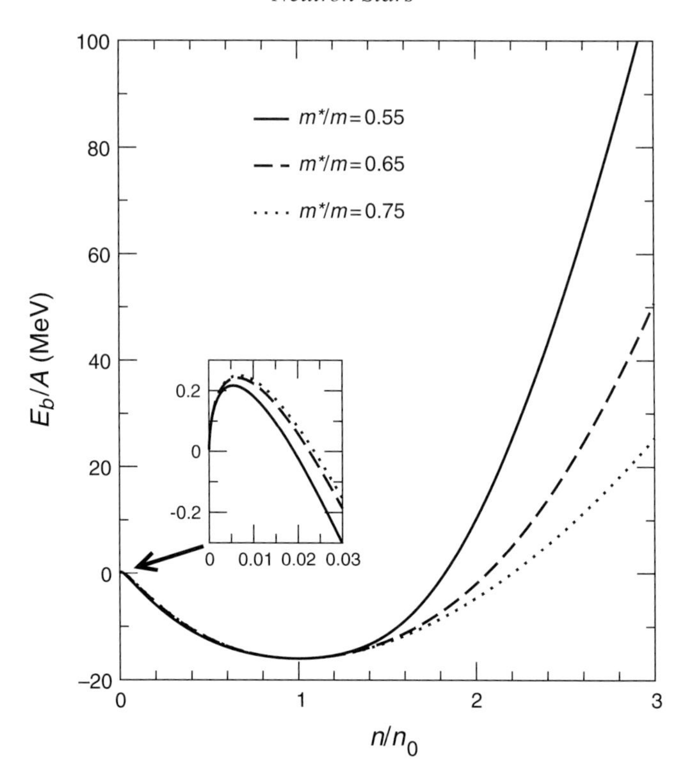

Figure 7.4 The EOS for nuclear matter with an incompressibility of $K_\infty = 240\,\text{MeV}$ and different effective masses at saturation density n_0.

If the energy per particle increases rapidly with increasing density, the nuclear EOS is said to be hard or stiff, otherwise it is said to be soft. There is no fixed criterion to determine when a nuclear EOS is hard or soft. However, one can relate one nuclear EOS to another one in terms of being harder or softer in comparison. Figure 7.4 shows the nuclear EOS in the form of the binding energy versus the baryon number density. The incompressibility is fixed to $K_\infty = 240\,\text{MeV}$ for all cases, while the effective mass at saturation density is varied from $m_N^*/m_N(n_0) = 0.55 \ldots 0.75$. One sees that the EOS differ only at high density, where the stiffness increases with decreasing effective masses. The inset shows the low-density behavior, where one sees that the binding energy increases for densities below $0.01n_0$, like a free Fermi gas, as interactions are negligible at very low densities.

7.3.2 Pure Neutron Matter

We turn our attention now to the case relevant for neutron stars, the case of pure neutron matter. We have seen that at the end of the outer crust, when reaching the neutron dripline, the Z/A ratio is of the order of 0.3, see Table 7.3. At the end of the inner crust, the charge of the unit cell is about $Z = 40$, while the number of nucleons reaches $A = 1,800$. The ratio of protons to all nucleons is in this case about 0.02, so that the contribution of protons and electrons is a small correction to the EOS of neutron star matter at the transition to the outer core.

For comparison, the lowest ratios of the charge to the mass number for known particle stable nuclei is $Z/A = 0.25$ for ^{8}He, 0.27 for ^{11}Li, and 0.29 for ^{14}Be. The latter two are known to be so extremely neutron-rich that they have a halo of a pair of neutrons surrounding the nuclear core. If one takes away one neutron, the remaining nucleus will be particle unstable. The delicate situation of these three constituents, two neutrons plus the nuclear core, is aptly classified as being a Borromean system. The coat of arms of the italian Borromeo family are three rings that are intertwined in such a fashion, so by removing one ring, all three rings fall apart.[7]

However, nuclear mass models are not able to describe such unusual nuclear systems. They can only describe masses of nuclei usually heavier than ^{16}O. Double magic nuclei with a low Z/A ratio are ^{132}Sn with $Z/A = 0.38$ and ^{208}Pb with $Z/A = 0.39$, which are usually considered in the fit procedure for the nuclear mass model parameters. We see that those Z/A values are closer to nuclear matter than to pure neutron matter. An extrapolation to pure neutron matter from nuclear mass models is therefore not well controlled and its results have to be taken with great care.

One can introduce expansion parameters for nuclear matter for nonvanishing isospin, that is, for nuclear matter with more neutrons than protons. Still, surface and Coulomb effects are ignored. Surface effects will be absent for bulk matter, Coulomb effects will be negligible as electrons compensate the charge of the protons in the outer core of neutron stars. We introduce the relative neutron to proton number asymmetry δ as

$$\delta = \frac{N - Z}{A}, \tag{7.55}$$

The first derivative of the binding energy with respect to δ vanishes, as isospin symmetric matter is the most stable configuration. The first nontrivial term is then the second derivative of the binding energy with respect to δ, the symmetry energy

$$S(n_B) = \frac{1}{2} \left. \frac{\partial^2 (E/A)}{\partial \delta^2} \right|_{\delta=0}, \tag{7.56}$$

[7] The symbol is much older and is known by the name of Valknut by the Norse people from Viking times.

which depends on the baryon number density n_B. The symmetry energy can be expanded in terms of the baryon number density to first order around the saturation density n_0

$$S(n_B) = J + L \cdot \frac{n_B - n_0}{3n_0}, \tag{7.57}$$

which results in two coefficients

$$J = S(n_0) \qquad L = 3n_0 \left. \frac{\partial S(n_B)}{\partial n_B} \right|_{n_0} . \tag{7.58}$$

One usually denotes J as the symmetry coefficient. The coefficient L is the slope parameter as it describes the density dependence to linear order of the symmetry energy. The factor 3 in its definition with respect to the baryon number density n_B originates from the historic expansion in terms of the Fermi momentum. The 2012 version of the FRDM, FRDM2012, derives the following values by a fit to all available atomic masses (Sagawa and Möller, 2017)

$$J = 32.3 \pm 0.5 \text{ MeV} \qquad L = 53.5 \pm 15 \text{ MeV} \qquad \text{(FRDM2012)}. \tag{7.59}$$

We realize that the symmetry coefficient is related to the coefficient a_{sym} of the Bethe–Weizsäcker mass formula, Eq. (7.13), which gives a value of $a_{\text{sym}} = 23\,\text{MeV}$, in obvious deviation from the value quoted for J. The reason is that FRDM2012 takes into account also a surface term that depends on the asymmetry δ, which is not considered in the Bethe–Weizsäcker mass formula. This discrepancy demonstrates again that the extrapolation from nuclear systems to bulk nuclear matter is precarious as it depends considerably on surface effects. The value of J gives us a measure of the binding energy difference of nuclear matter ($\delta = 0$) and pure neutron matter ($\delta = 1$). Taking just the quadratic term for the binding energy difference, pure neutron matter would be unbound by $E_b/A = -16\,\text{MeV} + J \approx +16\,\text{MeV}$.

The value of the slope parameter L is much less known compared to the one of J. However, L is more important for neutron stars, as it describes the density dependence of the symmetry energy and thereby fixes the pressure of pure neutron matter at saturation density. Recall that the derivative of the binding energy with respect to the baryon number density is proportional to the pressure, so indeed

$$P = n_0^2 \left. \frac{\partial(E/A)}{\partial n_B} \right|_{n_0} = \frac{L}{3} \cdot n_0. \tag{7.60}$$

The ratio of the pressure to the energy density of pure neutron matter at saturation density can then be estimated to be

$$\frac{P}{\varepsilon} = \frac{L/3 \cdot n_0}{(m_N - 16\text{ MeV} + J)n_0} = \frac{L/3}{923\text{ MeV} + J} \approx 2\%. \tag{7.61}$$

The pressure of pure neutron matter is much smaller than its energy density at n_0. Note that we ignored higher order terms in the expansion with respect to n_B and δ. On that level, the pressure of pure neutron matter at saturation density is determined by the slope parameter L. This makes sense as the pressure of normal nuclear matter vanishes at saturation density by definition, so the net pressure of pure neutron matter can only depend on the symmetry energy $S(n_B)$. With the ratio of the pressure to energy density at hand, we can estimate the thickness of the outer layer of a neutron star up to saturation density. Using Eq. (7.18), we find a thickness of $\Delta R/R \approx (3-6)\%$ for a compactness of $C = 0.2$–0.3. This shows that an essential part of a neutron star is composed of matter beyond saturation density. We note also that the pressure is very small at n_0, so a substantial increase of the pressure, a stiffening of the EOS, is needed to stabilize a neutron star as massive as $2M_\odot$. We expect, therefore, a considerable change of the neutron star EOS at densities beyond saturation density. We can turn the argumentation around and look for the critical compactness where $\Delta R/R \approx 1$. A neutron star with a maximum density of n_0 in the core would have a compactness of about $C = 1/50$. This is not too far from reality. As we will see, the corresponding neutron star using the state-of-the-art EOS up to n_0 has a mass of about $0.2M_\odot$ and a radius of about 20 km, which gives a compactness of $C = 0.015 \approx 1/70$.

The EOS of pure neutron matter can be tackled by different means than by considering extrapolations from nuclear matter. As it involves only the neutron–neutron interaction, it turns out that it can well be described by extrapolating from the two-body interaction to bulk matter. In general, the nucleon–nucleon two-body system can be classified in the total spin and the total isospin of the nucleon–nucleon pair. Protons and neutrons have spin 1/2 and can pair to either total spin zero or total spin one. As nucleons are fermions, it is known from quantum statistics that their total wavefunction must be antisymmetric, that is, it has to flip the sign when interchanging the nucleons in the pair. A state with two nucleons having the same spin are symmetric under the exchange of the nucleons, while two nucleons with different spins are antisymmetric. The same holds for the isospin. Hence, there remain only two possible states for the nucleon–nucleon pair. The two nucleons can only be paired by being symmetric in the spin and antisymmetric in the isospin or vice versa. Interchanging the two nucleons will flip the sign of the wavefunction as only one quantum number is antisymmetric. The state with total spin $S = 1$ and total isospin $I = 0$ is the deuteron. The state with total spin $S = 0$ and total isospin $I = 1$ is not bound, but it has a resonance close to threshold making this state nearly bound. Such a nearly bound state is called a quasi-bound state. A pair of two neutrons belongs to this latter state, as it has total isospin $I = 1$. The nucleon–nucleon two-body interaction for total angular momentum zero, the s-wave scattering interaction, can then be classified in these two channels, being either the $s = 1$ spin-triplet, $I = 0$ isospin-singlet state, or the $s = 0$ spin-singlet, $I = 1$ isospin-triplet state. The interaction can be quantified by the phase shift δ the wavefunction experiences for scattering at a central potential. The phase shift can be expanded in terms of the relative momentum k

$$k \cdot \cot\delta(k) = -\frac{1}{a_s} + \frac{1}{2} r_e \cdot k^2 + \cdots \tag{7.62}$$

Here, a_s is the scattering length in the s-wave and r_e is the effective range parameter. For the neutron–neutron (nn) scattering, one finds that the scattering length is very large: $a_{nn} = -18.5 \pm 0.3$ fm, while the effective range parameter is $r_e = 2.7$ fm. A large positive scattering length indicates a bound state. A large negative scattering length, a resonant state close to threshold. So we understand that this large negative scattering length originates from the quasi-bound state in the spin-singlet channel of the nucleon–nucleon interaction. The scale of the neutron–neutron scattering length is much larger than the size of the nucleon of about 1 fm. It is also larger than the typical interparticle distance for sufficiently high densities. The critical density can be estimated by setting $k_F = 1/(\pi a_{nn}) \approx 3$ MeV. This gives a density much lower than the neutron drip density of $n_{\rm drip} \approx 2 \times 10^{-3} n_0$, which marks the onset of the inner crust and the formation of a neutron liquid. Hence, the free neutrons in a neutron star do not 'see' this scattering length before they 'see' the neighboring neutrons. One says, that the scale of the neutron–neutron scattering length drops out and does not enter as a scale in a model of the pure neutron liquid. The properties of the neutron liquid then can only depend on the remaining scales at hand, which is just the Fermi momentum. The average energy per particle of the neutron can then only be proportional to the Fermi energy times a dimensionless factor. Pure neutron matter can then be described in terms of the unitary gas (UG), a generic gas of interacting fermions that depend only on the Fermi momentum. As neutrons are nonrelativistic, their energy per particle reads

$$\left(\frac{E_b}{A}\right)_{\rm (UG)} = \xi \cdot \left(\frac{E_b}{A}\right)_{\rm (Fermi\ gas)} = \xi \cdot \frac{3k_F^2}{10m_n}. \tag{7.63}$$

The dimensionless prefactor is a universal factor for all unitary Fermi gases and is called the Bertsch factor. It can be measured in atomic gases in the laboratory where the interaction of fermionic atoms, atoms with spin $s = 1/2$, is tuned by magnetic fields to ensure large scattering lengths, that is, the unitary Fermi gas case. The Bertsch factor has been determined in atomic traps of ^{6}Li to be $\xi = 0.370(5)(8)$, with the statistical and systematic error, respectively, given in brackets (Zürn et al., 2013). Eq. (7.63) represents a stunning result: the EOS of pure neutron matter is simply given by the expression for a free Fermi gas times a correction factor, which is known from laboratory measurements.

The reader might expect that there are corrections to the unitary Fermi gas limit as higher order terms in the expansion of the phase shift will become important at higher densities. The scale of the effective range parameter $r_e = 2.7\,\mathrm{fm}$ corresponds to a baryon density of about $10^{-4}\,\mathrm{fm}^{-3}$. Above that density, corrections from the effective range term have to be considered. Fortunately, these corrections can be estimated and will just lead to a slight increase of the value of ξ. Higher order contributions cancel each other so that the form of the unitary gas EOS holds up to a baryon density of $0.02\,\mathrm{fm}^{-3}$, which is model independent (Schwenk and Pethick, 2005).

The EOS of pure neutron matter for higher densities has been modeled in different ways, usually by starting to describe the two-body nucleon–nucleon interactions. In addition to nucleon–nucleon scattering data, the binding energy of the deuteron is reproduced in those ab initio approaches. Three-body interaction terms are implemented to describe the binding energy of three- and four-body nuclear systems, as the triton, ^{3}He and ^{4}He. Recent advances in the description of the nucleon forces start from a controlled expansion of the interaction terms, combining two concepts. The concept of an effective field theory sorts out the relevant degrees of freedom and all possible interaction operators are written down in an expansion series in the scales of the system at hand. The second concept is that of chiral symmetry, which has been known for a long time to give a successful description of scattering data. Combining these two concepts leads to chiral effective field theory (χEFT). As such, these models are not only able to describe two and higher order interaction terms on the same footing but are also able to delineate an error band in the final results.

The results for pure neutron matter EOS are shown in Figure 7.5 for different models. The model of Friedmann and Pandharipande (FP) and that of Akmal, Pandharipande, and Ravenhall (APR) are based on the older variational approaches with separate ansatze for two- and three-body forces. The other models shown use different many-body techniques, which have to be put on top of the basic chiral interaction terms included. The most recent one by Drischler et al. (2016), using the self-consistent Green's function theory and denoted by SCGF (2016) in the figure, reports an error band such that the EOS is well under control up to a density slightly above saturation density. One sees that the older variational models are within the range of the more modern calculations. It seems that it is sufficient to describe pure neutron matter by the nucleon–nucleon two- and three-body interactions. In fact, getting the neutron–neutron scattering data right is sufficient to describe pure neutron matter in the unitary gas limit, as seen earlier. In the figure, it is seen that the unitary gas limit gives a lower limit on the energy per baryon for pure neutron matter and that the model calculations are not too far above that limit. The striking conclusion is that the EOS of pure neutron matter, basically the EOS of the

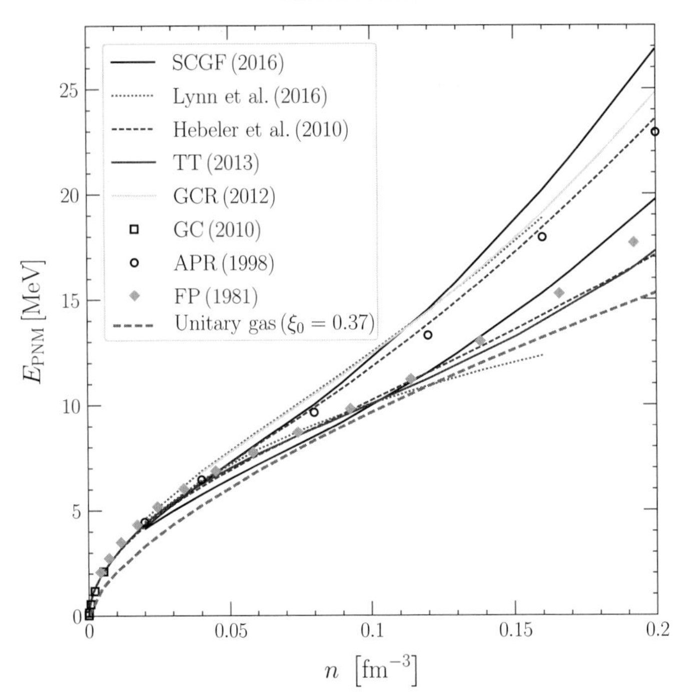

Figure 7.5 The energy per particle of pure neutron matter versus the baryon number density for different model calculations. The unitary Fermi gas limit shown for $\xi = 0.37$ is seen to be a lower limit for the model EOSs (taken from Tews et al., 2017). © AAS. Reproduced with permission. GC, Gezerlis & Carlson (2010); GCR, Gandolfi et al. (2012); TT, Togashi & Takano (2013)

outer core of neutron stars, is known up to baryon number densities slightly above saturation density, up to about $n \approx 0.2\,\mathrm{fm}^{-3}$.

Chiral EFT also makes a prediction for the asymmetry coefficient and the slope parameter (Krüger et al., 2013). However, they are defined in an expansion around the nuclear matter saturation point, that is, for an asymmetry of $\delta = 0$. For relating them to the properties of pure neutron matter, one has to make an ansatz for the dependence of the symmetry energy on the asymmetry δ to reach the point of $\delta = 1$. The simplest way is to assume a quadratic dependence of the symmetry energy $S(n_B)$ on δ:

$$\frac{E_b}{A}(n_B,\delta) = \frac{E_b}{A}(n_B,0) + S(n_B)\cdot\delta^2. \tag{7.64}$$

Then, the symmetry coefficient is the difference of the energy of pure neutron matter and nuclear matter and coincides with the definition used earlier, Eq. (7.56). The slope parameter is the same for pure neutron matter and for nuclear matter as it is independent on δ. The prediction of chiral EFT (within an improved ansatz for the symmetry energy) is then (Krüger et al., 2013)

$$J = 28.9 - 34.9\ \text{MeV} \qquad L = 43.0 - 66.0\ \text{MeV} \qquad \text{(Chiral EFT)}, \qquad (7.65)$$

which is compatible with the values quoted for FRDM2012, see Eq. (7.59).

What about the nuclear models discussed before for nuclear matter? These models are tuned to describe the properties of nuclei not the properties of few-body systems. In general, Skyrme and relativistic mean-field models cannot describe the binding energy of light nuclear systems, in particular not the one of the deuteron. Hence, the large neutron–neutron scattering length is not implemented. Therefore, these nuclear models cannot describe a priori the unitary gas and thereby not the EOS of pure neutron matter. However, there are Skyrme and relativistic mean-field models that can describe pure neutron matter. In those special cases, either the EOS of pure neutron matter has been included in the fit of the parameters of the model, as for the Skyrme parameter set SLy4 (Chabanat et al., 1997) discussed for the outer crust, or the coupling constants have been made density dependent to describe the results of Brueckner–Hartree–Fock calculations, which take into account two- and three-body forces, as for the relativistic mean-field parameter set DD2 (Typel et al., 2010). Relativistic mean-field models can be made compatible with the results of pure neutron matter by including new interaction terms (Todd-Rutel and Piekarewicz, 2005; Hornick et al., 2018).

7.3.3 Mass–Radius Relation

We are now in the position to discuss the full EOS for neutron stars and the resulting mass–radius relation.

Figure 7.6 shows the EOS for neutron star matter in the form of the pressure over the energy density in units of the energy density at the saturation point of nuclear matter, ε_0. The energy density range needed spans ten orders of magnitude from a mass density of $10^6\ \text{g cm}^{-3}$ from the outer crust to about $10^{16}\ \text{g cm}^{-3}$ in the center. The vertical dashed lines show the transition energy density from the outer to the inner crust at about $\varepsilon_c/\varepsilon_0 = 0.002$ and the one for the transition from the crust to the core at about $\varepsilon_c/\varepsilon_0 = 0.3$. Here we choose the relativistic mean-field model with the parameter set DD2 (Typel et al., 2010) for illustration. While the former critical energy density hardly changes within other model calculations, the latter one differs from model to model but stays around half saturation energy density. However, while the transition point is not well determined, the EOS hardly changes within the different models as it is mainly determined by the EOS of pure neutron matter. As one sees, the pressure is continuously increasing with increasing energy density. The EOS of the outer crust follows closely a power law. The onset

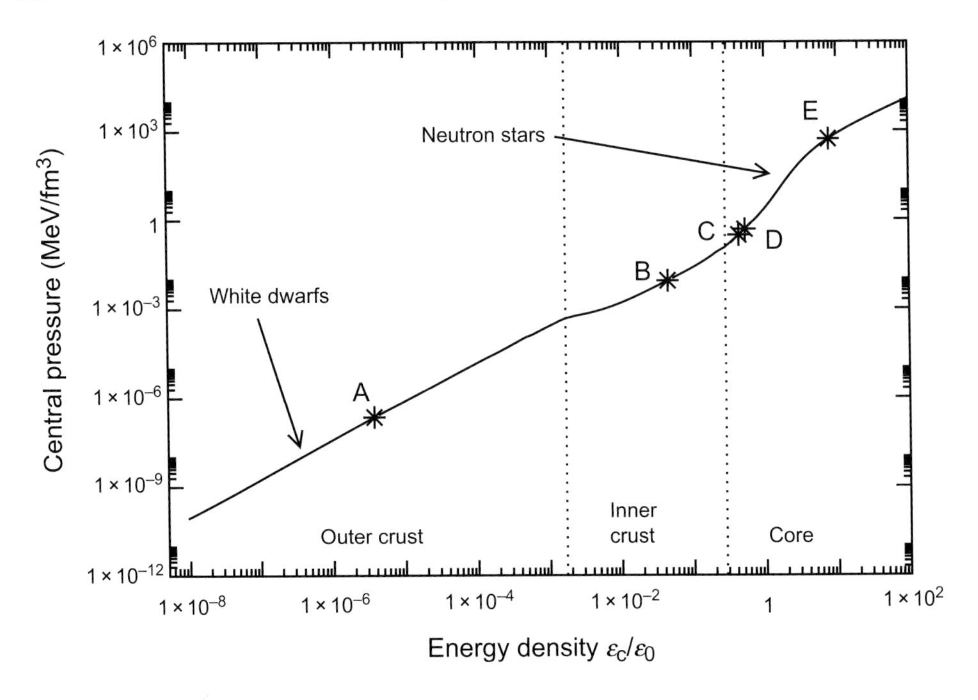

Figure 7.6 The EOS for neutron stars, pressure versus energy density, for the parameter set DD2. The vertical dotted lines delineate the corresponding regions inside neutron stars.

of the inner crust shows a sudden decrease of the slope due to the onset of the neutron liquid. In the core, the pressure increases strongly with the energy density. The labels A to E mark characteristic points for compact star configurations, to be discussed in more detail later. Here, point A indicates the maximum mass of white dwarfs at a central energy density of about $\varepsilon_c/\varepsilon_0 = 4 \times 10^{-6}$, which corresponds to a mass density of about 10^9 g cm^{-3}. Stable neutron stars appear between the points marked D and E, corresponding to energy densities of $\varepsilon_c/\varepsilon_0 = 0.5$–$8$. Note that the EOS is only known up to $\varepsilon_c/\varepsilon_0 = 1.3$ and that the upper limit on the energy density in the core of a neutron stars is model dependent.

For the energy densities encountered in the core of neutron stars, one has to distinguish now between the mass density and the energy density. The mass density ρ being defined as the nucleon mass times the baryon number density $\rho = m_N \cdot n_B$ will be smaller than the total energy density due to additional contributions to the energy density from interactions

$$\varepsilon = \rho + \text{ interaction terms.} \tag{7.66}$$

Figure 7.7 shows in a linear plot the baryon number density versus the energy density. One sees that the baryon number density starts to deviate from the relation

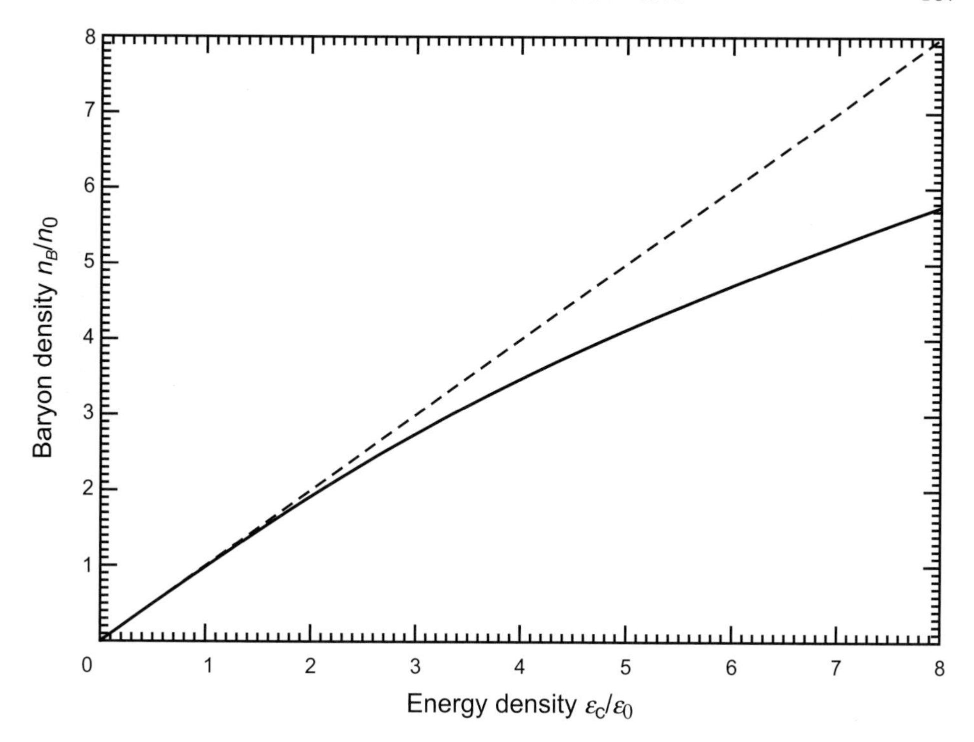

Figure 7.7 The baryon number density as a function of the energy density in units of the value of nuclear matter at the saturation point. The baryon number density (solid line) increases less rapidly than the energy density (dashed line).

$\rho \approx \varepsilon$ slightly above saturation density. At an energy density of $\varepsilon = 8\varepsilon_0$, the baryon number density is about 30% lower in comparison to $n_B = 5.7n_0$. As one can guess from the symbolic relation, Eq. (7.66), the quantitative difference between energy density and mass density will depend on modeling the interactions. However, as the interaction terms have to be repulsive to ensure the stability of neutron stars, the increase in the number density will be always lower in comparison to the one of the energy density.

In order to discuss the stable configurations of our compact star configuration, we take a look at the polytropic power defined as

$$\Gamma = \frac{\mathrm{d}\ln P}{\mathrm{d}\ln n_b} \tag{7.67}$$

and shown in Figure 7.8. We see a rich structure. Let us discuss them one by one from the left to the right along the curve. The first dip at the left-hand side stems from the transition from ^{56}Fe to ^{62}Ni in the outer crust. The transition from one nucleus to another in the outer crust is of first order, but the change in the

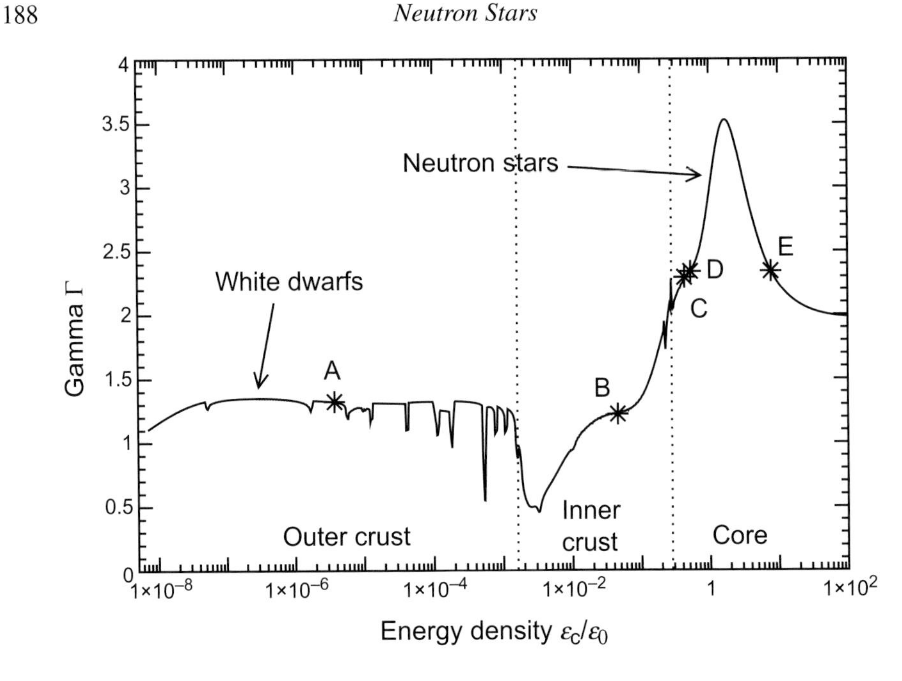

Figure 7.8 The polytropic power versus the energy density. The vertical dotted lines delineate the corresponding regions inside neutron stars.

energy density is quite small. In principle, Γ should drop to zero for a first-order phase transition. However, the transition is so sharp that it is smeared out due to the limited resolution of the EOS table. The next dip is the transition to ^{64}Ni just before point A is reached. At energy densities beyond point A, one sees several more pronounced phase dips due to the transition to heavier and more neutron-rich nuclei. Even if there is no dip, Γ is about 4/3, the lower limit of stability for polytropes. At the beginning of the inner crust, Γ drops down drastically to values of only about $\Gamma \approx 0.5$ due to the onset of the neutron liquid. If neutrons are started to be populated, they fill up the lowest levels in momentum first, producing nearly no contribution to the pressure. The baryon number density keeps increasing, however, so that a sudden drop in Γ occurs. Only at higher densities, neutrons are populated at higher momenta correlated with a sizable increase in the pressure. Γ starts to increase with increasing energy density. Just before the core is reached, a pronounced wiggle is visible in the curve. We realize that this is the transition from the lattice structure of the crust to the uniform liquid of the core. As the lattice vanishes abruptly, the transition is of first order. The change on energy density is seen to be larger compared to the transitions seen at lower densities, indicating that this transition is quite strong. It is also the region where the pasta structures discussed earlier are present. For the energy densities encountered in the core of neutron stars, Γ reaches its highest values. For the model used, Γ peaks at about $\Gamma = 3.5$. The high value of Γ ensures the stability of neutron star configurations that are stable within points D and E. Finally, Γ drops down to $\Gamma = 2$ at the highest densities shown, indicating an interaction dominated EOS. We stress that an EOS based on nucleon degrees of freedom will be unsuitable for these high densities, as the substructure of the nucleons will become important. Therefore, the high density limit is not the correct one. As we will discuss in the following chapter on quark stars, the correct high density limit is $\Gamma = 4/3$, the value for an ultrarelativistic gas of free particles.

7.3.4 Stability of Mass–Radius Configurations

The value of Γ is not sufficient to determine the stability of compact star configurations, as Γ is highly density dependent and the compact star configuration usually covers several orders of magnitude in density. For neutron stars, there will be always a density region that has a value of Γ below 4/3. Still, we know that neutron stars are stable. So we need to discuss a more general stability analysis. In the following, we take a look at linear perturbations of compact star configurations away from hydrostatic equilibrium. The differential equation turns out to correspond to a Sturm–Liouville eigenvalue equation. The eigenfunctions can be labeled by the number of nodes n with eigenfrequencies ω_n, which are ordered in energy as

$$\omega_0^2 < \omega_1^2 < \omega_2^2 < \cdots \tag{7.68}$$

Hence, if $\omega_0^2 > 0$, all modes must be real, so the compact star configuration in total is stable with respect to all possible radial oscillations. If $\omega_0^2 < 0$, then the lowest energy mode is imaginary, indicating an instability. Now one can solve the full Sturm–Liouville problem or one can resort to a trick: Set the trial value $\omega_n^2 = 0$. If the eigenfunction goes through zero N times, then there are N unstable modes. One can turn the argument around: If the eigenfunction does not go through zero, all modes are stable or at least marginally stable.

Another way to figure out the stability of compact star configurations is to look at the dependence of the total mass M and the radius R as a function of the central energy density ε_c. Stability demands first that the mass increases with increasing central energy density:

$$\frac{\mathrm{d}M}{\mathrm{d}\varepsilon_c} > 0. \tag{7.69}$$

So, extrema in the mass indicate a change in the stability of the compact star configuration. Hereby, the energy squared of a mode is changing its sign. Which energy squared changes its sign is determined by the slope of the radius as a function of the central energy density. If at the extremal point, the radius decreases with increasing central energy densities, an even mode changes. Vice versa, if the radius increases with increasing central energy density, an odd mode changes. In summary:

$$\left.\frac{dR}{d\varepsilon_c}\right|_{\text{extremum}} \begin{cases} < 0: & \text{the energy squarded of an even mode changes sign} \\ > 0: & \text{the energy squared of an odd mode changes sign.} \end{cases} \tag{7.70}$$

We will discuss these stability conditions by looking at the mass and radius of white dwarfs and neutron stars using the EOSs considered earlier from the outer crust to the core region of neutron stars.

Figures 7.9 and 7.10 show the mass and the radius as a function of the central energy density. The points marked A to E indicate extrema of the mass as a function of central energy density. At the lowest central energy density, we know that we are in the realm of white dwarfs, which we know to be stable. The first maximum with increasing central energy density at A marks the limiting mass of white dwarfs, the Chandrasekhar mass limit. The maximum mass is only about $1 M_\odot$. Note that this is a full calculation, taking into account a radial density profile and corrections from the lattice energy. Note also that we discuss here the outer crust in β-equilibrium, which consists mainly of ^{56}Fe with a charge to mass ratio of $Z/A = 26/56 \approx 0.46$. The Chandrasekhar mass is proportional to $(Z/A)^2$, so the maximum mass is reduced compared to the case of a carbon-oxygen white dwarf with $Z/A = 1/2$ (see the discussion of the Chandrasekhar mass limit following Eq. (5.40)). We see from Figure 7.10 that the radius decreases at point A, so an even mode changes sign. The mode that changes sign can only be the lowest one, the one for $n = 0$, so the configurations beyond point A are unstable. The next extremum in mass at point B is a minimum where the radius is now increasing with the central energy density. Hence, an odd mode changes sign, which in this case can only be the one for $n = 1$. So, another mode gets unstable, so in total the modes with $n = 0$ and $n = 1$ are unstable. At point C, we encounter a maximum in the mass where the radius increases with increasing central energy density. Again, an odd mode changes sign, which can only be the one for $n = 1$ again. Now, the mode for $n = 1$ turns stable again, so only the mode $n = 0$ remains to be unstable. At the point D, the mass shows a minimum, but now the radius decreases with increasing central energy density. Accordingly, an even mode changes sign, so ω_0^2 turns positive. Hence, all modes are stable now and point D marks the onset of another stable branch of compact star configurations, the stable neutron star branch. Point D indicates the minimum mass configuration of neutron stars, which is located at a mass of $M_{\text{min}} = 0.09 M_\odot$, with a radius of about $R \approx 250$ km. The central energy density is about $0.5\varepsilon_0$, which is about the energy density of the transition from the inner crust

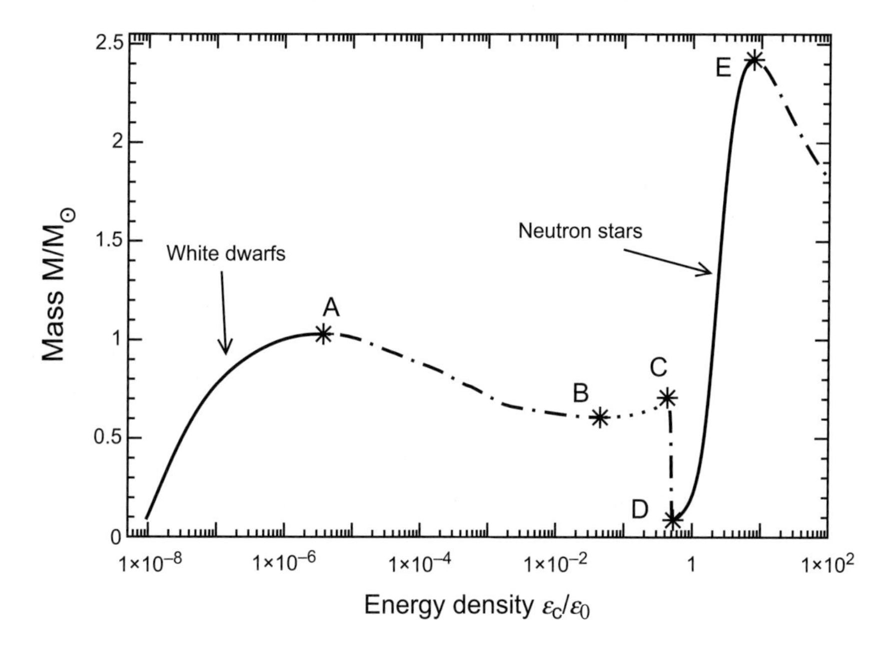

Figure 7.9 The mass of white dwarfs and neutron stars versus the central energy density.

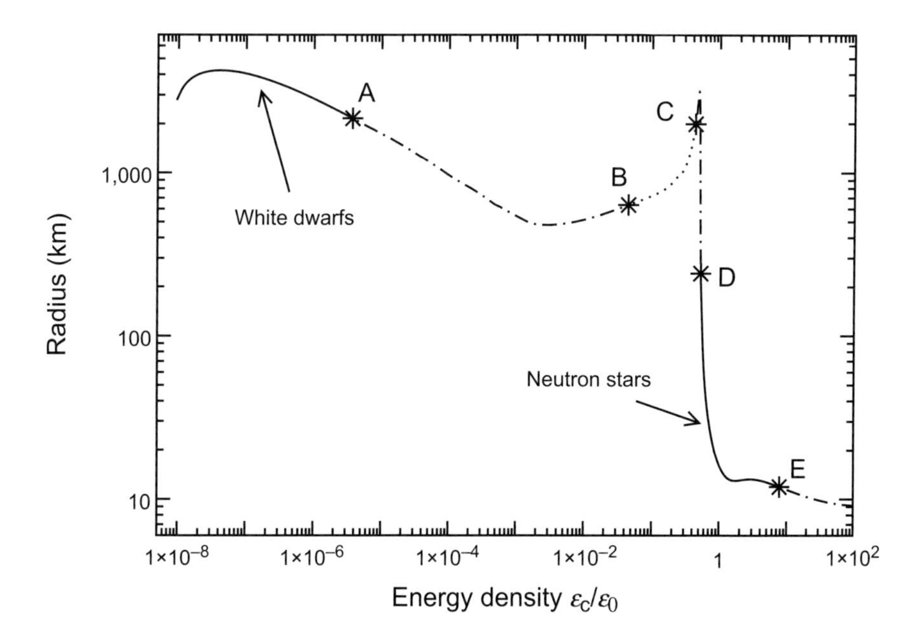

Figure 7.10 The radius of white dwarfs and neutron stars versus the central energy density.

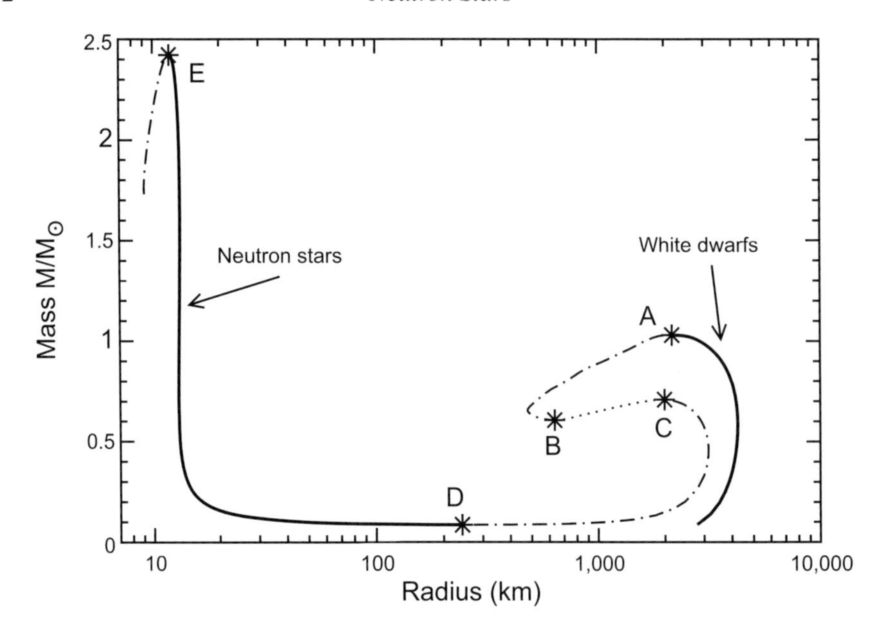

Figure 7.11 The mass–radius relation for white dwarfs and neutron stars.

to the outer core of neutron stars. The minimum mass configuration of neutron stars is therefore close to fully consisting of a lattice of nuclei with at most a small core of a neutron matter liquid.

Finally, point E marks the maximum mass configuration of neutron stars. Indeed, the radius at point E decreases with increasing central energy density, so ω_0^2 turns negative, that is, unstable as at point A. So in summary, there are two stable regions: one up to point A, the white dwarf branch, and the other one between points D and E, the neutron star branch, a posteriori justifying our labeling in the figures.

With a little moment of thought, one realizes that the considerations about the stability of compact star configurations can be recast for the mass–radius relation without the need for considering the mass and the radius as a function of the central energy density. This criterion for the stability of compact stars goes back to Bardeen et al. (1966). Consider a mass–radius curve as depicted in Figure 7.11. Start along the curve from the low energy density region, which will be usually at large radii. For low densities, we assume that we know that the configuration is stable. Then, looking along the curve with increasing energy density, a maximum in the mass is reached at point A where the curve continues in a counterclockwise direction. This behavior corresponds to a mode becoming unstable. We recall that the counterclockwise orientation is equivalent to a decreasing radius with increasing energy density. At point B, there is a minimum with a counterclockwise orientation indicating that another mode gets unstable. Here, the radius increases with

Table 7.4 *The conditions for modes getting stable or unstable for mass–radius relations of compact stars.*

Extremum\orientation	Clockwise	Counterclockwise
Maximum	Stable	Unstable
Minimum	Stable	Unstable

increasing energy density. At point C, the curve at this maximum shows a clockwise direction. The radius increases with increasing central energy density. Therefore, a mode gets stable. The minium at point D is also passed in a clockwise direction, so another mode gets stable. Here the radius decreases with increasing central energy density. Hence, the sequence following point D is stable. This recipe for figuring out stability is now reduced to the orientation of the curve going through an extremum in the mass–radius relation: A counterclockwise direction indicates that a mode gets unstable, a clockwise direction indicates that a mode gets stable. The four possible cases and its impact on the stability of modes is summarized in Table 7.4. We note that these stability criteria are generic for all compact star configurations, not only for white dwarfs and neutron stars.

Nowadays, one uses frequently EOSs of modern nuclear models, which are constructed for core-collapse supernova simulations and are also available for (nearly) zero temperature and β-equilibrium. These EOSs are available from online databases as the CompOSE webpage (see https://compose.obspm.fr).

7.4 Hyperon Matter: The Inner Core

So far, we have discussed neutron star matter in terms of nucleons and electrons only. But what about other particles? Can they be present in neutron stars too? The particle data group lists many more particles, particles observed in various nuclear and high-energy experiments (Tanabashi et al., 2018). Frankly, from our present knowledge, we do not know which particles are present in the core of neutron stars. But we can make some generic statements about the critical conditions for the presence of a particle in neutron star matter.

7.4.1 Onset of Particles in Neutron Star Matter

There are only two quantum numbers conserved for cold neutron star matter: the baryon number B and the charge Q. So, all we need is to sort the particles with respect to their baryon number and charge and look at the lowest energy states that will be populated first with increasing density. The lowest energy state for a baryon

Table 7.5 *Masses in MeV of leptons. Values from the particle data group (Tanabashi et al., 2018).*

Lepton	e^-	μ^-	τ^-
Mass	0.511	105.7	1776

number are the proton and the neutron, for charge Q it is the electron. Ok, so what are the next heavier particles?

For charge number, the next heavier particle is the muon μ^-, a lepton much heavier than the electron with a mass of $m_\tau = 106\,\text{MeV}$. We can treat it as a heavy electron, that is, as a free fermion as the electron. It belongs to the next family of particles of the standard model. The muon decays by the weak interaction on a timescale of $\tau_\mu = 2.2 \times 10^{-6}\,\text{s}$ to $\mu^- \to e^- + \bar{\nu}_e + \nu_\mu$, where $\bar{\nu}_e$ is the anti-electron-neutrino and ν_μ the myon-neutrino. Table 7.5 lists the masses of the three charged leptons known. Besides the electron and the muon, there is the tau lepton τ, which, however, is much heavier. As we will see, it is too heavy to appear in neutron star matter. Neutrinos also belong to the leptons, but they are not charged. So, they have the same quantum numbers as the photon (no charge and no baryon number). Therefore, it turns out that they are not present in cold neutron star matter (as the photon is not present at zero temperature). We note in passing that neutrinos are present in hot neutron star matter, as formed shortly after core-collapse supernovae or in neutron star merger (as is the photon).

There are other charged particles without a baryon number: the mesons. Mesons are particles that are subject to strong interactions. They are made of a quark–antiquark pair. The lightest ones are pions and kaons and their masses are listed in Table 7.6. We note that the lightest charged meson, the π^-, is heavier than the μ^-. Pions and mesons are stable with respect to strong interactions, but they decay via weak interactions to leptons. Neutron star matter is in weak equilibrium, which states that the timescale for weak interactions is much smaller than the typical timescale of neutron stars. The lifetime of pions and kaons is in the range of $10^{-8}\,\text{s}$ to $10^{-10}\,\text{s}$, certainly much smaller than the characteristic age of pulsars, which is in the range of hundreds of years and longer, see Section 6.5.3. In weak equilibrium, the weak interactions have sufficient time to equilibrate, meaning that the weak decay of particles is in balance with the weak production of particles. Once pions or kaons are present in weak equilibrium, the amount of decaying pions or kaons equals the amount of producing pions and kaons, so the net number of pions or kaons is unchanged. Indeed, the presence of pions and kaons has been considered in neutron star matter. The issue is not fully settled at present as the interaction of

Table 7.6 *Masses given in megaelectronvolts of mesons stable under strong interactions. Values from the particle data group (Tanabashi et al., 2018).*

Meson	$\pi^{\pm}$	π^{0}	$K^{\pm}$	K^{0}
Mass	139.6	135.0	493.7	497.6

pions and kaons in dense neutron star matter is not know. As pions and kaons are bosons, if they are present, they form a Bose condensate in the core of neutron stars.

For baryons, the next heaviest particle after the proton and the neutron is the Lambda Λ, with a mass of $m_\Lambda = 1116\,\text{MeV}$. It can be considered as a heavy neutron for neutron star matter as it has zero charge and baryon number one. When first discovered, the behavior of this new particle seemed so strange that it was called a strange particle and given later on the quantum number strangeness. For historical reasons, the strangeness number of the Λ is negative $S = -1$. In the quark picture, the neutron and the proton consist of three light quarks: The proton has two up and one down quark, the neutron has one up and two down quarks. In contrast, the Λ has a strange quark, it consists of one up, one down, and one strange quark. Hence, the strangeness quantum number is associated with the presence of the strange quark inside the hadron. Baryons with strange quarks are called hyperons to distinguish them to nucleons, which have no strange quark. The Λ hyperon decays to a nucleon and a pion within a timescale of $\tau_\Lambda = 2.6 \times 10^{-10}\,\text{s}$ via the weak interaction. Other heavier hyperons are the Sigma-hyperon Σ with $S = -1$ and masses of around $m_\Sigma = 1,190\,\text{MeV}$ and the Ξ^0 and Ξ^--hyperons with $S = -2$ and masses of $m_\Xi = 1,315\,\text{MeV}$ and 1,322 MeV, respectively. Finally, the heaviest hyperon is the Omega-hyperon Ω^-, with $S = -3$, so it consists of three strange quarks only. It is interesting to note that some of the hyperons have negative charge, so they can contribute to the overall charge neutrality of neutron star matter while carrying a nonvanishing baryon number. Table 7.7 lists all nucleons and hyperons that are stable under strong interactions, that is, they can only decay by weak interactions, so that the particle data group lists also a lifetime of the particle. The typical lifetime of those hyperons is $10^{-10}\,\text{s}$ (an exception is the Σ^0, which decays to a Λ and a photon by electromagnetic interactions and has a lifetime of about $10^{-19}\,\text{s}$). There are other, more massive baryons seen in experiments. However, they are resonances, that is, states that decay via strong interactions to the baryons listed in the Table 7.7.[8]

[8] For completeness, baryons with heavier quarks, as charmed baryons, are known that decay by weak interactions only, but their masses are above 1.8 GeV, so they are too heavy to appear in neutron star matter.

Table 7.7 *Masses given in megaelectronvolts of baryons stable under strong interactions. Values from the particle data group (Tanabashi et al., 2018).*

Baryon	p	n	Λ	Σ^+	Σ^0	Σ^-	Ξ^-	Ξ^0	Ω^-
Mass	938.3	939.6	1,116	1,189	1,193	1,197	1,315	1,322	1,672

In fact, all particles with a baryon number that are heavier than the nucleons are unstable and decay. So, if they are unstable, can we ignore them as they cannot be present in neutron star matter? The answer is no. Just think about the neutron that is, according to our discussion, present as a liquid in the inner crust and outer core of neutron stars. It is unstable in vacuum and decays via the weak interaction to a proton, an electron, and an anti-electron-neutrino on a timescale of $\tau_n = 880.2 \pm 1.0\,\mathrm{s}$. It is stabilized in neutron star matter because of the Pauli principle. The products of the decay involve fermions, most importantly the proton and the electron. If the available phase space of protons and electrons is blocked by the presence of other protons and electrons, the decay is Pauli-forbidden. Hence, the neutron cannot decay at sufficiently high densities of protons and electrons and forms a separate Fermi liquid in the inner crust and outer core of neutron stars. The critical condition for the onset of neutrons is that they become energetically favored in dense matter. In thermodynamics, the corresponding relation is just that of chemical equilibrium between neutrons, protons, and electrons. The neutron density is zero at the onset, so the Fermi momentum of neutrons is zero. Ignoring interactions for the neutrons, the critical condition for the onset of neutrons reads

$$\mu_n = m_n = \mu_p + \mu_e, \tag{7.71}$$

a relation which we have introduced in the discussion of dense matter in β-equilibrium in Chapter 3. We can generalize the condition for chemical equilibrium for a particle i with arbitrary baryon number B_i and charge Q_i to

$$\mu_i = B_i \cdot \mu_B + Q_i \cdot \mu_Q, \tag{7.72}$$

where μ_B is the baryon chemical potential or the neutron chemical potential and μ_Q is the charge chemical potential or the (negative) electron chemical potential. We can now use that relation for the critical condition for the onset of the muon. As the muon can be treated as a free particle, we arrive at

$$\mu_\mu = m_\mu = \mu_e. \tag{7.73}$$

If the electron chemical potential reaches the value of the muon mass, that is, if $\mu_e = 106\,\mathrm{MeV}$, muons will be populated in the core of neutron stars. Many models of the core of heavy neutron stars indeed predict such high electron chemical potentials, so the presence of muons in heavy neutron stars is likely.

The electron chemical potential and the corresponding electron fraction can be easily estimated for a free gas of neutrons, protons, and electrons. Charge neutrality demands that $n_e = n_p$, so the Fermi momenta of electrons and protons is equal, $k_{F,e} = k_{F,p}$. The condition of chemical equilibrium $\mu_n = \mu_p + \mu_e$ reads for a free gas

$$\sqrt{m_n^2 + k_{F,n}^2} = \sqrt{m_p^2 + k_{F,p}^2} + \sqrt{m_e^2 + k_{F,e}^2}. \tag{7.74}$$

Let us adopt the nonrelativistic approximation for nucleons, that is, $k_{F,N} \ll m_N$, and the relativistic one, that is, $k_{F,e} \gg m_e$ for electrons. Setting $k_{F,p} = k_{F,e}$, one finds

$$m_n + \frac{k_{F,n}^2}{2m_n} \approx m_p + \frac{k_{F,e}^2}{2m_p} + k_{F,e}. \tag{7.75}$$

For relativistic electrons, the electron chemical potential is just the electron Fermi momentum $\mu_e \approx k_{F,e}$. For $m_n \approx m_p$ and $k_{F,e} \ll m_p$, we arrive at a simple estimate for the electron chemical potential

$$\mu_e = \frac{k_{F,n}^2}{2m_n} \approx E_{F,n}^{(\mathrm{nr})} \approx 56\ \mathrm{MeV} \left(\frac{n}{n_0}\right)^{2/3}. \tag{7.76}$$

Hence, the electron chemical is about the nonrelativistic Fermi energy of neutrons, $E_{F,n}^{(\mathrm{nr})}$. The electron fraction can then be estimated to be

$$\frac{n_e}{n} \approx \frac{k_{F,e}^3}{k_{F,n}^3} \approx \left(\frac{k_{F,n}}{2m_n}\right)^3 = 5.1 \times 10^{-3} \frac{n}{n_0}, \tag{7.77}$$

which turns out to be rather small. Our estimates can be compared to an actual calculation of a free gas of nucleons and electrons where the results are shown in Table 7.8. We see that the electron chemical potential is indeed about the nonrelativistic neutron Fermi energy and reaches values of about 100 MeV at $3n_0$. This means that muons would appear around that density. The electron fraction is at most about 1% even at $3n_0$. For comparison, the baryon chemical potential is much larger than the electron chemical potential as it is given approximately by

$$\mu_B = \mu_n \approx m_n + \frac{k_{F,n}^2}{2m_n} = m_n + E_{F,n}^{(\mathrm{nr})}, \tag{7.78}$$

that is, the neutron mass plus the nonrelativistic neutron Fermi energy. Note that these estimates are for a noninteracting gas of particles. Interaction will change

Table 7.8 *The neutron Fermi momentum $k_{F,n}$, the baryon chemical potential μ_B, the nonrelativistic neutron Fermi energy $E_{F,n}^{(nr)}$, the electron chemical potential μ_e, and the electron (or proton) fraction n_e/n for a free gas of nucleons and electrons for different baryon number densities ($n_0 = 0.15\,fm^{-3}$ is used).*

n/n_0	k_n (MeV)	μ_B (MeV)	$E_{F,n}^{(\mathrm{nr})}$ (MeV)	μ_e (MeV)	n_e/n
0.5	257	974	35	35	2.6×10^{-3}
1.0	324	994	56	54	4.6×10^{-3}
2.0	409	1,024	89	82	8.2×10^{-3}
3.0	468	1,050	117	105	1.1×10^{-2}

those values. However, the results shown are in the right ballpark of model calculations including interactions: The electron chemical potential can be sizable, but the electron fraction stays small even at high densities.

We turn now our attention to the appearance of hyperons in dense netutron star matter. The critical condition for the onset of Λ hyperons, ignoring interactions for a moment, is given by

$$\mu_\Lambda = m_\Lambda = \mu_n. \tag{7.79}$$

If the neutron chemical potential reaches 1116 MeV, Λs will appear in neutron star matter when ignoring interactions. The energy to add a Λ hyperon is then equal to its mass, so it can be added to the medium while maintaining energy conservation. Such values of the neutron chemical potential are commonly reached in models of the neutron star core. We can estimate the critical density for the onset of the Λ by just considering a free gas of pure neutron matter. The critical condition for Λs to appear reads

$$m_\Lambda = \mu_n = \sqrt{m_n^2 + k_{F,c}^2}, \tag{7.80}$$

where $k_{F,c}$ is the critical Fermi momentum of the neutrons, which is then determined to be

$$k_{F,c} = \sqrt{m_\Lambda^2 - m_n^2} = 602 \text{ MeV}. \tag{7.81}$$

The corresponding critical baryon density for pure neutron matter turns out to be

$$n_c = \frac{1}{3\pi^2} k_{F,c}^3 = 0.96 \text{ fm}^{-3} = 6.4 n_0. \tag{7.82}$$

So, Λs would appear in pure neutron matter at a baryon density of $n_c = 6.4 n_0$ when ignoring interactions.

Table 7.9 *Critical density n_c for the appearance of exotic particles (other than neutrons, protons, and electrons) in neutron star matter for free gas of particles. No further particles appear up to a density of* $20n_0$.

Particle	n_c (fm^{-3})	n_c/n_0	μ_Q (MeV)	μ_B (MeV)
μ^-	0.457	3.0	−106	1,050
Σ^-	0.616	4.1	−125	1,072
Λ	1.26	8.4	−123	1,116

Let us discuss the composition of neutron star matter for noninteracting particles in more detail. We take into account all the stable particles known up to a mass of 1.8 GeV: the baryons from Table 7.7, the leptons from Table 7.5, and the mesons from Table 7.6. We assume that we have weak equilibrium, so only charge and baryon number are conserved quantities. Then, we calculate for increasing baryon number density the baryon and the charge chemical potential and check for the appearance of new particles besides neutrons, protons, and electrons, that is, when the mass of the particles equals its chemical potential, $m_i = \mu_i$, using the equation of chemical equilibrium (7.72). The resulting critical densities for new particles to appear in neutron star matter is shown in Table 7.9. One sees that the muon appears at $3n_0$. The electron chemical potential at that density is 106 MeV and equals the muon mass, as it should be at the onset of the appearance of the muons. The value of the critical density is in line with our earlier estimates, see Table 7.8. The Λ hyperons appear at a density of $8.4n_0$ where the baryon chemical potential reaches a value of 1,116 MeV, which is equal to the Λ mass. The critical density is larger than our estimate as we neglected the presence of other particles besides neutrons. We note, however, that the Σ^- hyperon appears before the Λ hyperon at a density of $4.1n_0$. Looking at the chemical potentials at that density, we see that the chemical potential for the negatively charged baryon Σ^-, the sum of the baryonchemical potential, and the chemical potential of the electron $\mu(\Sigma^-) = \mu_B + \mu_e = (125 + 1072)$ MeV $= 1,197$ MeV is just equal to the mass of the Σ^-. Hence, the more massive Σ^- appears at a lower density than the Λ by virtue of the large electron chemical potential that overcompensates the larger mass of the Σ^- compared to that of the Λ.

No other particle considered in our model appears in the calculation up to densities of $20n_0$. Even the light mesons do not appear. The negatively charged pions will appear when the electron chemical potential reaches the mass of charged pions. However, with the appearance of the negatively charged Σ^-, there is no need to increase the density of electrons further to ensure charge neutrality, so the electron chemical potential does not increase any further but starts to decrease with increasing baryon number density. The maximum electron chemical potential is reached with the onset of the Σ^-, which is $\mu_e^{\max} = 125\,\text{MeV}$, which is less than the pion mass so that pions do not appear. Charged kaons are even heavier, so they do not appear in neutron star matter for a free gas of particles either.

The situation depicted will change when interactions are considered for neutron star matter. We know that interactions have to be included in the description of neutron stars for pure neutron matter, otherwise the maximum mass will be only $M = 0.71 M_\odot$ in contradiction to the observation of pulsar masses of up to $2M_\odot$. In fact, taking into account the additional particles appearing for a free gas in Table 7.9 will produce a softer EOS than that for a gas of neutrons, protons, and electrons only. For example, when the Σ^- appears, an electron with a high Fermi momentum and a neutron with a high Fermi momentum are converted to a Σ^- with a nearly vanishing momentum just above the critical density. Less momentum in the systems results in smaller increase of the pressure with baryon number density. The energy density, controlled by the masses of the particles, will increase. The overall effect is that the matter has less pressure to counterbalance gravity, causing a decrease in the maximum mass. For the free gas EOS, the effect is small, the maximum mass changes to $0.69 M_\odot$ for the full free gas EOS. We will see a much more pronounced effect in the next subsection, where we discuss the impact of interactions on the appearance of particles in neutron star matter.

Contrary to the muon, interactions are important for the baryons to consider their presence in neutron star matter. Interactions for Λs have been studied, either by scattering experiments or by the observations of bound systems of hyperons and nucleons, which are called hypernuclei.

7.4.2 Hypernuclei

Hypernuclei were detected first in cosmic ray experiments in 1953 by Danysz and Pniewski (1953). Since then, hypernuclei have been produced and studied in the laboratory from mass number $A = 3$, the hypertriton ${}^3_\Lambda\text{H}$, consisting of a deuteron plus a Λ, to mass number $A = 208$ for the hypernucleus ${}^{208}_{\Lambda}\text{Pb}$. The notation for hypernuclei is ${}^A_\Lambda Z$, where one puts the chemical symbol for the charge Z. The Λ is counted in for the mass number of the nucleus, so A corresponds to the number of baryons of the total bound system. Λ hypernuclei are stable under strong interactions but decay by weak interactions, so their lifetime is of the order of the lifetime of the Λ, which is $\tau_\Lambda = 2.6 \times 10^{-10}\,\text{s}$. That lifetime corresponds to a decay length of $c\tau_\Lambda = 7.8\,\text{cm}$, so tracks of hypernuclei visible to the naked eye have been observed in emulsion experiments. Modern experiments use beams of kaons or pions to produce hypernuclei, thereby replacing a neutron of the original nucleus with a Λ:

$$K^- + {}^AZ \to {}^A_\Lambda Z + \pi^- \qquad \pi^+ + {}^AZ \to {}^A_\Lambda Z + K^+. \tag{7.83}$$

The outgoing pion and kaon carry information on the binding energy of the Λ inside the nucleus.

Figure 7.12 shows a summary of the presently known single particle energies of hypernuclei from spectroscopy. Note that there are in addition hypernuclei known down to mass number $A = 3$ from emulsion experiments that are not shown in the plot. The single particle energies can be simply described by the quantum numbers of the shell model without a spin-orbit splitting, which are s_Λ, p_Λ, d_Λ, f_Λ, up to even g_Λ. The lines show a shell model calculation for solving the Schrödinger equation with a Wood–Saxon potential, which constitutes a textbook showcase par excellence. The only parameter fixing the Λ interaction is the potential of the Λ in bulk matter, which is the limit $A \to \infty$. The figure shows the single particle energies as a function of $A^{-2/3}$, which serves two purposes. First, a dependence on $A^{-2/3}$ corresponds to a dependence on the radius of the nucleus R as R^{-2}, from the empirical relation that $R = r_0 \cdot A^{1/3}$, so it reflects effects from the surface of the nucleus. Second, the bulk limit $A \to \infty$ can be read off from the binding

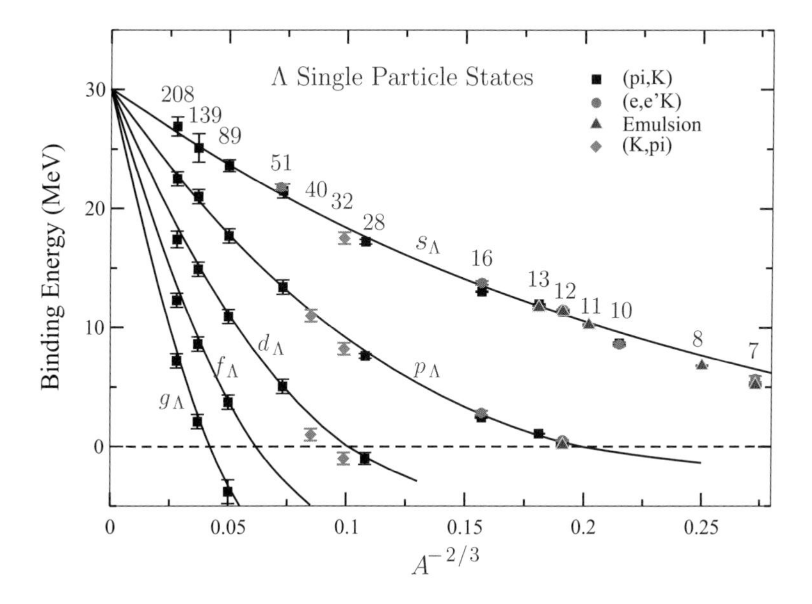

Figure 7.12 The Λ single particle energies of Λ hypernuclei with different mass numbers A. In bulk nuclear matter, that is, for $A \to \infty$, the Λ feels an attractive potential of 30 MeV. The numbers refer to the mass number of the hypernucleus. Reprinted figure with permission from Gal et al. (2016). Copyright (2016) by the American Physical Society

energy axis at the left side of the plot. The binding energy of the Λ in bulk matter is determined to be 30 MeV, so the Λ feels an attractive potential in bulk nuclear matter of

$$U_\Lambda(n = n_0) = U_{\Lambda,0} = -30 \text{ MeV} \tag{7.84}$$

at saturation density n_0. A refined fit to the single-particle energies reveals that there is a nontrivial density dependence of the Λ potential as a function of the baryon number density parametrized in the form

$$U(n) = -a \cdot \left(\frac{n}{n_0}\right) + b \cdot \left(\frac{n}{n_0}\right)^\gamma \tag{7.85}$$

such that the potential exhibits a minimum close to saturation density (Millener et al., 1988). Hence, hypernuclear data show that the Λ potential turns less attractive and eventually repulsive at densities beyond saturation density. The density dependence of the Λ potential can be rewritten as

$$U_\Lambda(n) = U_{\Lambda,0}\left[\frac{\gamma}{\gamma-1}\left(\frac{n}{n_0}\right) - \frac{1}{\gamma-1}\cdot\left(\frac{n}{n_0}\right)^\gamma\right] \tag{7.86}$$

by demanding a minimum at $n = n_0$. The second term in the brackets is the term dominating at high densities for $\gamma > 1$. The precise value of γ cannot be fixed from the data on Λ hypernuclei. The expectation is that interactions grow as n^2 for high densities, which corresponds to $\gamma = 2$. Then, the Λ potential would switch from attraction to repulsion at $n = 2n_0$.

The experimental situation for heavier hyperons than the Λ is much less certain, see Gal et al. (2016) for a review on hypernuclei. For Σ hyperons, it is established now that the Σ potential in nuclear matter is strongly repulsive with a likely value of $U_{\Sigma,0} = +30 \pm 20$ MeV. No bound Σ hypernucleus has been detected, with the exception of the quasi-bound hypernucleus ${}^{4}_{\Sigma}\text{He}$. For the Ξ hyperons, there are a few old emulsion data suggesting bound Ξ hypernuclear states. Scattering of a K^- on a nucleus and emission of a K^+ deposits two units of strangeness, thereby enabling the formation of a Ξ hyperon inside a nucleus. The corresponding measurements hint at an attractive potential for the Ξ inside the nucleus. The Ξ potential in nuclear matter is much less attractive compared to the Λ, likely to be about half of it, that is, $U_{\Xi,0} \approx -15$ MeV. Nothing is known experimentally about Ω hypernuclei. It is interesting to note that the potential of the nucleons, Λ, and Ξ in nuclear matter seem to scale with the number of nonstrange quarks. The nucleon has three nonstrange quarks with a potential of -50 MeV at saturation density. The Λ has two nonstrange quarks and its potential is -30 MeV, which is about two-thirds that of the nucleon. The Ξ has one nonstrange quark where a potential of -15 MeV amounts to about a third of the one of the nucleon or half of that of the

Λ. This pattern or rule can be referred to as the quark-counting rule for hypernuclear potentials.

The potential between hyperons can be experimentally accessed by observing double hypernuclei. There exists one firmly established double hypernucleus, the ${}^{6}_{\Lambda\Lambda}$He, which consists of a ^{4}He core surrounded by two Λs. There exists also a few more double hypernuclear events. However, the interpretation of the double hypernucleus and its binding energy is not uniquely established, unfortunately. What one knows is that the total binding energy of the double hypernucleus ${}^{6}_{\Lambda\Lambda}$He shows some additional binding energy that has to originate from the interaction between the two Λs. Hence, the interaction between two Λs is known to be attractive. The amount of the additional bond energy is small, however, so that the $\Lambda - \Lambda$ interaction appears to be weaker than the one between nucleons. The other hyperon–hyperon interactions are not known experimentally. In the future, experiments at the Japan Proton Accelearator Research Complex (J-PARC) in Japan and at the PANDA experiment at the FAIR in Germany will produce and study double hypernuclei to determine the hyperon–hyperon interaction. The ALICE experiment at CERN's Large Hadron Collider and the CBM experiment at FAIR has been and will be measuring correlations between hyperons produced in relativistic heavy-ion collisions, thereby extracting the hyperon–hyperon interaction strength, see e.g. Acharya (2019). For a list of future experiments involving hyperons, see, for example, Gal et al. (2016).

7.4.3 Hyperons in Dense Matter

Let us see how interactions can change the critical density for the appearance of Λs in dense neutron star matter. Assuming pure neutron matter, the critical condition reads now

$$m_\Lambda + U_\Lambda(n) = \sqrt{k_{F,c}^2 + m_n^2} + U_n(n). \tag{7.87}$$

The critical potential difference between neutrons and Λs

$$\Delta U_{\text{crit}}(n) = U_n(n) - U_\Lambda(n) = m_\Lambda - \sqrt{k_{F,c}^2 + m_n^2} \tag{7.88}$$

can shift the critical density to lower values compared to the free case. The potential difference needed to have $n_c/n_0 = 1, 2, 3$ corresponding to $k_{F,c} = 324$, $410, 470$ MeV has to be $\Delta U(n) = 123, 92, 66$ MeV, respectively. Let us adopt the form of the potential of Eq. (7.86) for nucleons and Λs in the following. We set for definiteness $\gamma = 2$ and use the experimentally determined potentials at saturation density of $U_n(n_0) = -50$ MeV and $U_\Lambda(n_0) = -30$ MeV for neutrons and Λs, respectively. The potential difference of neutrons and Λs is then given by

$$\Delta U(n) = \left(U_{n,0} - U_{\Lambda,0}\right)\left\{2\left(\frac{n}{n_0}\right) - \left(\frac{n}{n_0}\right)^2\right\} \tag{7.89}$$

and assumes values of $\Delta U(n) = -20, 0, +60\,\text{MeV}$ for $n/n_0 = 1, 2, 3$. As the neutron potential rises more steeply at high densities, eventually neutrons will be disfavored compared to Λs. We see that at $n = 3n_0$ the potential difference is close to the value needed to fulfill the critical condition for Λs to appear, Eq. (7.88). Hence, Λs will appear in neutron matter slightly above $n = 3n_0$ for our choice of the potential. Note that for higher nonlinear terms in the potential, that is, for $\gamma > 2$, the potential difference $\Delta U(n)$ will be rising even more rapidly with density shifting the critical density to a value even below $3n_0$. The potential difference will then be $\Delta U(n) = -20, +20, +180\,\text{MeV}$ for $\gamma = 3$, for example. So far, we ignored the impact of the symmetry energy on the chemical potential of neutrons. For pure neutron matter, the parameter δ for the expression of E/A with the symmetry energy, see Eq. (7.64), is $\delta = 1$, so the energy is shifted up by the symmetry energy $S(n_B)$, moving the onset of the appearance of Λs to an even lower density. There is, however, a lower limit on the critical density for hyperons to appear. The change of the sign of $\Delta U(n)$ is necessary for the hyperon onset, which occurs at $n = 2n_0$ for $\gamma = 2$, so the critical density cannot be much smaller than $n \approx 2n_0$. Hence, the expectation is that hyperons will appear in dense neutron star matter at a critical density somewhere between $n_c = (2 - 3)n_0$. From our discussion of the hypernuclear potentials, the Σ is disfavored to appear before the Λ as its potential is repulsive. The Ξ hyperon potential is likely to be attractive, which would favor its appearance in neutron star matter compared to the free gas case.

Indeed, many nuclear models including hyperons find that hyperons appear in the density range of $n_c = (2-3)n_0$ in neutron star matter, see, for example, Schaffner-Bielich (2008). The onset of new particles appearing in dense neutron star matter marks the beginning of the inner core of neutron stars. We will denote in the following the region in the core consisting of neutrons, protons, and leptons (electrons plus possibly muons) as the outer core and the region where other particles, as the Λ, appear as the inner core of neutron stars.

7.4.4 The Hyperon Puzzle

The presence of hyperons in the inner core of neutron stars turns out to be a major problem for modern models of neutron star matter at high density. As we discussed in the last subsection, the appearance of hyperons at $n_c = (2 - 3)n_0$ can be generically traced back to the smaller potential of the Λ compared to the one of the neutron. Thereby, the Λ feels less repulsion at high density compared to the neutron and becomes energetically favored at a density of $n_c = (2-3)n_0$. The smaller repulsion can be attributed to the quark-counting rule: As the Λ has only two nonstrange quarks, the potential depth is only two-thirds that of the neutron.

The problem with Λ matter in the core of neutron stars is that they become increasingly favored at densities beyond the critical density n_c as the difference in the potentials for neutrons and Λs increases with density, see Eq. (7.89). This feature is different compared to protons, which appear at the start of the outer core in neutron stars. Protons and neutrons feel the same potential in nuclear matter. For neutron matter, the symmetry energy has to be considered that favors equal amounts of neutrons and protons over pure neutron matter. However, the corresponding isospin dependent potential is considerably smaller compared to the isoscalar one in nuclear matter. Recall that the symmetry energy is about 32 MeV while nuclear scalar and vector potentials can be hundreds of megaelectronvolts. Hence, protons are not getting as abundant in the outer core compared to the Λs once Λs appear at n_c in the inner core. The substantial fraction of Λs present at high densities will turn the neutron star to a compact star more aptly called a hyperon star. One can also consider such a neutron star as a giant hypernucleus floating in free space.

Once a substantial fraction of hyperons is present in the inner core of neutron stars, the EOS will be substantially modified. The contribution of the hyperons to the EOS from interactions will be considerably smaller compared to that of neutrons or nucleons. The interaction strength for hyperons will be two-thirds that of the nucleons, reflecting the quark-counting rule. We know from our general discussion of compacts stars with an interaction dominated EOS, that the maximum mass scales with the interaction strength y as

$$M_{\max} \propto y \cdot M_{\text{Landau}}$$

for $y \gg 1$, see Eq. (4.100). Hence, if the interaction strength y is reduced by two-thirds for hyperons in dense matter, so is the maximum mass reduced by two-thirds for hyperon stars compared to the case of a neutron star consisting of nucleons only. Modern nuclear models based on nucleons only hardly achieve maximum masses of $3M_\odot$, but should get a maximum mass of at least $2M_\odot$ to comply with pulsar mass measurements. If, say, the maximum mass of a compact star for a nuclear model without hyperons is about $(2.0-3.0)M_\odot$, then the maximum mass for the nuclear model with hyperons will be only about $(1.3-2.0)M_\odot$ according to their reduced interaction strength. Except for the upper end of this range, the maximum mass of the hyperon star will be in violation of the observed pulsar masses of $2M_\odot$. This constitutes the hyperon puzzle: Once hyperons are taken into account in the nuclear models for neutron star matter, the maximum mass of the neutron star configuration will be (in nearly all cases) below the observed pulsar mass of $2M_\odot$. Recall, that for Ξs the interaction with nucleons would be even only a third that of the nucleon–nucleon interactions according to the quark-counting rule and as indicated from hypernuclear data. The presence of Ξs in neutron star matter would shift the maximum mass to even lower values.

One could think of several ways out of this dilemma:

- One excludes hyperons in the nuclear models by hand. However, constructing a model of the nuclear interaction by ignoring experimental data from hypernuclei is sweeping the intrinsic failure of the nuclear model under the carpet.
- One pushes up the critical density for the onset of hyperon formation in neutron star matter beyond the maximum density in neutron stars. As we have seen, the onset of hyperon formation is related to the density dependence of the potential difference between neutrons and hyperons. Pushing up the critical density implies that the repulsion felt by the hyperons in dense matter is not much smaller than the one felt by neutrons, which would be in contrast to the weaker Λ–nucleon interaction seen in hypernuclei.
- There is an additional repulsion between hyperons at high densities so that the fraction of hyperons is suppressed. However, this repulsion has to compensate the weaker repulsion between nucleons and hyperons. If Ξs are also present, this compensation would be even more difficult.

Hence, the hyperon puzzle is a real puzzle. At present, there is no accepted solution to the problem (for a review we refer to Chatterjee and Vidaña, 2016). One solution, however, is a particular striking one: Hyperons appear but before they can destabilize the neutron star a new phase appears at high density with a stiff EOS supporting a $2M_\odot$ compact star. That new phase would be not based on hadronic degrees of freedom, nucleons, and hyperons, but on a new degree of freedom in the form of the constituents of hadrons, that is, quarks, forming a quark matter core. We will turn our attention to this possible phase in the coming chapter.

7.5 Structure of Neutron Stars

Finally, we discuss the overall structure of a neutron star developed so far. Figure 7.13 shows an illustrative example of the structure of neutron star for a neutron star mass of $2M_\odot$ as a function of the radius r. The lines mark the boundaries of the different parts of the neutron star. The corresponding energy density in units of ε_0, the energy density of nuclear matter at saturation density n_0, can be read off from the y-axis, which is drawn not to scale. The actual values are taken from a calculation using the parameter set DD2, which we have used before in this section for the mass–radius diagram, for example. The color coding marks the different compositions. We read off from the figure that

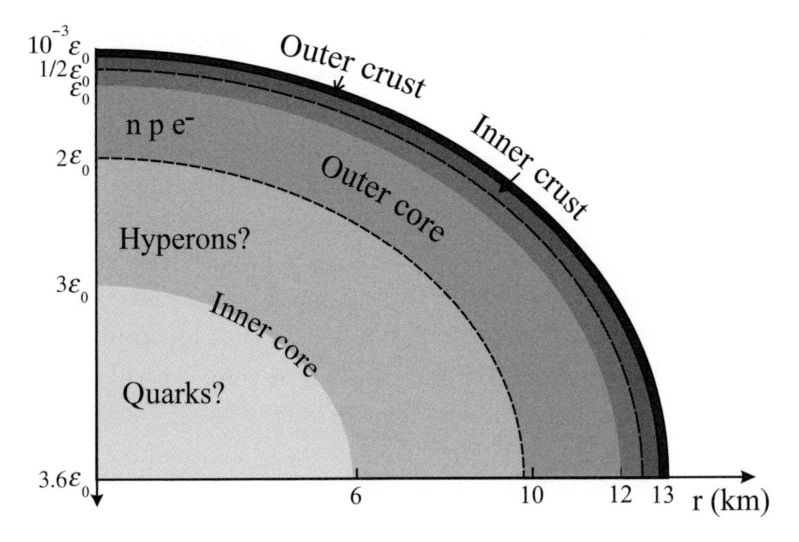

Figure 7.13 The structure of neutron star as a function of the radius r and the energy density ε for a neutron star mass of $2M_\odot$ for the parameter set DD2. The lines mark the boundaries of the different parts of the neutron star (outer and inner crust, outer and inner core).

- The outer crust, consisting of a lattice of nuclei in a gas of electrons, starts at the total radius of the neutron star of 13.1 km and extends to about $10^{-3}\varepsilon_0$, which corresponds to a radius of 12.9 km.
- The inner crust, a lattice of nuclei immersed in a liquid of neutrons and electrons, continues up to an energy density of about $0.5\varepsilon_0$, corresponding to a radius of 12.5 km.
- The outer core reaches the energy density at saturation density ε_0 at a radius of 12.0 km. Up to this energy density, we know the EOS. The outer core extends to $2\varepsilon_0$ at a radius of 9.8 km. It consists of neutron matter with an admixture of protons and electrons.
- The inner core fills most of the neutron star. Hyperons might be present at $(2 - 3)\varepsilon_0$. Quark matter might appear at $3\varepsilon_0$, which corresponds to a radius of 5.9 km. The maximum energy density in the core is $3.6\varepsilon_0$.

We see that the inner 12 km of the $2M_\odot$ neutron star have an energy density in excess of that of saturation density ε_0. Only for the outer 1.1 km we know the EOS. We quote here also the values for a calculation of a $1.4M_\odot$ neutron star with the parameter set DD2: The total radius is 13.2 km and an energy density of ε_0 is reached at a radius of 11.2 km. So in this case, a $1.4M_\odot$ neutron star has a known EOS for the outer 2 km while the inner 11.2 km have an higher energy density. The energy density reaches a value of $2.5\varepsilon_0$ at maximum in the center of the $1.4M_\odot$ neutron star.

The central energy density quoted is a lower limit to the central energy density of neutron stars. The measurement of gravitational waves from a neutron star merger GW170817 by the LIGO and Virgo collaboration implies a maximum radius of a neutron star of about 13 km, see Annala et al. (2018); Most et al. (2018); De et al. (2018); Abbott et al. (2018b), and the discussion in Chapter 10. As we get a radius of about 13 km for the parameter set DD2, the values quoted for the central energy density are at the limit and for the highest radius allowed by GW170817. Neutron stars with smaller radii than 13 km will have a higher central energy density. There is an upper bound on the maximum energy density reached for neutron stars, too, by demanding causality. For the stiffest possible EOS with $c_s^2 = 1$ and for a maximum mass of at least $2M_\odot$, one finds that the maximum energy density possible for neutron stars is about $13.5\varepsilon_0$, see Eq. (4.86).

Note that in the actual calculation, neither hyperons nor quark matter have been considered. So the color coding for the inner core are for illustrative purposes. Recall that we do not know at present what is going on in the inner core of the neutron star. But it could harbor some interesting physics and some surprises, which we will discuss in the coming two chapters.

Exercises

(7.1) What is the gravitational binding energy of a neutron star with $R = 10\,\text{km}$ and $M = 2M_\odot$? Compare to the energy emitted in 10 billion years of the Sun ($L_\odot = 4 \times 10^{33}\,\text{erg s}^{-1}$) and of a galaxy ($L_{\text{galaxy}} \sim 10^{11} L_\odot$).

(7.2) Show that the condition of stability of the pasta phase $\varepsilon_{\text{surf}} = 2\varepsilon_{\text{Coul}}$, see Eq. (7.33), results from the minimization of the total energy per volume.

(7.3) Show that the pressure of pure neutron matter at saturation density is given by $P = n_0 \cdot L/3$, where L is the slope parameter as defined in Eq. (7.58).

(7.4) Show that the energy per particle for a Fermi gas as given by $E/A = 3k_F^2/(10m)$ does not depend on the degeneracy factor of the fermions.

(7.5) Show that the Fermi momenta of nuclear matter and pure neutron matter at the same baryon number density differ by a factor $2^{1/3}$.

(7.6) Consider a hypothetical particle with baryon number two, a dibaryon. What is the critical condition for its appearance in neutron star matter if it is uncharged? Assume that the mass of the dibaryon is 2 GeV and calculate the critical density for a free neutron gas.

(7.7) Derive the potential difference between neutrons and Λ hyperons, Eq. (7.88), from Eq. (7.85), assuming that there is a minimum in the potential at saturation density. Find the critical density for the onset of Λ hyperons numerically or graphically for $\gamma = 2$ and 3 using Eq. (7.88).

8

Quark Stars

In this chapter, we discuss the properties of quark stars. What we have in mind is to describe matter based on quark degrees of freedom, not hadronic degrees of freedom, nucleons, and hyperons, as done in the last chapter.

Historically, the transition to quark matter in neutron stars and the possible existence of quark stars was discussed first by Ivanenko and Kurdgelaidze (1965). A first attempt to describe the properties of quark stars was made by Itoh (1970). It is interesting to note that both works appeared before the advent of quantum chromodynamics, which was developed in the mid-70s.

As far as we know today, quarks are the elementary particles, the basic constituents of hadrons. The interaction between quarks is described by the quantum field theory called quantum chromodynamics or QCD for short. Let us have a look at QCD first as a starting point for discussing the properties of quark matter at high densities suitable for quark star configurations.

8.1 Quantum Chromodynamics

The underlying theory of strong interaction is based on quark and gluon degrees of freedom. In the quark picture, the baryons (Greek for heavy), as for the nucleons and the hyperons, consist of three quarks. The mesons (Greek for middle), as for pions and kaons, consist of a pair, a quark with an antiquark. Baryons and mesons are called hadrons (Greek for thick). In contrast leptons (Greek for light) are not made of quarks but are elementary particles as the quarks. Note that the Greek origin in the naming scheme has historical origins as it reflects the masses of the particles known in the early days of the development of QCD. Today, heavy mesons and heavy leptons are known too. In addition, there exist several candidates today of hadrons with four, tetraquarks, and five quarks, pentaquarks, as well as states consisting of gluons only, glueballs, which are all hadrons too. Hadrons generically stand for composite particles that are interacting by the strong force.

The interaction between quarks is mediated by the exchange of gluons. The corresponding quantum field theory is the theory of quantum chromodynamics or QCD for short, whereby the quarks and the gluons carry a color charge (the origin of chromo is from the Greek word for color). The gluons are interacting with each other, as they carry a color charge, making the theory a nonabelian one, meaning that there are nonlinear interactions in the theory. The nonlinear character of QCD gives rise to new phenomena, in particular the one of confinement: Quarks are confined in our world within hadrons, that is, baryons and mesons, such that the overall color of the constituents, quarks and gluons, cancels out and the outside world observes only color neutral hadronic states. Thereby, the interaction of QCD is short-range with an intrinsic scale given approximately by the size of hadrons, which has been measured to be about 1 fm. Using the uncertainty relation, this scale corresponds to an energy scale of about 200 MeV, the typical scale usually associated with QCD. The gluons are massless, while the quarks have nonvanishing masses. The presently known masses from the particle data group are listed in Table 8.1. One sees that the lightest quarks, the up and down quarks, have masses of just a few megaelectronvolts. Neutrons and protons are made of these quarks. According to the quark model, the proton consists of two up and one down quark, the neutron of one up and two down quarks. To get the observed charges of the proton and the neutron, the quarks have to have fractional charges. The up quark has a charge of +2/3, the down quark one of $-1/3$. As three quarks combine to a baryon, the baryon number of all quarks has to be 1/3. The next heavier quark is the strange quark. Baryons with a strange quark (and light quarks) are the hyperons discussed in the previous chapter. The mass of the strange quark is about 100 MeV and its charge is $-1/3$, as for the down quark. The hyperons decay by weak interactions such that the strange quark is transformed to a down quark. The mass of the strange quark is below the scale of QCD of 200 MeV so that it is grouped with the up and down quark as a light quark. The masses of the heavier charm, the bottom, and the top quark are well above the QCD scale, so these quarks are considered as heavy quarks.

8.1.1 Quark and Gluon Condensates

The predictions of the theory of QCD have been tested in numerous experiments. High-energy scattering experiments on protons reveal that there are three pointless particles inside the proton. The mass spectrum of hadrons can be explained by QCD, most impressively by solving the QCD Lagrangian with lattice gauge calculations on supercomputers. However, the phenomenon of confinement is still an unresolved puzzle. Another feature of QCD is that the masses of the quarks of the nucleons do not add up to the mass of the nucleon. The masses of three up or

Table 8.1 *Charges and masses of the six quarks of the standard model. Reprinted table from Tanabashi et al. (2018). Copyright (2018) by the American Physical Society.*

Quark flavor	Up	Down	Strange	Charm	Bottom	Top
Charge	+2/3	$-1/3$	$-1/3$	+2/3	$-1/3$	+2/3
Mass	$2.2^{+0.5}_{-0.4}$ MeV	$4.7^{+0.5}_{-0.3}$ MeV	95^{+9}_{-3} MeV	$1.275^{+0.025}_{-0.035}$ GeV	$4.18^{+0.04}_{-0.03}$ GeV	173.0 ± 0.4 GeV

down quarks give just a mass of several megaelectronvolts, orders of magnitude below the mass of the nucleon of about 1 GeV. For hyperons, the situation does not improve, as the mass of the strange quark of about 100 MeV is still an order of magnitude below the masses of hyperons, which range between 1,116 MeV for the Λ with one strange quark to 1,672 MeV for the Ω^- with three strange quarks. The masses of nucleons and hyperons have to originate then mainly from the gluonic contribution. This observation leads to two important concepts. One relates to the effect that masses of light quarks can be ignored as a first lowest order approximation. The QCD Lagrangian without a quark mass term exhibits a symmetry: chiral symmetry.[1] Left- and right-handed quarks cannot be distinguished in a massless theory. The other one relates to the masses of the nucleons and hyperons being dominated by the contribution of gluons. The gluons form a gluon condensate that permeates spacetime. Inside the hadron, the gluon condensate constitutes a background in energy density. As gravity couples to energy, the energy density of the gluon condensate gives the essential contribution that appears as the mass of the hadron to an outside observer.

Another way to visualize a hadron in terms of quarks and gluons was advocated by a group at the Massachusetts Institute for Technology (MIT) in Boston, USA, in the 1970s. Consider a collection of quarks appropriate for the hadron under consideration, so the quarks provide the relevant quantum number of the hadron, as baryon number, charge, or isospin, and (quark) flavor (strangeness, charm, beauty, ...) as well as spin and parity. To keep the quarks together, a boundary condition was imposed at the surface of the hadron, ensuring that quarks could not escape, which served as an effective way to model confinement. The quarks are therefore confined in a sphere of a certain radius. The stability of the construction was ensured by balancing the pressure originating from the kinetic energy of the quarks with an outside pressure acting on the surface of the hadron. The outside pressure is pointing inward, thereby the hadron experiences a negative pressure from the vacuum outside to arrive at a vanishing total pressure at the boundary of the hadron. The so constructed sphere of quarks was named the bag, the outside pressure, an essential input parameter, the bag parameter B. This model was refined with additional phenomenological parameters, a Casimir or zero-point energy term and an interaction term, as well as with a nonvanishing quark mass for the strange quark. This so-called MIT bag model was able to describe successfully the hadron mass spectrum for the lowest lying meson and baryon states (Chodos et al., 1974; DeGrand et al., 1975) except for the pion, see later. It was also extended to describe hadron states with charm quarks (Hasenfratz et al., 1980). However, it turned out that the bag parameter had to be readjusted to describe the charmed hadron states,

[1] The term chiral originates from the Greek word for handedness.

from the original value of $B^{1/4} = 145-235\,\mathrm{MeV}$. This is the range usually considered for the bag parameter B. Note that the corresponding energy density of the bag parameter amounts to $B = 57.5-400\,\mathrm{MeV\,fm^{-3}}$ or in terms of the energy density of nuclear matter at saturation density to 0.4–2.8ε_0. It should not be too surprising that these scales are similar to the ones of nuclear matter, as QCD is the underlying theory of the nuclear interaction as well as that for the hadron mass spectrum.

The reader might worry that there is a negative pressure involved in the MIT bag model and how this relates to the energy density of the gluon condensate mentioned earlier. In fact, thermodynamic consistency demands that a constant vacuum energy density involves a constant negative vacuum pressure of equal magnitude. Just consider the first law of thermodynamics:

$$\varepsilon = -P + T \cdot s + \mu \cdot n$$

and consider the vacuum case of vanishing temperature T and number density n. One arrives at the relation

$$\varepsilon_{\mathrm{vac}} = -P_{\mathrm{vac}} = B.$$

The bag parameter B can therefore be associated with the energy density of the gluon condensate in vacuum.

There is another condensate that one can associate with the quarks. The quark condensate emerges from a nonvanishing vacuum expectation value of the quark fields. The quarks inside hadrons couple to the quark condensate, thereby attaining a constituent quark mass, which is about one-third of the nucleon mass, that is, about 300 MeV, for the light quarks. The quark condensate acts as a mass term for the quarks inside hadrons. Therefore, the quark condensates break spontaneously the chiral symmetry of QCD dynamically, so that left- and right-handed quarks couple to each other. The spontaneous breaking of the chiral symmetry implies a massless Goldstone boson as a consequence of the Goldstone theorem. As there is also an explicit breaking of chiral symmetry due to the small mass for the light quarks, the Goldstone boson is not exactly massless. Therefore, it is called a pseudo-Goldstone boson. It can be associated with the pion, which is the lightest known hadronic state of QCD. We note in passing that the MIT bag model cannot explain the pion mass, as it does not consider the effects of spontaneous chiral symmetry breaking. However, the hadron mass spectrum with light quarks can very well be described by effective chiral models with hadronic degrees of freedom.

As known from solid-state physics, condensates melt at a critical temperature. The same feature is present for the gluon and the quark condensate. The melting of the former condensate can be associated with the transition from the confined states of hadrons to a deconfined state of quarks and gluons, the quark–gluon plasma. The melting of the latter condensate is related to chiral symmetry restoration, so the masses of light quarks change from their constituent quark mass to that of the bare or current quark mass, as listed in Table 8.1. The critical temperature of QCD has been determined by solving the QCD Lagrangian on supercomputers using lattice gauge theory. Since the last few years, results have been available of physical quark masses that find that the transitions to deconfinement and to the chirally restored phase occur at about the same temperature of $T_c \approx 150\,\text{MeV}$. The QCD phase transition is a rapid crossover, so it is neither of first nor of second order. The critical temperature quoted is defined by the peak in the first derivative of the quark and gluon condensates with respect to temperature, that is, the point of the most rapid change with temperature.

In principle, the condensates should also melt at high densities. At which densities the melting is happening is still unknown and a topic of intense research. In particular, the chiral phase transition at high densities has been predicted in many effective models to be of first order. Unfortunately, lattice gauge calculations cannot solve QCD at high densities at present. So, the location as well as the order of a QCD phase transition at high density cannot be determined by first principle calculations of solving the QCD Lagrangian directly.

8.1.2 Asymptotic Freedom

The principal feature of QCD at high temperature, high density, or high energy in general is asymptotic freedom. The coupling strength between quarks and gluons turns out to depend strongly on the characteristic energy scale of the physical processes involved. With increasing energy, the coupling constant decreases so that the coupling constant vanishes at infinitely high energies. Then, the quarks do not interact with each other, they are asymptotically free. This feature of QCD emerges from the self-interaction between the gluons and has been predicted on theoretical grounds from the QCD Lagrangian. The coupling strength of QCD is usually quoted in terms of the QCD fine-structure constant

$$\alpha_s = \frac{g^2}{4\pi}, \tag{8.1}$$

where g is the coupling constant between quarks and gluons. The notation is in obvious analogy to the fine-structure constant of quantum electrodynamics, which is defined as $\alpha_{\text{EM}} = e^2/(4\pi)$, where the coupling constant e is the electromagnetic charge. The QCD coupling strength has been determined from several experiments covering the energy scale of the τ-lepton mass of $m_\tau = 1.776\,\text{GeV}$ to the electroweak scale of about 1 TeV. Figure 8.1 shows the measured values of α_s as a function of the energy scale of the experimental reaction probing

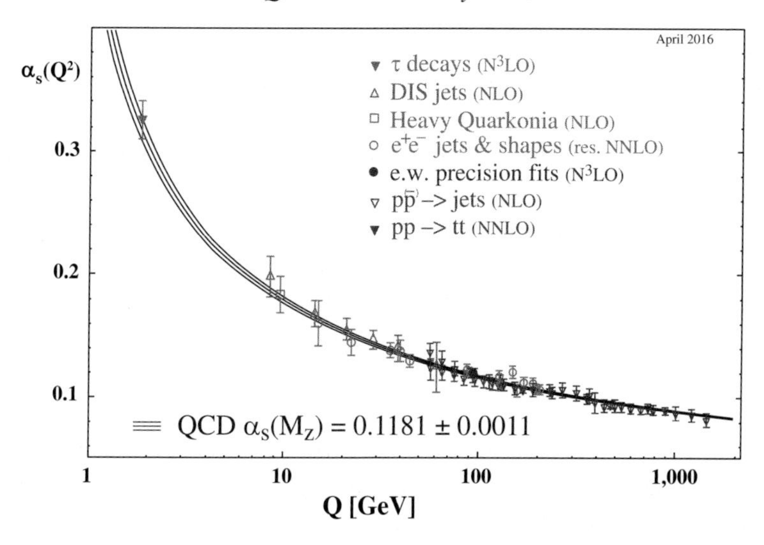

Figure 8.1 The QCD coupling strength α_s as a function of the energy scale. Reprinted figure with permission from Tanabashi et al. (2018). Copyright (2018) by the American Physical Society

QCD. The value of α_s determined at the lowest energy from the decay of the τ-lepton is $\alpha_s(1.8\,\text{GeV}) = 0.3$. At the highest energy, α_s decreases to about $\alpha_s(100\,\text{GeV}) = 0.1$. For comparison, the fine-structure constant of quantum electrodynamics is about $\alpha_{\text{EM}} \approx 1/137$, so two orders of magnitude smaller. Hence, QCD interactions are much stronger than electromagnetic ones. The line drawn through the experimental data points is the prediction of QCD, which is fixed by just one measured value of α_s at a specific energy scale. One sees that the coupling strength is strongly energy dependent and decreases with increasing energy. One says the coupling constant is 'running' (which is still a misnomer as it is not a constant at all). As α_s is small at high energies, interactions between quarks can be described by a perturbative expansion in the coupling constant, that is, by perturbative QCD (pQCD). In principle, the QCD Lagrangian has besides the quark masses no intrinsic scale. Truncating the perturbative series at a certain order introduces not only an uncertainty but also a scale dependence, the (unphysical) renormalization scale, which becomes smaller with increasing order of the truncated terms. The error in the pQCD calculations is then estimated by choosing a fiducial physical scale and varying the renormalization scale between half and twice its value. So far, the predictions of pQCD have been successfully checked experimentally in various scattering experiments at high energies.

The running of the strong coupling constant is encoded in the β-function, its logarithmic derivative with respect to the renormalization scale squared.

The β-function can be computed perturbatively within QCD in a series of powers of the strong coupling constant g as:

$$\frac{\mathrm{d}\alpha_s}{\mathrm{d}\ln\bar{\Lambda}^2} = \beta(\alpha_s) = -b_0 \cdot \alpha_s^2 - b_1 \cdot \alpha_s^3 - \cdots \tag{8.2}$$

Note the minus sign on the right-hand side that expresses the feature of asymptotic freedom of QCD, that is that the coupling constant decreases with increasing scale. The coefficients b_0, b_1, ... depend on the renormalization scheme used and on the number of flavors N_f considered. It is standard nowadays to use the modified minimal subtraction ($\overline{\text{MS}}$) renormalization scheme, where the renormalization scale is customarily denoted as $\bar{\Lambda}$. The coefficients in the $\overline{\text{MS}}$ scheme are given as:

$$b_0 = \frac{33 - 2N_f}{12\pi} \qquad b_1 = \frac{153 - 19N_f}{24\pi^2}. \tag{8.3}$$

To first order, the differential equation for $\alpha_s(\bar{\Lambda})$ can be solved analytically to give

$$\alpha_s(\bar{\Lambda}) = \frac{1}{b_0 \ln\left(\bar{\Lambda}^2/\Lambda^2_{\overline{\text{MS}}}\right)} = \frac{4\pi}{(11 - \frac{2}{3}N_f)\ln\left(\bar{\Lambda}^2/\Lambda^2_{\overline{\text{MS}}}\right)}, \tag{8.4}$$

where $\Lambda_{\overline{\text{MS}}}$ is an integration constant to be fixed by experiment. A fit to the data shown in Figure 8.1 by associating $\bar{\Lambda}$ with the center-of-mass energy Q gives for $N_f = 3$ that $\Lambda_{\overline{\text{MS}}} = 332 \pm 17\,\text{MeV}$ where also higher orders in the β-function are taken into account (Tanabashi et al., 2018). One sees, that the strong coupling constant decreases logarithmically with the energy scale one is probing QCD.

8.1.3 Free Quark Matter at High Temperature

The concept of asymptotic freedom has also been applied for describing the equation of state (EOS) of the quark–gluon plasma at high temperatures and compared to lattice gauge calculations of QCD. A prediction of asymptotic freedom in general is that the EOS should reach asymptotically that of a free gas of particles, that is, the Stefan–Boltzmann limit with the appropriate degrees of freedom. For noninteracting particles, the pressure as a function of temperature for vanishing chemical potentials can be expressed as an integral over the distribution function of a free gas. In the limit of high temperatures, effects from quark masses for which $m \ll T$ holds can be ignored.

The pressure integral for free particles reads

$$P(T) = \gamma \int \frac{\mathrm{d}^3k}{(2\pi)^3} \frac{k^2}{E(k)} f(T) = \frac{\gamma}{2\pi^2} \int_0^\infty \mathrm{d}k \frac{k^4}{E(k)} f(T), \tag{8.5}$$

where $f(T)$ is the distribution function, γ the degeneracy factor, and $E = \sqrt{k^2 + m^2}$ the energy of the particle with mass m. The distribution function for fermions is given by the Fermi–Dirac distribution function, the one for bosons by the Bose–Einstein distribution function, which differ by a sign, with a plus sign for fermions and a minus sign for bosons:

$$f(T) = \frac{1}{e^{E/T} \pm 1}. \tag{8.6}$$

For massive particles and in the limit of $m \gg T$, the distribution function approaches that of the Boltzmann distribution function. The contribution of massive particles to the pressure is then exponentially suppressed. As a rule of thumb, particles are exponentially suppressed for temperatures lower than $T < m/3$ and start making a significant contribution to the pressure for temperatures higher than $T > m/3$. Hence, the main contribution to the pressure comes from massless particles or massive particles in the limit $m \ll T$. The pressure integral can be solved analytically for the case $m = 0$, where $E(k) = k$. It gives

$$P(T) = \frac{\gamma}{2\pi^2} \int_0^\infty \mathrm{d}k \frac{k^3}{e^{k/T} \pm 1} = \frac{\pi^2}{90} T^4 \begin{cases} \gamma \text{ for bosons} \\ \frac{7}{8}\gamma \text{ for fermions.} \end{cases} \tag{8.7}$$

The Stefan–Boltzmann limit is defined as the sum of the pressures of all particles in the limit $m \ll T$:

$$P_{\mathrm{SB}} = \frac{\pi^2}{90} T^4 \left(\sum_{\text{bosons}} \gamma_{i,bosons} + \frac{7}{8} \sum_{\text{fermions}} \gamma_{i,fermions} \right), \tag{8.8}$$

which is determined by the sum of the degeneracy factors, with a prefactor of $7/8$ for the fermionic contribution, which we denote as the relativistic degrees of freedom in the following.

Let us count the degeneracy factors for the quark–gluon plasma. Gluons are massless bosons and carry a spin $s = 1$ as the photon. Gluons can have two different kinds of polarizations similar to the photon. The color degrees of freedom are $N_c^2 - 1$, where N_c is the number of colors. For QCD, there are three colors, so the three quarks inside a baryon combine to a color-charge neutral object. Hence, there are eight gluons with different color charges. The degeneracy factor for gluons is then in total

$$\gamma_{\text{gluon}} = 2 \cdot (N_c^2 - 1) = 2 \cdot 8 = 16. \tag{8.9}$$

Quarks have a spin $s = 1/2$ and can carry N_c charges. The contribution of anti-quarks gives another factor of two. Also, there are different flavors, in total six different flavors for the up, down, strange, charm, bottom, and top quarks. The contribution of quarks to the pressure depends now on the temperature chosen.

For a temperature of around $T = T_c$, the up and down quark masses fulfill the condition $m \ll T$, see Table 8.1. The strange quark mass m_s is lower than T_c but not much. For the other quarks, the heavy quarks, the mass is much larger than T_c, so $m \gg T$ holds and their contribution to the pressure can be safely ignored at temperatures around T_c. The degeneracy factor for quarks can then recast in terms of the number of flavors with $m \ll T$ and reads

$$\gamma_{\text{quarks}} = 2 \cdot (2s+1) \cdot N_c \cdot N_f = 2 \cdot 2 \cdot 3 \cdot N_f = \begin{cases} 24 & (N_f = 2) \\ 36 & (N_f = 3). \end{cases} \tag{8.10}$$

The contribution of the quarks to the Stefan–Boltzmann limit has to be multiplied by the factor $7/8$ so that the relativistic degrees of freedom from quarks is 21 without ($N_f = 2$) and 31.5 with the contribution from the strange quarks ($N_f = 3$). Summing up, the Stefan–Boltzmann limit for the quark–gluon plasma is given by

$$P_{\text{SB}} = \frac{\pi^2}{90} T^4 \left(16 + 10.5 \cdot N_f\right) = \frac{\pi^2}{90} T^4 \begin{cases} 37 & (N_f = 2) \\ 47.5 & (N_f = 3) \end{cases} \tag{8.11}$$

for temperatures around T_c. According to our rule of thumb, the contribution of the strange quarks to the pressure will become important for temperatures of $T > m_s/3 \approx 30\,\text{MeV}$, so we expect that the pressure of the quark–gluon plasma approaches the limit for three quark flavors $N_f = 3$ for temperatures around T_c. The next heavier quark, the charm quark, will then add to the pressure at temperatures of $T > m_c/3 \approx 400\,\text{MeV}$.

We can also have a look at the relativistic degrees of freedom well below the critical temperature in the hadron gas. Looking at the known hadron masses, see for example, the tables in the last section, the only particle that has a mass below T_c is the pion. The pion is a boson with spin zero and isospin one, that is, it has three differently charged states, π^+, π^-, and π^0, with similar masses. According to our rule of thumb, the pions contribute to the pressure significantly for a temperature of more than $T \approx m_\pi/3 \approx 50\,\text{MeV}$. Hence, there are only three relativistic degrees of freedom for the hadron gas. Compared to the relativistic degrees of freedom of a quark–gluon plasma, this is an order of magnitude less. Hence, we expect a substantial increase in the pressure when going from the hadron gas to the quark–gluon plasma.

8.1.4 Lattice Data at High Temperature

Lattice gauge calculations show, however, that the limit of asymptotic freedom is not reached, even at temperatures well above the critical temperature where the condensates melt. So, there is still some residual interaction left in the quark–gluon

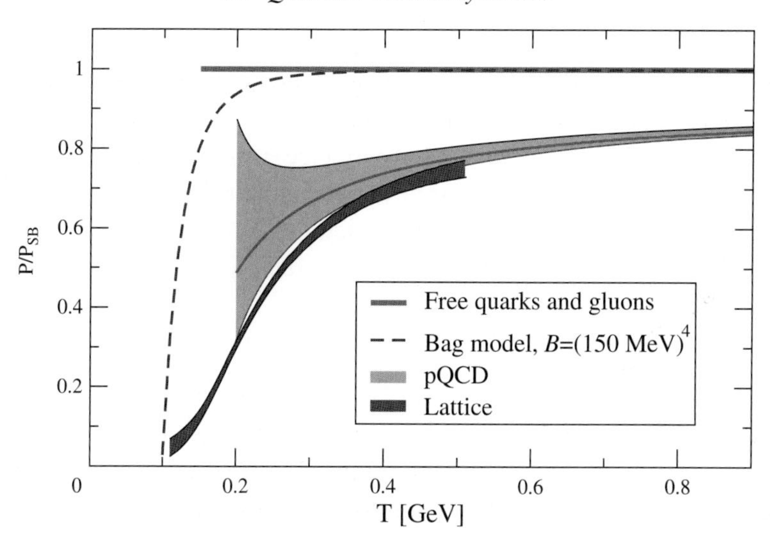

Figure 8.2 The pressure of matter consisting of quarks and gluons versus the temperature from lattice QCD calculations compared to calculations from pQCD and the MIT bag model (taken from Fraga et al., 2014). © AAS. Reproduced with permission

plasma. pQCD can describe the deviation from the Stefan–Boltzmann with corrections from interactions. The individual contributions of pQCD exhibit alternating signs, so the perturbative series only converges at temperatures of several times the critical temperature T_c. The contribution of gluons at low momenta, so-called soft gluons, need to be resummed nonperturbativley, which then allows for an improved description of the lattice data at temperatures closer to T_c.

Figure 8.2 shows the pressure relative to the Stefan–Boltzmann limit as a function of temperature for different calculations. Results of lattice QCD calculations are shown for $N_f = 2 + 1$ flavors with physical quark masses. One sees that the pressure increases drastically between $T = 100\,\text{MeV}$ and $300\,\text{MeV}$, signaling the transition to the quark–gluon plasma. The transition is a rapid cross-over, so neither a first- nor a second-order phase transition. One defines the temperature of the steepest increase as the pseudo-critical temperature, often denoted as the critical temperature in short but misleading terms, of QCD which is $T_c \approx 150\,\text{MeV}$. The energy density in the cross-over region of temperatures of $T = 145-163\,\text{MeV}$ is found to be $\varepsilon = 0.18-0.5\,\text{GeV fm}^{-3}$ (Bazavov et al., 2014) which corresponds to $1.3-3.6\varepsilon_0$. The pQCD results are state-of-the-art resummed calculations at order $\alpha_s^{5/2}$ and are reasonably close to the lattice QCD results. While the error band in the prediction of pQCD calculations is sizable at a temperature of $T = 200\,\text{MeV}$, it becomes smaller and smaller at higher temperatures. For both cases, one sees that the pressure remains significantly below the Stefan–Boltzmann limit for free quarks and gluons, indicating a substantial contribution from interactions. The dashed line shows the result of the bag model, which assumes a noninteracting gas of quarks and gluons with a vacuum term. One sees that the bag model shows a sudden increase in the pressure. However, the results of the bag model are close to the Stefan–Boltzmann limit at temperatures not too far from the critical temperature of QCD in contrast to the lattice QCD results. Hence, we conclude that interactions and not vacuum terms are important for describing the EOS of the quark–gluon plasma.

Equipped with a basic knowledge of QCD, we start now calculating the EOS of cold quark matter. It will serve as input for determining the properties of compact stars made of quarks as the elementary degrees of freedom: quark stars.

8.2 Free Strange Quark Matter

As we learned in the last section, quarks are asymptotically free, which means that they can be described as free particles at asymptotically large energy scales. That energy scale can be anything, not only collisional energy. Tuning the number density or the baryon chemical potential to sufficiently high values, quark matter is probed at high energy scales too. Just imagine the Fermi energy of quarks as the corresponding energy scale, at which quarks interact with each other.

Let us start by considering a gas of noninteracting quarks at vanishing temperature and nonvanishing baryon chemical potential μ_B in the following. As quarks carry 1/3 of the baryon number of a baryon, we define the quark chemical potential μ_q as just 1/3 that of the baryon chemical potential. Taking into account also an electrochemical potential for considering charge neutral strange quark matter, we recall that the chemical potential reads in general

$$\mu_i = B_i \cdot \mu_B + Z_i \cdot \mu_Z, \tag{8.12}$$

where B_i is the baryon number and Z_i the charge of particle species i. Here, μ_Z denotes the charge chemical potential, so electrons and muons have a chemical potential of $\mu_e = \mu_\mu = -\mu_Z$. Then, the chemical potential is given by

$$\mu_u = \frac{1}{3}\mu_B + \frac{2}{3}\mu_Z \qquad \mu_d = \frac{1}{3}\mu_B - \frac{1}{3}\mu_Z \qquad \mu_s = \frac{1}{3}\mu_B - \frac{1}{3}\mu_Z \tag{8.13}$$

for the up, down, and strange quarks, respectively, see Table 8.1. We introduce the quark (number) chemical potential via the defining relation $\mu_q = \mu_B/3$, so it represents the quark number as a conserved quantity. Then, the chemical potentials of the quarks can be rewritten in terms of the quark chemical potential μ_q and the electron chemical potential μ_e as

$$\mu_u = \mu_q - \frac{2}{3}\mu_e \qquad \mu_d = \mu_q + \frac{1}{3}\mu_e \qquad \mu_s = \mu_q + \frac{1}{3}\mu_e. \tag{8.14}$$

We note that the chemical potential for down and strange quarks is the same, as they carry the same baryon number and charge. Strangeness is not conserved in quark star matter as we know from our discussion of hyperons in neutron star matter. The nonconservation of strangeness is part of what we will call weak equilibrium in the following, which is defined by demanding that all weak interactions are in equilibrium. The weak equilibrium includes then also the transformation of up and down quarks via the weak reaction

$$u + e^- \longleftrightarrow d + \nu_e, \tag{8.15}$$

which is the quark analog of β-equilibrium for nucleons. Hence, the chemical potentials of quarks and electrons in weak equilibrium are related by

$$\mu_d = \mu_u + \mu_e \text{ and } \mu_s = \mu_d. \tag{8.16}$$

Note that in the way we set up the chemical potential of quarks in Eq. (8.14), the conditions of weak equilibrium are automatically fulfilled.

The grand canonical potential $\Omega = -P \cdot V$ is the sum of the contributions of all quark species

$$\frac{\Omega}{V} = -P = -\sum_i P_i = -\frac{1}{4\pi^2}\sum_i \mu_i^4. \tag{8.17}$$

The number density of quark species i is

$$n_i = \frac{\partial P}{\partial \mu_i}. \tag{8.18}$$

The total energy density is determined via the thermodynamic relation

$$\varepsilon = -P + \sum_i \mu_i \cdot n_i. \tag{8.19}$$

The pressure for massless particles at zero temperature is given by

$$P_i = \frac{\gamma}{3}\int \frac{\mathrm{d}^3k}{(2\pi)^3}\, k = \frac{\gamma}{6\pi^2}\int_0^{k_{F,i}} \mathrm{d}k\, k^3 = \frac{\gamma}{24\pi^2} k_{F,i}^4, \tag{8.20}$$

where $i = u, d, s$. The chemical potential is equal to the Fermi energy or the Fermi momentum $\mu_i = E_{F,i}(k_{F,i}) = k_{F,i}$. The degeneracy factor for quarks is the same as computed earlier, except one has to take out the contribution of antiquarks, which is absent at vanishing temperature. Then, the degeneracy factor for quarks at zero temperature is given by $\gamma_i = (2s+1)N_c = 2 \cdot 3 = 6$ and is the same for all flavors. Hence, the total pressure for massless quarks reads

$$P = P_u + P_d + P_s = \frac{1}{4\pi^2} \sum_i k_{F,i}^4 = \frac{1}{4\pi^2} \left(\mu_u^4 + \mu_d^4 + \mu_s^4 \right) . \tag{8.21}$$

The number density of the quark species i is

$$n_i = \frac{\gamma_i}{6\pi^2} k_{F,i}^3 = \frac{1}{\pi^2} \mu_i^3 , \tag{8.22}$$

which can be either derived by using the thermodynamic relation, Eq. (8.18), or by integration over the three-dimensional Fermi sphere up to the Fermi momentum. Finally, the energy density turns out to be simply that of a relativistic free gas of particles

$$\varepsilon = 3P \quad \text{or} \quad P = \frac{1}{3}\varepsilon . \tag{8.23}$$

As a last step, we need to consider the condition of charge neutrality by adding leptons to quark matter. The total charge Z of quarks alone is

$$Z = \sum_i Z_i \cdot n_i = \frac{2}{3} n_u - \frac{1}{3} n_d - \frac{1}{3} n_s . \tag{8.24}$$

We realize that there is a trivial solution for charge neutral strange quark matter for equal number densities of up, down, and strange quarks, so no leptons are needed to ensure overall charge neutrality:

$$n_u = n_d = n_s \quad \text{and} \quad n_e = n_\mu = 0 . \tag{8.25}$$

In this case, the Fermi momenta of all quark flavors are the same, so the total baryon number density is

$$n_B = \frac{1}{3} (n_u + n_d + n_s) = \frac{1}{\pi^2} k_{F,q}^3 = \frac{1}{\pi^2} \mu_q^3 , \tag{8.26}$$

where $k_{F,q}$ is the quark Fermi momentum and μ_q the quark chemical potential. These relations are consistent only for a vanishing charge chemical potential $\mu_Z = 0$. Then, the chemical potentials of all quark species are the same in weak equilibrium:

$$\mu_q = \mu_u = \mu_d = \mu_s \quad \text{with} \quad \mu_Z = -\mu_e = -\mu_\mu = 0 . \tag{8.27}$$

We can compute now the baryon number density as a function of the quark chemical potential as

$$n_B = \frac{1}{\pi^2} \mu_q^3 = 0.36\,\text{fm}^{-3} \cdot \left(\frac{\mu_q}{300\,\text{MeV}} \right)^3 . \tag{8.28}$$

In models of neutron stars, we have encountered baryon chemical potentials of around 1 GeV and more, see for example, Table 7.9. For the corresponding quark chemical potentials of about one-third of these values of $\mu_q = 300$, 400, and 500 MeV, one arrives at baryon number densities of $n_B = 0.36$, 0.84, and 1.6 fm^{-3} or of 2.4, 5.6, and $11 n_0$, respectively. The corresponding energy densities are

$$\varepsilon = 3P = \sum_i \frac{\gamma_i}{8\pi^2}\mu_i^4 = \frac{9}{4\pi^2}\mu_q^4 = 240\,\text{MeV fm}^{-3} \cdot \left(\frac{\mu_q}{300\,\text{MeV}}\right)^4, \qquad (8.29)$$

where we used Eq. (8.21). For quark chemical potentials of $\mu_q = 300$, 400, and 500 MeV, one computes energy densities of $\varepsilon = 240$, 760, and 1,900 MeV fm^{-3} or of 1.7, 5.4, and $14\varepsilon_0$, respectively. Certainly, one would expect that quark matter is present for energy densities larger than the saturation energy density of nuclear matter, so values much smaller than $\mu_q = 300$ MeV are disfavored. Note that the baryon number densities and energy densities for quark matter in units of that for nuclear matter, n_0 and ε_0, are different. In particular, the energy density grows faster than linear with the number density as $\varepsilon \propto n_B^{4/3}$. For comparison, the average mass density of a typical neutron star is

$$\bar{\rho} = \frac{3M}{4\pi R^3} = 270\,\text{MeV fm}^{-3} \cdot \left(\frac{M}{2M_\odot}\right)\left(\frac{10\,\text{km}}{R}\right)^3, \qquad (8.30)$$

which is about $2\varepsilon_0$ for a compact star with a mass of $M = 2M_\odot$ and a radius of 10 km. Hence, we expect that the quark chemical potential relevant for quark stars will be of the order of several hundred megaelectronvolts. The range of $\mu_q =$ 300–500 MeV covers a sufficiently wide range of number and energy densities.

Of course, we neglected the nonvanishing quark masses. While the up and down quark masses are much smaller than the quark chemical potential $m_u \approx m_d \ll \mu_q = 300$ MeV, the strange quark mass is not that much smaller. To investigate the effect of the strange quark mass, one has to start now from the charge neutrality condition

$$n_e = \frac{2}{3}n_u - \frac{1}{3}n_d - \frac{1}{3}n_s. \qquad (8.31)$$

It will turn out a posteriori that the distribution of muons can be safely ignored. The energy–momentum relation for strange quarks changes to

$$\mu_s = E_{F,s} = \sqrt{k_{F,s}^2 + m_s^2}, \qquad (8.32)$$

which has to be used to get a relation between the quark and the electron chemical potentials from the charge neutrality condition. Assuming that $\mu_q \gg m_s \gg \mu_e \gg m_e$, one finds that the electron chemical potential can be estimated to be

$$\mu_e \approx \frac{m_s^2}{4\mu_q} = (5 - 8)\,\text{MeV} \tag{8.33}$$

for $\mu_q = 300\text{–}500\,\text{MeV}$, which is much smaller than the quark chemical potential. The reader is asked to verify the relation in the exercises. The corresponding ratio of the electron to baryon number densities turns out to be

$$\frac{n_e}{n_B} \approx \frac{1}{3}\left(\frac{m_s}{2\mu_q}\right)^6 = 10^{-5} - 10^{-7} \tag{8.34}$$

for $\mu_q = 300\text{–}500\,\text{MeV}$, so the contribution of electrons can be safely ignored for strange quark matter in weak equilibrium.

We have seen that the scales for quark star matter and neutron star matter are not so different, as the underlying theory is QCD in both cases. This similarity will also show up in the properties of quark stars whose gross properties will be not so different compared to neutron stars. However, there will be a crucial difference between quark stars and neutron stars, as strange quark matter allows for the rise of a new class of compact stars: selfbound stars.

8.3 Selfbound Stars

We consider now additional input from QCD for the description of quark matter. In this section, we take a look at the impact of the presence of a quark or gluon condensate on the properties of quark matter. So, let us start with the pressure of a free gas of quarks where we add a vacuum pressure from the quark or gluon condensate. We have seen before that the vacuum pressure is equal to minus the vacuum energy density, so we can write:

$$P = P_{\text{quarks}} + P_{\text{vac}} = P_{\text{quarks}} - \varepsilon_{\text{vac}}. \tag{8.35}$$

What we need now is the relation between the pressure and the energy density of quarks. Assuming massless noninteracting quarks, that relation would be the one of a relativistic gas. The total energy density can then be written as

$$\varepsilon = \varepsilon_{\text{quarks}} + \varepsilon_{vac} = 3P_{\text{quarks}} + \varepsilon_{vac}. \tag{8.36}$$

Putting this relation into Eq. (8.35), one arrives at

$$P = \frac{1}{3}\left(\varepsilon - \varepsilon_{vac}\right) - \varepsilon_{\text{vac}} = \frac{1}{3}\left(\varepsilon - 4\varepsilon_{vac}\right). \tag{8.37}$$

If one sets the vacuum energy density equal to the bag constant $B = \varepsilon_{\text{vac}}$, this EOS is just the one of the MIT bag model introduced in Eq. (4.65). For the MIT bag model, we know from scaling analysis that the solutions of the Tolman–Oppenheimer–Volkoff (TOV) equation scale with the vacuum energy density. In particular, the maximum mass and the corresponding radius scales with $\varepsilon_{\text{vac}}^{-1/2}$, see Eq. (4.66). Moreover, the whole mass–radius relation scales with $\varepsilon_{\text{vac}}^{-1/2}$. For the range of the MIT bag constant of $B = 145-235\,\text{MeV}\,\text{fm}^{-3}$, one finds maximum masses of $M_{\text{max}} = 2.01-1.57M_{\odot}$, with radii of $R_{\text{crit}} = 10.9-8.56\,\text{km}$, respectively. These values are close to the ones found for models of neutron stars made of nucleonic matter. Hence, a determination of the maximum mass and its radius of a compact star from, for example, pulsar data will not allow us to distinguish between an ordinary neutron star and a quark star made of strange quark matter with a nonvanishing vacuum term.

The EOS of the MIT bag model has the interesting property that the pressure vanishes at a nonvanishing value of the energy density. This generic feature defines a new class of EOSs, the one of selfbound matter. Associated with selfbound matter, the corresponding compact star configurations are called selfbound stars in general.

Selfbound stars: Compact star configurations with an EOS where the energy density has a nonvanishing value at zero pressure.

Hence, our conclusion can be put into the form that one cannot distinguish between an ordinary star and a selfbound quark star from the maximum mass and the radius.

We note that neutron star matter has a vanishing energy density at zero pressure. Also, free fermion gases have a vanishing energy density at zero pressure. However, we also know at least one example realized in nature of selfbound matter: nuclear matter. Nuclear matter in its ground-state has a vanishing pressure at saturation density, which corresponds to a nonvanishing energy density of $\varepsilon_0 = 140\,\text{MeV}\,\text{fm}^{-3}$. So, spheres of nuclear matter, that is, nuclei, are stable without the pull of gravity present for compact stars. However, there is a maximum number of baryons that allow for stable nuclei. The heaviest stable nucleus is ^{208}Pb with a mass number of $A = 208$. Then, the repulsive Coulomb force will make heavier nuclei unstable and they will decay eventually to lighter nuclei. Astrophysical objects as compact stars need to be uncharged to be stable. Charge-neutral matter of nucleons and leptons consists mainly of neutrons. We know from state-of-the-art calculations of neutron matter that it is unbound, a result that emerges also from the fact that the symmetry energy of nuclear matter is much larger than the binding energy of nuclear matter. Nuclei just made of neutrons have therefore not been observed in nature. Neutron stars are bound by virtue of the attractive forces of gravity, counteracting the repulsive nuclear forces in hydrostatic equilibrium.

The situation is different for strange quark matter. Quark matter consisting of up, down, and strange quarks in weak equilibrium has essentially no electrons. So, if strange quark matter in bulk is selfbound, then spheres consisting of strange quark matter are bound by virtue of the vanishing total pressure. The gravitational attraction is not needed for hydrostatic equilibrium. In fact, gravity only sets a limit on the maximum mass of the quark matter sphere. For smaller blobs of strange quark matter, gravity can be ignored, and they are all stable. The energy density of these bound spheres of strange quark matter is constant, as it is fixed by the vacuum energy density. Hence, selfbound stars have the mass–radius relation of a sphere with constant energy density, so

$$M \propto R^{-3} \quad \text{(selfbound stars)}. \tag{8.38}$$

The size of the spheres can be arbitrarily small, so the mass–radius relation for selfbound stars starts at the origin, that is, selfbound stars have arbitrarily small radii. Of course, there is a microscopic limit on the size of strange quark matter spheres when surface effects are becoming important. For a surface energy on the typical scale of QCD, one expects that surface effects are important for quark spheres with radii in the range of a few Fermi, that is, 18 orders of magnitude smaller than the typical radius of quark stars of a few kilometers. These microscopic quark spheres made of strange quark matter are called strangelets. Their existence is hypothetical as their stability hinges on the binding energy of strange quark matter compared to nuclear matter. We know that nuclei are stable, they do not decay to strangelets. However, the decay of nuclei to strangelets involves the simultaneous transformation of strange quarks to down quarks via weak interactions that can take longer than the age of the universe of about 14 billion years. There exists heavy nuclei that decay by weak interactions such that their lifetimes are longer than the age of the universe. For example, the nucleus ^{147}Sm used for radiometric dating of the oldest material in our solar system has a half-life of about 106 billion years. So, it could be that strangelets are more stable than nuclei and coexist with ordinary matter. On the other hand, quark matter made of up and down quarks only will decay to nuclei or to nucleons instantly by strong interactions, as they have the same quantum numbers of baryon number, charge, and (zero) strangeness.

The concept of exotic matter being more stable than ordinary matter was introduced by Bodmer (1971). He speculated that there exists absolutely stable matter that has a higher binding energy than nuclear matter. Then nuclear matter would not be the true ground state of matter and would be metastable. The timescale of the decay of nuclear matter is much longer than the age of the universe, as there is a barrier blocking it so that the existence of absolutely stable matter is not in contradiction to observation. In addition to nucleonic matter and baryonic matter, he discussed also strange quark matter as an option for absolutely stable matter. The scientific discussion on absolutely stable strange quark matter got ignited, however, many years later with the work of Witten (1984). He studied the cosmological QCD phase transition and speculated that astronomically large spheres of absolutely stable strange quark matter could have formed if that phase transition was of first order. In an appendix, he discusses the properties of compact stars made of absolutely stable strange quark matter, which were denoted as strange stars thereafter. The notion that there exists absolutely stable strange quark matter is nowadays known as the Bodmer–Witten hypothesis.

So, there are two conditions that need to be fulfilled for strange matter:

(1) Droplets of quark matter without strange quarks are less stable than nuclei, so nuclei cannot decay by strong interactions to quark droplets.
(2) Quark matter with strange quarks is more stable than nuclear matter, such that it cannot decay to nuclei.

The first condition can be written in the form that the binding energy per baryon number (or mass number) of two-flavor quark matter droplets of up and down quarks is higher than the total energy of the most stable nucleus, ^{56}Fe. We add a presumed correction energy from the surface term for the quark matter blob of 4 MeV (Farhi and Jaffe, 1984) so that we get as the two-flavor constraint for bulk matter:

$$\frac{E_{\text{bulk}}}{A}(N_f = 2) > m_N - \frac{E_b}{A}(^{56}\text{Fe}) + \frac{E_{\text{surf}}}{A}(N_f = 2) \approx 934\,\text{MeV}. \tag{8.39}$$

The second condition states that three-flavor quark matter in bulk is more stable than ^{56}Fe, so the three-flavor constraint reads

$$\frac{E_{\text{bulk}}}{A}(N_f = 3) < m_N - \frac{E_b}{A}(^{56}\text{Fe}) \approx 930\,\text{MeV}. \tag{8.40}$$

We can adopt those constraints for the MIT bag model with three massless quark flavors. The binding energy in bulk can be set to the baryon chemical potential for vanishing pressure by virtue of the Hugenholtz–van Hove theorem, see Eq. (7.47) and below the equation. For $N_f = 2$, the baryon chemical potential is determined to be $\mu_B = \mu_u + 2\mu_d$ (so μ_e cancels out in the expression). The charge neutrality condition is given by $n_d = 2n_u$, which is equivalent to $\mu_d = 2^{1/3}\mu_u$, so $\mu_B = (1 + 2^{4/3})\mu_u$. The pressure of quarks is equal to the bag constant in equilibrium, so

$$B = P = \frac{1}{4\pi^2}\left(\mu_u^4 + \mu_d^4\right) = \frac{1}{4\pi^2}\left(1 + 2^{4/3}\right)\mu_u^4 = \frac{1}{4\pi^2}\left(1 + 2^{4/3}\right)^{-3}\mu_B^4. \tag{8.41}$$

Finally, the two-flavor condition gives the following constraint

$$\frac{E_{\text{bulk}}}{A}(N_f = 2) = \mu_B(N_f = 2) = \left(\left(1 + 2^{4/3}\right)^3 4\pi^2 B\right)^{1/4} > 934\,\text{MeV}, \tag{8.42}$$

which is fulfilled for $B^{1/4} > 145$ MeV. The case $N_f = 3$ can be handled more straightforwardly, as we can use Eq. (8.29) with the condition $\varepsilon = 4B$ in equilibrium. Then, the second condition can be written as

$$\frac{E_{\text{bulk}}}{A}(N_f = 3) = \mu_B(N_f = 3) = \left(3^3 \cdot 4\pi^2 B\right)^{1/4} < 930\,\text{MeV}, \tag{8.43}$$

which results in the three-flavor condition $B^{1/4} < 163\,\text{MeV}$. We note that the binding energy for strange matter can be as large as about 100 MeV for $B^{1/4} = 145\,\text{MeV}$.

As of this writing, the existence of the hypothetical strange matter being more stable than nuclear matter has not been ruled out unambiguously. Strangelets are less stable with decreasing baryon number due to finite-size effects, as surface and curvature terms are becoming important even if Coulomb effects are small. Searches for short-lived strangelets in terrestrial experiments as relativistic heavy-ion collisions have been negative so far (see Abelev et al., 2007, and references therein). Stable small strangelets cannot exist, as they would have been produced a long time ago by cosmic rays hitting the Moon or Earth's atmosphere (Busza et al., 2000; Ellis et al., 2008). However, astronomically large strange matter in the form of strange stars could have similar maximum masses and radii as ordinary neutron stars, as we have seen for the MIT bag model. Then, pulsar mass measurements alone cannot rule out the hypothetical existence of strange stars. The mass–radius relation for strange stars of $M \propto R^{-3}$ distinguishes them from neutron stars for small masses. While the radius for neutron stars gets larger with lower masses, that of strange stars gets smaller. However, strange stars can have a crust of a lattice of nuclei surrounded by electrons similar to the outer crust of neutrons stars. There are no free neutrons that can be eaten up by the strange quark matter. Due to the nonvanishing strange quark mass, there is a slight surplus of positive charge for strange quark matter, so charged nuclei are separated from the strange matter core by a Coulomb barrier. Strange stars with a crust have a total radius that is increasing with decreasing mass, as for neutron stars, see Section 8.5.

The maximum mass of strange stars within the MIT bag model can be calculated by using Eq. (4.66). We note that the maximum mass of a strange star will be $M_{\text{max}} = 2.01 M_\odot$ for a bag constant of $B^{1/4} = 145\,\text{MeV}$, with smaller masses for larger values of B. The mass limit from the pulsar mass measurement of PSR J0740+6620 is $2.14^{+0.10}_{-0.09} M_\odot$, see Table 6.1, so the bag constant has to be smaller than $B < 145\,\text{MeV}$ to be compatible with observations. However, the two-flavor constraints demand that $B > 145\,\text{MeV}$ for absolutely stable strange quark matter and for selfbound strange star configurations to exist. Hence, we conclude that the simple MIT bag model does not give strange star solutions compatible with the modern mass constraint from pulsar mass measurements. So, one needs to extend the MIT bag model, for example, by including interactions, to allow for acceptable strange star configurations. Effects of interactions for the properties of quark stars will be studied in the next section. For the classic papers on strange stars, see Haensel et al. (1986) and Alcock et al. (1986), and for a review see Weber (2005).

8.4 Interacting Quark Matter

In Section 8.1, we discussed the perturbative interactions between quarks and the resulting EOS as a function of temperature for vanishing chemical potential, see Figure 8.2. The pressure gets reduced compared to the case of free noninteracting quarks at the same temperature. In the following, we want to explore the impact of perturbative interactions as a function of chemical potential at vanishing temperature to apply it to strange star configurations. We assume $N_f = 3$ massless quarks in the following.

The perturbative series of QCD for nonvanishing chemical potential depends on the QCD fine structure constant α_s, that is, on even powers of the strong coupling constant. Note that this is different compared to the finite temperature case, where the perturbative series depends on all powers of the strong coupling constant. The odd powers emerge from resumming diagrams involving gluon diagrams that are not present at vanishing temperature. By truncating the perturbative series at a given order, logarithmic terms appear that depend on the renormalization scale $\bar{\Lambda}$. The pressure as a function of the quark chemical potential μ within pQCD up to order $\mathcal{O}(\alpha_s^2)$ turns out to be of the form

$$P(\mu) = \frac{N_f}{4\pi^2}\mu^4 \left\{1 - 2\left(\frac{\alpha_s}{\pi}\right) - \left[G + N_f \ln\frac{\alpha_s}{\pi} + \left(11 - \frac{2}{3}N_f\right)\ln\frac{\bar{\Lambda}}{\mu}\right]\left(\frac{\alpha_s}{\pi}\right)^2\right\}, \tag{8.44}$$

with $G = 10.376 - 0.536N_f + N_f \ln N_f$. The original calculation goes back to the work of Freedman and McLerran (1977), Baluni (1978), and Toimela (1985). The modern form of the pressure as written earlier is given, for example, in Fraga et al. (2001), but we use an updated numerical value taken from the improved calculation of Gorda et al. (2018).

Let us have a look term by term at the expression of the pressure of Eq. (8.44). The first term is the contribution of a free gas of massless quarks, the Stefan–Boltzmann limit:

$$P_{\mathrm{SB}} = \frac{N_f}{4\pi^2}\mu^4. \tag{8.45}$$

The second term gives the first correction to the pressure of free quarks at order α_s. For $\alpha_s = 0.3$–0.5, the pressure is reduced by about $2\alpha_s/\pi \approx 20$–30%, respectively. The next terms are corrections proportional to α_s^2 and to $\alpha_s^2 \ln \alpha_s$. The last term depends explicitly on the renormalization scale $\bar{\Lambda}$. The prefactor in front of the logarithm of that term looks familiar and bears resemblance to the first coefficient of the running of the strong coupling constant described by the β-function $\beta(\alpha_s)$, see Eqs. (8.2) and (8.3). This similarity is no accident. In fact, the prefactor has to have exactly the form given in order to have the pressure independent of the unphysical renormalization scale $\bar{\Lambda}$ up to order α_s^2. Let us see how this works by looking at the derivative of the pressure with respect to the logarithm of the renormalization scale $\bar{\Lambda}$ squared

$$\frac{\mathrm{d}(P/P_{\mathrm{SB}})}{\mathrm{d}\ln\bar{\Lambda}^2} = -\frac{2}{\pi}\frac{\mathrm{d}\alpha_s}{\mathrm{d}\ln\bar{\Lambda}^2} + \left(11 - \frac{2}{3}N_f\right)\frac{\mathrm{d}\ln\bar{\Lambda}}{\mathrm{d}\ln\bar{\Lambda}^2}\left(\frac{\alpha_s}{\pi}\right)^2 + \mathcal{O}(\alpha_s^3) \tag{8.46}$$

$$= -\frac{2}{\pi}b_o\alpha_s^2 + \left(11 - \frac{2}{3}N_f\right)\frac{1}{2}\left(\frac{\alpha_s}{\pi}\right)^2 + \mathcal{O}(\alpha_s^3) \tag{8.47}$$

$$= -\frac{2}{\pi}b_o\alpha_s^2 + 4\pi b_0\cdot\frac{1}{2}\left(\frac{\alpha_s}{\pi}\right)^2 + \mathcal{O}(\alpha_s^3) \tag{8.48}$$

$$= \mathcal{O}(\alpha_s^3), \tag{8.49}$$

where we used Eq. (8.2) in the second line and Eq. (8.3) in the third line. Hence, we see that the pressure depends on the renormalization scale only at higher order, at order α_s^3. One can turn the argument around and think of the dependence on $\bar{\Lambda}$ as a measure of higher order effects in the perturbative expansion. By varying $\bar{\Lambda}$, one can check for higher order effects and the reliability of the perturbative expansion. Usually, one draws a band of uncertainty for the result of pQCD by varying $\bar{\Lambda}$ within a factor of two. The fiducial scale to be chosen for $\bar{\Lambda}$ is given by the characteristic energy at which the quarks and gluons are interacting. In the case considered here, at zero temperature and high density, the typical energy of two quarks interacting with each other would be twice the quark chemical potential. So, we choose $\bar{\Lambda} = 2\mu_q$ and vary it between $\bar{\Lambda} = \mu_q$ and $\bar{\Lambda} = 4\mu_q$ to check the perturbative expansion.

Figure 8.3 shows the pressure of quark matter relative to the Stefan–Boltzmann limit as a function of the baryon chemical potential. The results of the pQCD calculation are shown in a band, where the renormalization scale $\bar{\Lambda}$ is varied by a factor two and a factor 1/2 away from the fiducial scale $\bar{\Lambda} = 2\mu_q = 2\mu_B/3$. One sees that even at $\mu_B = 6\,\mathrm{GeV}$ the pressure of interacting quark matter is about 20% below the Stefan–Boltzmann limit. The width of the band is small at $\mu_B = 6\,\mathrm{GeV}$, at the order of a few percent, while it increases considerably for lower baryon chemical potentials. At a baryon chemical potential of $\mu_B = 3\,\mathrm{GeV}$, the interaction reduces the pressure in the range of 20–40% compared to a free gas. For an even lower baryon chemical potential, the pressure drastically decreases with decreasing baryon chemical potential and eventually even vanishes at a nonvanishing value of the quark chemical potential. Note that the chemical potential where the pressure vanishes is equivalent to the energy per particle due to the Hugenholtz–van Hove theorem, see Eq. (7.47). Therefore, if the pressure vanishes below a chemical potential of $\mu_{B,c} = 930\,\mathrm{MeV}$, quark matter will be absolutely stable. We see that this critical chemical potential is within the band shown.

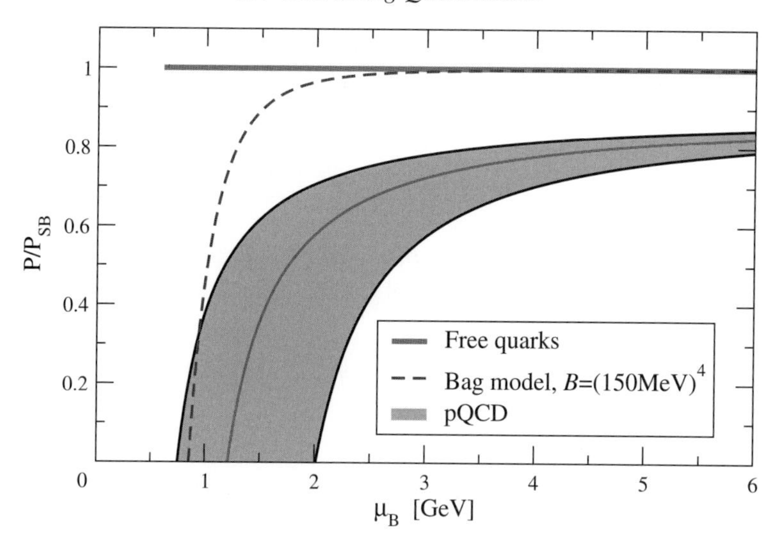

Figure 8.3 The pressure of quarks and gluons relative to the Stefan–Boltzmann limit at zero temperature versus the baryon chemical potential for pQCD and the MIT bag model (taken from Fraga et al., 2014). © AAS. Reproduced with permission

Of course, at a certain point, the perturbative series breaks down. From the band, we can estimate that the result of pQCD can be trusted for baryon chemical potential starting from about 3 GeV and higher. This means that the result of pQCD cannot be used for the chemical potentials encountered in compact stars of around 1–1.5 GeV. So, whether or not absolutely stable strange quark matter exists cannot be answered by pQCD nor what the EOS of quark matter relevant for quark stars is. However, let us use the EOS as an effective model to describe quark matter for quark stars, keeping in mind that it is not justified to do so within pQCD. There is an important lesson to be learned that is a general feature of EOSs for compact stars and that we will outline later.

For comparison, the result of the bag model for a bag constant of $B^{1/4} = 150\,\text{MeV}$ is also shown in Figure 8.3. The pressure rises rapidly with increasing baryon chemical potential getting close to the Stefan–Boltzmann limit at a baryon chemical potential of about $\mu_B = 2\,\text{GeV}$. For higher baryon chemical potentials, the pressure of the bag model is basically indistinguishable from a free gas of quarks. These results of the bag model are in contradiction to the prediction of pQCD. Hence, the bag model should not be used to describe quark matter at high densities. There is, however, a way to modify the bag model to make it compatible with the results from pQCD, at least in the range of baryon chemical potentials of up to a few gigaelectronvolts. If one introduces an effective model with a correction factor for the effective degrees of freedom such that the pressure is reduced compared to the Stefan–Boltzmann limit, one arrives at a pressure as a function of the quark chemical potential μ_q of the form

$$P(\mu_q) = \frac{N_f}{4\pi^2} a_{\text{eff}} \cdot \mu^4 - B_{\text{eff}}, \tag{8.50}$$

with the effective parameters a_{eff} and B_{eff}. The pressure of the effective model is within a few percent of the results from pQCD for the choices of about $a_{\text{eff}} = 0.63$ and $B_{\text{eff}} = 200\,\text{MeV}$ (Fraga et al., 2001).

As input to the TOV equation, we need the pressure as a function of energy density. Using the thermodynamic relations

$$n = \frac{\partial P}{\partial \mu_q} \quad \text{and} \quad \varepsilon = -P + n \cdot \mu_q \tag{8.51}$$

one can calculate thermodynamically consistently the EOS of pQCD for quark stars. Note that the calculation of the number density n and the energy density ε involves a derivative, so that the derivative of the strong coupling constant appears in the expressions. One can think that these terms are of a higher order in α_s and can be neglected. However, they have to be kept to ensure thermodynamic consistency. It is only the thermodynamic potential $\Omega/V = -P$ that is expanded in powers of α_s, not every thermodynamic quantity derived from it. We can use the effective model to calculate an approximation to the EOS of pQCD analytically. One finds that the EOS reads

$$P = \frac{1}{3}(\varepsilon - 4B_{\text{eff}}), \tag{8.52}$$

which turns out to be independent of the parameter a_{eff}. The EOS does not depend on the effective degrees of freedom and one recovers the EOS of the original bag model with an effective bag constant. The independence on the parameter a_{eff} also means that the corrections from interactions in terms of a constant strong coupling strength α_s drop out for the EOS in the form $p = p(\varepsilon)$. Hence, corrections from pQCD are small as long as α_s is not depending strongly on the chemical potential.

The lesson to be learned is that the EOS in the form $p = p(\mu)$ can be quite different, as we have seen in Figure 8.3 for pQCD and the bag model, but can lead to a similar EOS in the form $p = p(\varepsilon)$, see Eq. (8.52). Hence, the mass–radius relation of a compact star can be similar for different forms of the pressure as a function of the chemical potential that is introduced to conserve particle number. It points to the generic feature that gravity is blind with respect to the composition of matter, which is just a consequence of the strong equivalence principle.

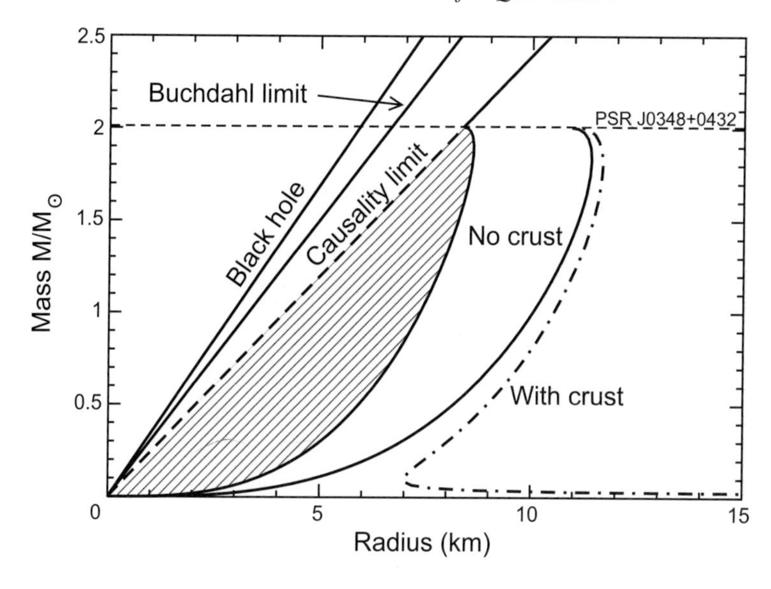

Figure 8.4 The mass–radius relation of quark stars with and without an outer crust of a lattice of nuclei. The line for a compactness of $C = 1/2$ is the static black hole limit, the one at $C = 4/9$ the Buchdahl limit for compact stars. The causality limit is shown by the mass–radius curve for the EOS with $c_s^2 = 1$ and the line for a compactness of $C = 0.354$.

8.5 Mass–Radius Relations for Quark Stars

Figure 8.4 shows the mass–radius relation of quark stars using the bag model EOS in the form of Eq. (8.52). We have chosen a value of $B_{\rm eff} = 145\,{\rm MeV}$ so that the maximum mass is at $2M_\odot$ to be compatible with the $2M_\odot$ pulsar mass measurement. Here, the result from the MIT bag model can be used, see Eq. (4.66), where the maximum mass scales as $B^{-1/2}$. As the pressure vanishes at a nonvanishing value of the energy density of $\varepsilon_0 = 4B_{\rm eff}$, the quark star configurations are selfbound and gravity is not needed to ensure hydrostatic equilibrium. The mass–radius relation follows then a curve that reflects a constant overall energy density so that $M/R^3 = $ const. or $M \propto R^{-3}$. Therefore, the mass–radius curve starts at the origin. It reaches a maximum mass when the pull of gravity becomes as important as the Fermi pressure of the quarks, which is balanced by the vacuum pressure $B_{\rm eff}$.

If a crust is taken into account for strange stars, the mass–radius curve has a similar maximum mass, but the curve is shifted toward higher radii by a few hundred meters. At a mass of about $0.1M_\odot$ and a radius of about 7 km, the mass–radius curve deviates strongly from the case with a crust as the radius starts to increase rapidly with decreasing mass. A halo of a crust forms for such low mass configurations as the gravitational pull of the quark matter core weakens so that the crust is less compressed compared to the more massive quark matter core.

On the left side of the plot, various limits are shown. The region to the left of the compactness limit for nonrotating black holes of $C = GM/R = 1/2$ is forbidden by general relativity for nonrotating spherical configurations. The Buchdahl limit shows the maximum compactness allowed by general relativity for static spheres of fluids with a compactness of $C = 4/9$, see Eq. (2.159). For fluids that are causal in the sense that the speed of sound is at most the speed of light, $c_s^2 \leq 1$, the maximum compactness possible is $C = 0.354$, see Eq. (4.80), which is shown by the dashed line that continues as a solid line above $2M_\odot$. No causal EOS can produce a configuration that is to the left of this line.

One can exclude an even larger region in the mass–radius diagram if one takes also into account that the maximum mass has to be at least as large as about $2M_\odot$ from the pulsar mass measurement of PSR J0348+0432, see Table 6.1. The mass–radius curve for the causal EOS with $c_s^2 = 1$ shown is tuned to achieve a maximum mass of $2M_\odot$. The maximum of this causal mass–radius curve is located at a radius of about 8.3 km and lies, as it has to be, on the causality line of $C = 0.354$. The hatched area between the causal mass–radius curve and the causality line is then excluded in addition to the area left of the causality line. Note that the exclusion limit from causality continues above $2M_\odot$ along the causality line, which is plotted then as a solid line. If a more massive pulsar is found in the future, the causal mass–radius curve will be shifted up and toward the right, increasing the area that can be excluded from causality.

Table 8.2 summarizes the properties of selfbound stars, as strange stars, versus that of neutron stars. It is important to note that the property of the EOS for selfbound stars, a vanishing pressure at a nonzero value of the energy density, alone determines that the resulting compact star configurations are bound by interaction not by gravity and that the mass–radius relation can start from the origin. Neutron star matter, on the other hand, has always a nonvanishing pressure at a nonvanishing energy density, so neutron stars are bound by the gravitational pull balancing the

Table 8.2 *A comparison of the properties of selfbound stars and neutron stars.*

	Selfbound stars	Neutron stars
EOS	Vanishing pressure at nonzero energy density	Vanishing pressure only at vanishing energy density
Stability	Bound by interactions	Bound by gravity
Mass–radius relation	Starts at the origin (without a crust)	Starts at large radii for small masses
Minimum mass	Arbitrarily small masses and radii possible	$M \approx 0.1 M_\odot$ at $R \approx 200$ km

nonzero pressure of matter. For neutron stars in particular, there is a minimum mass of $M_{\min} \approx 0.1 M_{\odot}$ at a radius of about $R \approx 200\,\text{km}$. Note that in both cases, no detailed statement can be made on the maximum mass, only that a maximum mass exists.

Exercises

(8.1) Estimate the number density at which nucleons start to overlap in a geometrical picture. Assume that nucleons are hard spheres with a radius of the charge radius of the proton of $r_p = 0.84\,\text{fm}$. Assume that the highest average density is that of close-packing of equal spheres with a volume fraction of $\pi/(3\sqrt{2})$.

(8.2) When the work on the properties of quark stars appeared in Itoh (1970), QCD was not known yet. So, a gas of free quarks with a mass of 10 GeV was used as this was the limiting accelerator energy at that time. Calculate the corresponding maximum mass using the scaling solution of Eq. (4.56). Take into account that the degeneracy factor for quarks would be $g = 6$ for up, down, and strange quarks without color and compare your result to the one of the original paper ($M_{\max} \approx 3.6 \times 10^{-3} M_{\odot}$).

(8.3) Estimate the effects from electrons for quark matter consisting of up, down, and strange quarks in weak equilibrium. Assume that the up and down quarks are massless and that the strange quark mass m_s is nonvanishing. Assume also that the quarks are noninteracting particles (fermions). Use the condition of weak equilibrium $\mu_u + \mu_e = \mu_d = \mu_s$ and the condition of charge neutrality

$$n_e = \frac{2}{3}n_u - \frac{1}{3}n_d - \frac{1}{3}n_s$$

as well as the relation between the chemical potential and the Fermi momentum for a free gas of fermions $\mu_i = \sqrt{k_F^2 + m_i^2}$ with the appropriate masses. Derive the relation for the electron chemical potential for the limit $\mu_q \gg m_s \gg \mu_e \gg m_e$, which reads

$$\mu_e \approx \frac{m_s^2}{4\mu_q},$$

where μ_q is the quark chemical potential. Verify the estimates of μ_e for typical quark chemical potentials of $\mu_q = 300\text{–}500$ MeV and the corresponding ratios of the electron to baryon number densities given in Eqs. (8.33) and (8.34). Check a posteriori that the approximations made in the derivation are justified.

(8.4) Calculate the properties of quark matter consisting of massless up and down quarks in weak equilibrium. Estimate the electron chemical potential in terms of the up quark chemical potential to linear order from the condition of charge neutrality. Note that the quarks have an additional degree of freedom compared to electrons, which is the color degree of freedom. Hence, the degeneracy factor of quarks is three-times larger than that of electrons. Show that the electron density is much smaller than the quark density, so $n_e \ll n_d \approx 2n_u$. Derive the relation for the electron chemical potential $\mu_e \approx (2^{1/3} - 1)\mu_u$ and use it to estimate the electron-to-baryon number ratio.

(8.5) Estimate the critical baryon number density for the onset of strange quarks in up and down quark matter in weak equilibrium using the results of the previous exercise by showing that $n_{B,c} \approx m_s^3/(2\pi^2)$.

(8.6) Estimate the critical baryon number density for the onset of charm quarks in quark matter with strange quarks in generalized weak equilibrium, that is, assuming that the charm quantum number is not conserved. (Answer: about $180n_0$).

(8.7) Show that for a pressure of the form

$$P(\mu_q) = a \cdot \mu_q^4 - B$$

the EOS is

$$P = \frac{1}{3}(\varepsilon - 4B),$$

that is, independent on the parameter a by using the thermodynamic relations given in Eq. (8.51). Show that the energy per particle for vanishing pressure depends on the parameter a so that the binding energy can be tuned independently from the mass–radius relation.

9

Hybrid Stars

In this chapter, we combine the findings we have for the low-density equation of state (EOS) from a pure gas of neutrons and for the high-density EOS from perturbative quantum chromodynamics (pQCD). Both regions in density are well constrained by our present knowledge of the strong interaction force. The combination of the EOS for neutron matter at low density with the one of quark matter at high density will result in an EOS based on two different pictures of quasi-particles, hadrons versus quarks, in the dense medium. Therefore, such an EOS is called a hybrid EOS and the resulting compact star configurations are called hybrid stars.

The transition from neutron matter, or hadronic matter, to quark matter can be a smooth one. And so far we discussed only smooth functions of the pressure versus the energy density when considering compact stars. However, a true phase transition, that is, a second- or a first-order phase transition, will cause either a kink in the EOS or a jump in the energy density. At present, it cannot be ruled out that the transition from hadrons to quarks in dense matter is of first or second order. We have seen that there is rapid change of the order parameters related to chiral symmetry and deconfinement at high temperatures and low density from lattice quantum chromodynamics (QCD). It is not completely clear at present whether not that rapid crossover transition turns into a first-order phase transition at high density and low temperature, relevant for hybrid stars, or not.

We will discuss in this section also the impact of a non-monotonous EOS on the properties of compact stars. The possibility of a phase transition in dense matter will open new features of the mass–radius relation, which can be formulated quite generically and model independently.

9.1 Combining Neutron and Quark Matter

In this subsection, we combine our findings for the EOS of neutron matter at low density as outlined in the chapter on neutron stars, Chapter 7, with the one of quark matter as outlined in the section on quark stars, Chapter 8. Let us summarize first what we learned so far about the EOS in these two limiting density regimes.

At low densities, the EOS can be described in terms of nucleons and nuclei where the effective models are well constrained by our knowledge of the nuclear interactions. Up to an energy density of about $10^{-3}\varepsilon_0$, where $\varepsilon_0 = 140\,\mathrm{MeV\,fm}^{-3}$ is the energy density of nuclear matter, the EOS for neutron stars is described by a lattice of nuclei immersed in a gas of electrons. The corresponding region for a neutron star is the outer crust. From $10^{-3}\varepsilon_0$ up to about $\varepsilon_0/2$, neutron star matter is in the form of a lattice of nuclei surrounded by electrons and a liquid of neutrons. From that energy density on up to about ε_0, neutron star matter is in its purest form, a liquid of pure neutron matter. The EOS of pure neutron matter has been reliably computed with an uncertainty reaching about 20% for an energy density slightly above ε_0. Pure neutron matter is unbound by $E/A = +16\,\mathrm{MeV}$, so the baryon chemical potential at saturation energy density for neutron matter is about $\mu_B = m_N + 16\,\mathrm{MeV} = 955\,\mathrm{MeV}$.

On the high end of the energy density we have to consider quark matter. At present, we can only solve QCD bulk matter for zero temperature at extremely high density within pQCD. The uncertainty in the pressure grows with decreasing quark chemical potential μ_q. At about $\mu_q = 900\,\mathrm{MeV}$, which corresponds to a baryon chemical potential of $\mu_B = 2.7\,\mathrm{GeV}$, the uncertainty reaches 20%. At such a quark chemical potential, the energy density for free quark matter can be calculated from Eq. (8.29) to be about a whooping $140\varepsilon_0$. Corrections from interactions will reduce that value somewhat. We read from Figure 8.3 that the pressure is about 1/2–3/4 that of the Stefan–Boltzmann limit for a free gas of quarks at $\mu_B = 2.7\,\mathrm{GeV}$, so the energy density is in the range of 70–$100\varepsilon_0$ at the lower end of the EOS from pQCD.

In summary, there is a humongous gap between the two different regions in energy density known so far, from about ε_0 at the upper end of the neutron matter EOS to about 70–100 ε_0 for the lower end of the quark matter EOS. Note that the difference in the energy densities is about two orders of magnitude while the difference in the baryon chemical potential is just $\mu_B = 955\,\mathrm{MeV}$ at the upper end of the low-density EOS to $\mu_B = 2.7\,\mathrm{GeV}$ at the lower end of the high-density EOS, a seemingly innocuous factor of barely three.

Figure 9.1 shows the pressure of neutron star matter and quark matter as a function of the quark chemical potential. The quark chemical potential is shifted by the energy (per quark number) of iron nuclei, which is the lowest chemical potential for the neutron star crust with $\mu_{\mathrm{iron}} = E/A(^{56}\mathrm{Fe}) = 930\,\mathrm{MeV}$. The reason to shift the chemical potential is that the baryon number or quark number density vanishes below this chemical potential. So, nonvanishing densities are only reached for chemical potentials above that value. Note also that the plot is on a double logarithmic scale.

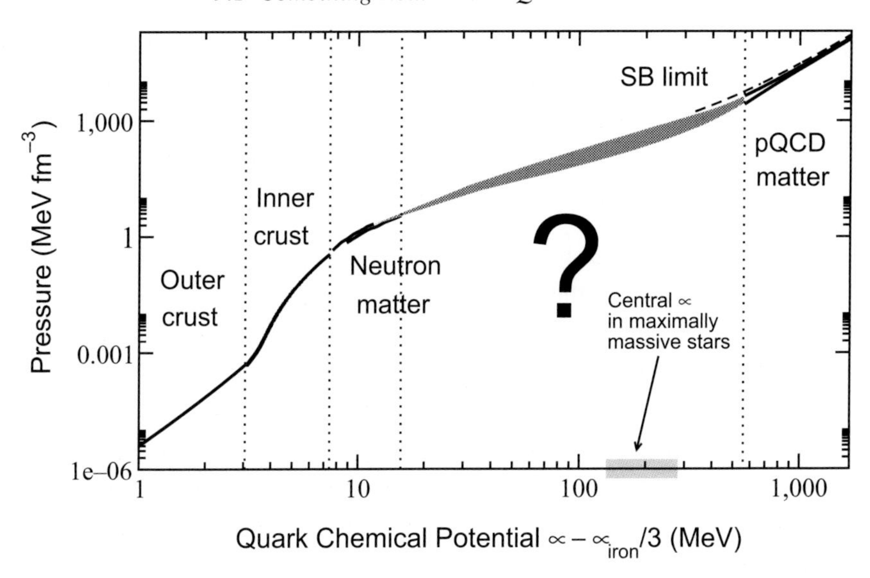

Figure 9.1 The EOS for neutron stars (taken from Kurkela et al., 2014). © AAS. Reproduced with permission. pQCD, perturbative QCD; SB, Stefan–Boltzman limit

The different regions of a neutron star are delineated by vertical lines in Figure 9.1. The outer crust ends at a baryon chemical potential being equal the neutron mass, which corresponds to a quark chemical potential above the one of ^{56}Fe of about 3 MeV. The inner crust extends to a shifted quark chemical potential of 7.4 MeV. The upper end of the pure neutron matter EOS at saturation density is located at a shifted quark chemical potential of about 16 MeV. The EOSs for the outer and inner crusts are shown by the solid line. Those of pure neutron matter EOS are depicted by two solid lines, corresponding to the upper and lower limit of the presently known uncertainty band. At the upper end of the quark chemical potential, the uncertainty band from pQCD is also shown by two lines, starting at a shifted quark chemical potential of 560 MeV. Plotting the shifted quark chemical potentials, one sees that the regions of the low- and high-density EOS are separated by more than an order of magnitude in the shifted quark chemical potential relating to a correspondingly large separation in the baryon number or quark number density.

The EOS can be interpolated in a model independent way by using a set of curves matching the two limiting EOSs. The hatched area shows possible smooth interpolations between the neutron matter EOS and the pQCD EOS, which was done by using piecewise polytropes, a widely used method, which we will discuss in the following in more detail. The shaded region along the x-axis marks the range of central energy densities reached for the maximum mass configuration using the interpolated EOSs. Clearly, the central energy densities for neutron stars is well within the unknown region of the EOS, well above the neutron matter EOS and well below the perturbative EOS in chemical potentials.

9.1.1 Piecewise Polytropes

The simplest way to interpolate between the two limiting EOSs at low and high densities is by polytropes, the same type of EOSs we have already encountered in the discussion of white dwarfs. Various other parameterizations for neutron star EOSs are possible and have been discussed in detail by Lindblom (2018), as for example, a parameterization in terms of the speed of sound.

The use of piecewise polytropes for the neutron star EOSs started with the work of Read et al. (2009), which showed that a representative set of neutron star EOSs reported in the literature can be well approximated by piecewise polytropes. The relation of mass and radius measurements to the neutron star EOS has been explored in terms of piecewise polytropes in Özel and Psaltis (2009) and Özel et al. (2010). Piecewise polytropes starting from a low-density EOS have been used to constrain the mass–radius relation of neutron stars by Steiner et al. (2010). Their method has been extended for interpolating between neutron matter and pQCD results using piecewise polytropes by Kurkela et al. (2014).

We recall that polytropes have the generic form

$$P(n) = K \cdot n^{\Gamma}, \tag{9.1}$$

where n is the number density, K and Γ are parameters. To ensure a smooth matching at both ends of the interpolation, one needs at least two polytropes. The two parameters of the polytrope are matched to the EOS by ensuring continuity in the pressure (equal pressure) and in the slope of the pressure as a function of chemical potential (i.e., equal number density). So, let us consider a polytrope of this form and allow for an arbitrary pressure P_0 at the matching number density n_0 to ensure a smooth transition. Now, the full thermodynamic potential is the pressure as a function of the chemical potential. So, we have to recast the generalized polytrope in a form so that it depends on the chemical potential, where the integration constant is fixed according to the matching point at μ_0 with the pressure $P_0 = P(\mu_0)$ and the number density $n_0 = n(\mu_0)$. We derive now the energy density as a function of the number density, which is the correct thermodynamic potential to be used. Then, we obtain the expression for the chemical potential. We are making use of our discussion of polytropes before by resorting to the thermodynamic relation of Eq. (3.114):

$$P = n^2 \cdot \frac{\mathrm{d}(\varepsilon/n)}{\mathrm{d}n}. \tag{9.2}$$

The differential equation for the energy density reads

$$\frac{\mathrm{d}\,(\varepsilon/n)}{\mathrm{d}n} = \frac{P}{n^2} = K \cdot n^{\Gamma-2}, \tag{9.3}$$

which can be integrated to result in

$$\varepsilon(n) = \frac{K}{\Gamma - 1} \cdot n^{\Gamma} + c \cdot n, \tag{9.4}$$

where c is an integration constant to be fixed to the matching point. So far we have just repeated the discussion of polytropes for white dwarfs. The polytropes considered for white dwarfs are matched to a low-density gas of free nucleons so that for white dwarfs, $c = m_N$, assuming that $\Gamma > 1$. For fitting to a general point of another EOS, the constant c has to be fixed differently. The chemical potential can be computed from the energy density to be

$$\mu = \frac{\mathrm{d}\varepsilon}{\mathrm{d}n} = c + \frac{K \cdot \Gamma}{\Gamma - 1} \cdot n^{\Gamma-1}. \tag{9.5}$$

Now we can fix the integration constant, as we see that we have to set

$$c = \mu_0 - \frac{K \cdot \Gamma}{\Gamma - 1} n_0^{\Gamma-1} \tag{9.6}$$

for matching. The number density can be transformed in terms of the chemical potential by inverting Eq. (9.5):

$$n = \left[(\mu - c)\frac{\Gamma - 1}{K \cdot \Gamma}\right]^{1/(\Gamma-1)} = \left[(\mu - \mu_0)\frac{\Gamma - 1}{K \cdot \Gamma} + n_0^{\Gamma-1}\right]^{1/(\Gamma-1)} \tag{9.7}$$

and we can rewrite the pressure in terms of the chemical potential:

$$P(\mu) = K \cdot \left[(\mu - \mu_0)\frac{\Gamma - 1}{K \cdot \Gamma} + n_0^{\Gamma-1}\right]^{\Gamma/(\Gamma-1)}. \tag{9.8}$$

The constant K is fixed by the requirement that $P_0 = P(\mu_0) = K n_0^{\Gamma}$ if n_0 is given. In that way, the pressure, the chemical potential, and the number density are equal at the matching point, ensuring a smooth transition. Note that Γ can be varied arbitrarily as long as $\Gamma \neq 1$, so a set of curves can be generated with different values of Γ. The EOS for generalized polytropes can be recast in the form suitable for input into the Tolman–Oppenheimer–Volkoff (TOV) equations as:

$$\varepsilon = \frac{1}{\Gamma - 1} P + \left(\mu_0 n_0 - \frac{\Gamma}{\Gamma - 1} P_0\right)\left(\frac{P}{P_0}\right)^{1/\Gamma} \tag{9.9}$$

for a given matching point at μ_0 with a continuous pressure $P_0 = P(\mu_0)$ and number density $n_0 = n(\mu_0)$.

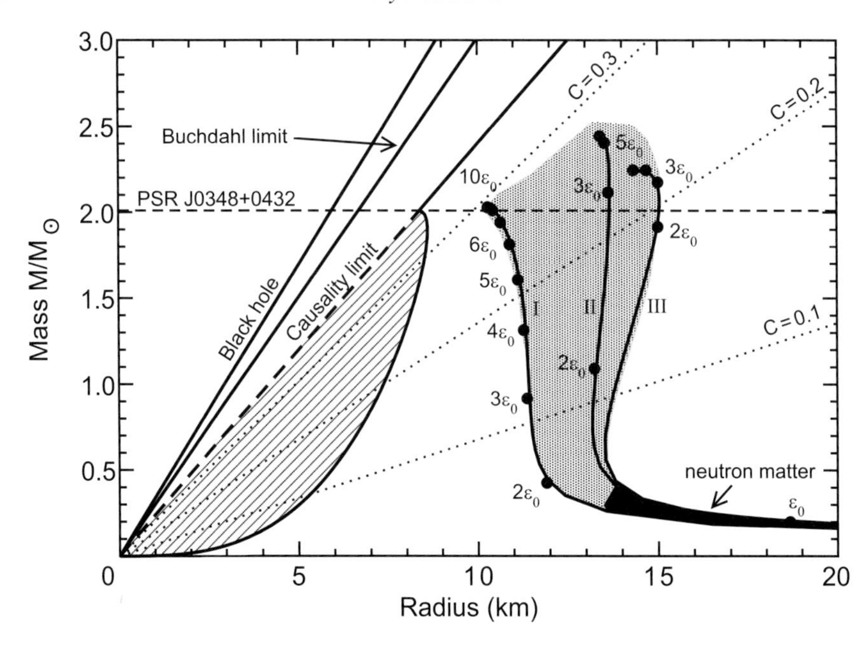

Figure 9.2 The mass–radius relation for neutron stars. (data taken from Kurkela et al., 2014)

Starting from a low-density matching point, several polytropes with different values of Γ can be stitched together by the method outlined earlier. For the final polytrope, Γ has to be chosen to match smoothly into the uncertainty band of the EOS of pQCD. The whole setup of piecewise polytropes then has to be checked for causality, that is, that the speed of sound does not exceed the speed of light $c_s^2 \leq 1$. The EOS is thereby bounded in its stiffness. On the other hand, the maximum in the mass–radius curve has to reach at least $2M_\odot$ to be compatible with the pulsar mass measurement of PSR J0348+0432. Thereby, the EOS cannot be too soft. The result will be an uncertainty band in the mass–radius diagram where the right border of the band is given by the causality constraint and the left border by the pulsar mass constraint.

Figure 9.2 shows the resulting mass–radius diagram for hybrid stars when the EOS has been interpolated by the use of two polytropes. As for the mass–radius relation of quark stars, see Figure 8.4, several limits on the mass–radius diagram are also included in the figure. The line denoted as black holes stands for the limiting compactness of $C = 1/2$ for static black holes. General relativity forbids any mass–radius configuration to the left of this line for static configurations. The Buchdahl limit corresponds to the compactness of $C = 4/9$ for static spheres of an incompressible fluid. The causality line originates from the causal EOS $p = \varepsilon - \varepsilon_0$, with $c_s^2 = 1$ and a maximum mass of $2M_\odot$. The maximum compactness due to causality is $C = 0.354$ in general, see Eq. (4.80), and is shown as a solid line for masses larger than $2M_\odot$. No mass–radius configuration can be to the left of the combined causality line, shown by the solid line, without violating causality and the $2M_\odot$ pulsar mass constraint. The mass–radius relation for neutron stars starts at small masses and large radii. The dark shaded region marks the uncertainty band from the pure neutron matter EOS, which ends around radii of $R = 13-14$ km and a mass of about $0.2M_\odot$. The gray shaded region outlines the uncertainty band in of mass and radius combinations of the interpolating piecewise polytropes, which are compatible with causality and reach at least $2M_\odot$. The dotted lines depict values of constant compactness for $C = 0.1$, 0.2, and 0.3. One sees that all the neutron star configurations above $1M_\odot$ have a compactness between $C = 0.1$ and 0.3.

There are three representative piecewise polytropes shown within the gray shaded region. Along the mass–radius curves, the central energy densities are labeled in units of the energy density of nuclear matter ε_0. The mass–radius relation labeled 'I' at the left boundary is the most compact one. It reaches a central energy density of about $10\varepsilon_0$ for the maximum mass configuration. The mass–radius relation label 'II' in the middle of the gray shaded region has the highest maximum mass of the three representative cases of about $2.4M_\odot$, with a maximum central energy density of about $5.6\varepsilon_0$. The mass–radius relation labeled 'III' has the largest radius and corresponds therefore to the stiffest EOS. The central energy density for the maximum mass configuration is about $5\varepsilon_0$.

We can read off the figure now the range of radius and central energy densities for typical masses of a neutron star. For a neutron star with a mass of $1.4M_\odot$, the radius is limited to $R = 11-14.5$ km with central energy densities of $\varepsilon = 1.6$ to about $4\varepsilon_0$. For a neutron star with a mass of $2M_\odot$, the radius is in the range of $R = 10-15$ km with central energy densities of about $\varepsilon = 2-8\varepsilon_0$.

9.2 Phase Transitions in Dense Matter

So far, we have considered a smooth, interpolating EOS. In the following, we consider the possibility of a first-order phase transition in dense matter and its impact on the mass–radius diagram of hybrid stars.

9.2.1 The QCD Phase Diagram

As outlined earlier, the EOS for neutron star matter between pure neutron matter and pure perturbatively interacting quark matter is unknown at present. One knows from QCD that hadrons, consisting of confined quarks, form at low density and that quarks are noninteracting at high density due to asymptotic freedom. At some intermediate density, a transition from hadronic degrees of freedom to quark degrees of freedom has to happen. In principle, this can be a smooth crossover or a first-order phase transition. From our discussion of QCD, the phases at low and high density can be described in terms of the quark and gluon condensates, which vanish at high energy scales. The low density phase is then in the chirally broken and confined phase, the high density phase in the chirally restored and deconfined phase. The two order parameters for chiral symmetry and confinement must not be coupled to each other. In fact, chiral symmetry is the appropriate one at nonvanishing density, while confinement is only well defined at vanishing density for pure gluon theory. The conjecture for a first-order phase transition in QCD at high density is then that there is chiral phase transition. As one knows from lattice QCD the phase transition is a crossover at vanishing density, the first-order chiral phase transition line has to end somewhere at low density in a critical end point, which is of second order.

A QCD phase diagram is depicted in Figure 9.3 as a function of temperature and quark chemical potential. The hadron phase is located at low chemical potentials, the quark–gluon plasma at high chemical potentials and temperature. At low temperatures and high chemical potentials, the quarks can combine and form diquark pairs, which give rise to the phenomenon of color superconductivity, a supercon-

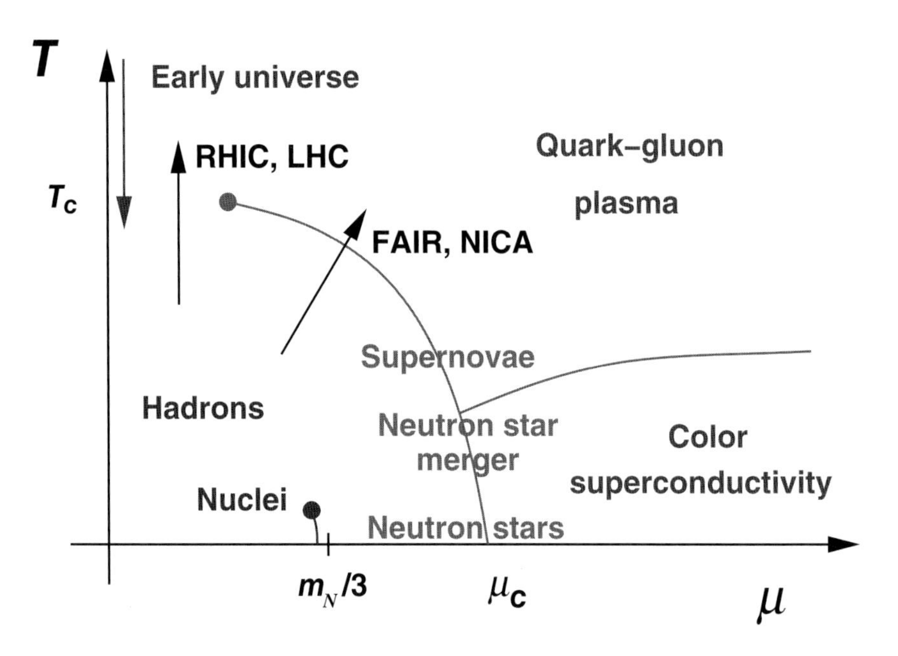

Figure 9.3 Cartoon of the QCD phase diagram of temperature versus quark chemical potential with conjectured phase transition lines. FAIR, Facility for Antiproton and Ion Research; LHC, Large Hadron Collider; NICA, Nucleon Based Ion Collider Facility; RHIC, Relativistic Heavy-Ion Collider.

ducting phase in color charges of QCD in analogy to the superconducting phase in solid-state physics for electromagnetism. The three phases could be separated by conjectured phase transition lines. At high temperatures and low chemical potentials, the phase transition line ends in a critical end point. The phase structure bears resemblance to that of water, where the gas, liquid, and solid phase corresponds to the hadron gas, the quark–gluon plasma, and the color-superconducting phase, respectively. At the meeting point of the three phases is the triple point, where all three phases are in equilibrium.

What we know at present from the QCD phase diagram is located at only two small regions: lattice QCD data at small chemical potentials and temperatures around the pseudo-critical temperature of $T_c \approx 150\,\text{MeV}$ and nuclear data at zero temperature and quark chemical potentials of $\mu_q \approx m_N/3$. For the latter region, one knows from the nuclear EOS that there is a liquid–gas phase transition from a gas of nuclei to bulk nuclear matter. That phase transition is of first-order and ends in a critical end point at a temperature of $T = 17.9(4)\,\text{MeV}$ and a density of $n = 0.06(1)\,\text{fm}^{-3}$, values that have been determined from nuclear reactions (Elliott et al., 2013).

The QCD phase diagram can be explored in terrestrial experiments by means of heavy-ion collisions. The heavy-ion colliders at Brookhaven National Laboratory (Relativistic Heavy-Ion Collider [RHIC]) and at CERN (European Organization for Nuclear Research) (Large Hadron Collider [LHC]) probe the QCD phase diagram at high temperatures and small chemical potentials. The future heavy-ion facilites at GSI in Darmstadt (Facility for Antiproton and Ion Research [FAIR]) and at the Joint Institute for Nuclear Research in Dubna (Nucleon Based Ion Collider Facility [NICA]) will explore the QCD phase diagram at moderate temperatures and chemical potentials. Hot and dense QCD matter can be also studied via investigating the early universe, which passes through the QCD phase diagram along the temperature axis at zero chemical potential, reaching a temperature of T_c at a time of about 10^{-5} s after the big bang. The hot and dense matter created in core-collapse supernovae and neutron star mergers is located at moderate temperatures and moderate to high chemical potentials.

9.2.2 Maxwell Construction

In the following, we describe the construction of a first-order phase transition applicable for cold neutron star matter. So, we set the temperature to be zero. We start by describing the EOS in terms of only one chemical potential, the baryon chemical potential. Let us consider two phases, I and II, which are in thermodynamic equilibrium with a common separating interface inside a neutron star. First, hydrostatic equilibrium demands that the pressure has to be equal on both sides of the interface, so that $P_{\rm I} = P_{\rm II}$. If the pressures are not equal, there is a pressure gradient and the phase with the higher pressure will expand and the other one will shrink until the pressures are equal. Second, the chemical potential has to be equal on both sides of the interface, so $\mu_B^{(\rm I)} = \mu_B^{(\rm II)}$. If the chemical potentials are not equal, there will be a gradient in the chemical potentials. Particles in the phase with the higher chemical potential have a higher energy at the Fermi surface. They will travel through the interface to the phase with the lower chemical potential, which is energetically favored until the change in the concentration will cause the chemical potentials to be equal. So, in summary, two phases are in thermodynamic equilibrium if

$$P_{\rm I}\left(\mu_B^{(\rm I)}\right) = P_{\rm II}\left(\mu_B^{(\rm II)}\right) \qquad \text{with} \qquad \mu_B^{(\rm I)} = \mu_B^{(\rm II)}. \tag{9.10}$$

Figure 9.4 depicts schematically a first-order phase transition and its standard Maxwell construction using the condition of phase equilibrium of Eq. (9.10). Figure 9.4a shows the EOS as the pressure versus the energy density. Phase I at low density reaches a maximum in the pressure at point B. Then the pressure decreases with increasing energy density from point B to point C. As the derivative of the pressure with respect to the energy density is the speed of sound squared, the region from B to C has a negative speed of sound squared, so the speed of sound would be imaginary, indicating an instability. From point C on, the pressure rises again with the energy density, where matter is now in the high density phase II. To find the point of phase equilibrium, one needs to take a look at the pressure as a function of chemical potential, which is illustrated in Figure 9.4a. Phases I and II corresponds to two lines in the plot that intersect at points A of phase I and point D of phase II. At this point, the pressure and the chemical potentials in both phases are equal. The phase equilibrium is shown in Figure 9.4a by the horizontal line, that is, a line of constant pressure, connecting point A in phase I with point D in phase II.

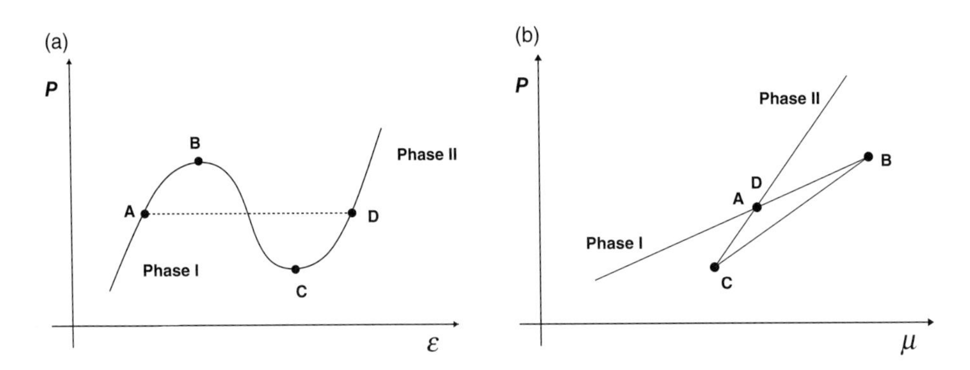

Figure 9.4 A first-order phase transition in terms of the pressure as a function of (a) energy density and (b) the chemical potential.

The grand canonical potential $\Omega(\mu, V)$ is minus the pressure as a function of the chemical potential times the volume. In bulk matter, only the dependence on the chemical potential remains as $\Omega/V = -P(\mu)$. Thermodynamic equilibrium corresponds to minimizing the grand canonical potential, which is equivalent to maximizing the pressure at a given chemical potential. Following this procedure for the curves pictured in Figure 9.4b, phase I is the thermodynamically favored one for low chemical potentials up to point A. Then, from point D, which is at the same point as point A, up to higher chemical potentials, phase II is the thermodynamically favored one. We see now that the region from point A to point B is the unfavored overdense region of phase I and the region from point C to D is the unfavored under-dense region of phase II. We note that it is possible that point C is located at negative pressure. In fact, this is the case for the liquid–gas phase transition of nuclear matter, which is a strongly first-order phase transition. From thermodynamic relations, the derivative of the pressure with respect to the chemical potential is the number density. We can read from the plot that the slope of the curve in phase I is smaller than the one of phase II at the crossing point, indicating a jump in the number density during a phase transition. We also realize that the phase transition has to go from a low- to a high-density phase with increasing chemical potential, as dictated by maximizing the pressure as a function of chemical potential. It would not work the other way around.

Following the stable sequence of the pressure as a function of energy density in the left plot of Figure 9.4a, it emerges that the coexistence region of phase I and phase II between points A and D corresponds to a region of constant pressure and a jump in the energy density. For compact stars, a region with constant pressure corresponding to a vanishing pressure gradient cannot withstand the pull of gravity. Therefore, hydrostatic equilibrium will shrink the region between point A and D inside a compact star to a zone with vanishing extension. Hence, as a function of the radial coordinate inside a compact star, the energy density will jump from the phase at point A to the one at point D instantly at a critical radius. There is no mixed phase possible inside a compact star according to the Maxwell construction.

9.2.3 Gibbs Construction

We know that in neutron star matter two conserved quantum numbers and correspondingly two chemical potentials are involved. The extension of the condition of phase equilibrium from one to two chemical potentials is straightforward.

It is interesting to note that the very first papers on phase transitions in the core of neutron stars did not take into account that besides the baryon chemical potential, the charge chemical potential also has to be taken into account. For simplicity, just the energy density versus the baryon number density was considered for the phase transition to quark matter, see, for example, Baym and Chin (1976). Phase constructions with two independent chemical potentials have been adopted for considering the transition to the quark–gluon plasma in relativistic heavy-ion collisions, see Greiner et al. (1987, 1988). As matter is not in weak equilibrium in heavy-ion collisions, the chemical potential for the quantum number strangeness has been considered in addition to the baryon chemical potential to describe a first-order phase transition at high temperature and density. They found that a mixed phase with separate regions of positive and negative strangeness number density occurs. That the charge chemical potential has to be considered for phase transitions to quark matter in neutron stars in addition to the baryon chemical potential was first discussed in detail by Glendenning (1992). He showed that a proper description of the phase transition with two independent chemical potentials allows for the presence of a large mixed phase inside neutron stars. And that in this mixed phase, separate regions with positive and negative charge densities occur.

Hydrostatic equilibrium will still demand that the pressures in both phases are the same, so $P_{\rm I} = P_{\rm II}$. Chemical equilibrium now demands that all chemical potentials are equal in both phases, so the baryon chemical potential μ_B and the charge chemical potential μ_Q have to be the same across the phase boundary. In summary, the so-called Gibbs construction generalizing the Maxwell construction reads

$$P_{\rm I}\left(\mu_B^{(\rm I)},\mu_Q^{(\rm I)}\right) = P_{\rm II}\left(\mu_B^{(\rm II)},\mu_Q^{(\rm II)}\right) \quad \text{with} \quad \mu_B^{(\rm I)} = \mu_B^{(\rm II)}, \quad \mu_Q^{(\rm I)} = \mu_Q^{(\rm II)}. \tag{9.11}$$

The pressure in one phase is now a function of two chemical potentials, so it represents a surface in the space of baryon and charge chemical potentials, not a line as in the Maxwell case. The crossing of the two surfaces will result in a line that depends on the two chemical potential and not a point, as in the Maxwell case. Figure 9.5 depicts the situation for the Gibbs construction by showing the surfaces of the pressure of the two phases versus the neutron (baryon) chemical and the electron (charge) chemical potential. The label 'HP' stands for the low-density hadron phase, phase I, the label 'MP' for the mixed phase, and the label 'QP' for the quark phase, phase II. One sees that the two surfaces of the pressure for the two phases intersect in one line. For a given baryon chemical potential, the charge chemical potential is fixed by total charge neutrality. The line denoted as 'HP' shows the combination of the baryon and charge chemical potential on the pressure surface for phase I, where phase I has a vanishing total charge. At the crossing point marked with '1', the pressure in phase II becomes equal to that of phase I. However, for the combination of the baryon and charge chemical potential, phase II will not be uncharged. The line marked 'QP' shows the line for vanishing charge for phase II, which hits the phase coexistence line at the point marked '2' not at '1'. As for the Maxwell construction, one could think about jumping from

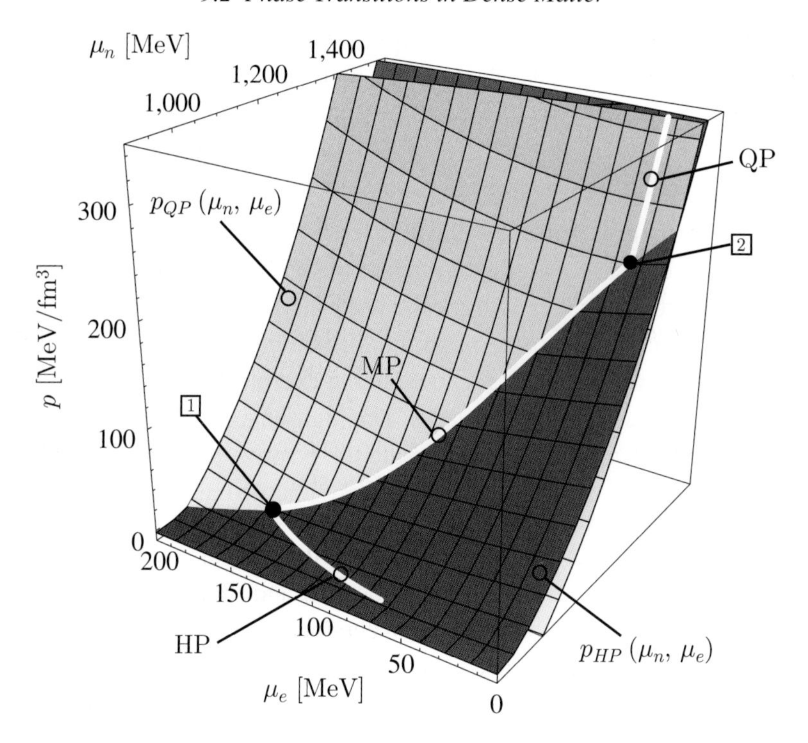

Figure 9.5 The Gibbs construction for nuclear matter and quark matter. Reprinted from Schertler et al. (2000), copyright (2000), with permission from Elsevier.

phase I at point '1' to phase II at point '2' to ensure locally charge neutral phases. This construction would result in a jump of both chemical potentials.

There is another way to construct the phase transition by allowing for locally charged bubbles of one phase to be immersed in the other phase with the opposite charge. The charge density and the volume fraction χ of the two phases have to be combined in such a way that the charge vanishes globally, on scales larger than the size of the charged bubbles:

$$n_Q^{(\text{total})}(\mu_B, \mu_Q) = (1 - \chi) \cdot n_Q^{(\text{I})}(\mu_B, \mu_Q) + \chi \cdot n_Q^{(\text{II})}(\mu_B, \mu_Q) = 0. \tag{9.12}$$

A similar relation holds for the total energy density, which is

$$\varepsilon^{(\text{total})}(\mu_B, \mu_Q) = (1 - \chi) \cdot \varepsilon^{(\text{I})}(\mu_B, \mu_Q) + \chi \cdot \varepsilon^{(\text{II})}(\mu_B, \mu_Q). \tag{9.13}$$

In Eq. (9.12), the charge densities of the two phases have to be evaluated at the same baryon and charge chemical potentials along the phase coexistence line. The so-constructed mixed phase starts at point '1' with small bubbles of highly charged phase II in the slightly charged medium of phase I with the opposite charge. Along the phase coexistence line, the volume fraction of phase II increases but also the charge density will decrease toward point '2' where the charge density of phase II vanishes. At point '2' the situation is reversed as phase I occupies a small volume fraction but is highly charged being inside a medium of slightly charged matter of phase II with the opposite charge.

We have left open the explicit sign of the charges of the two phases in the discussion. For nuclear matter as phase I, it will be favorable to have a positive charge, as this is closer to the case of nuclear matter with equal amounts of protons and neutrons. Nuclear matter is the case with the lowest energy by virtue of the asymmetry energy. On the other hand, quark matter with up, down, and strange quarks is charge neutral already for low-charge chemical potentials, see for example, Eq. (8.33) for the case of free quarks. The high-charge chemical potential of the nuclear matter phase needed to compensate the positive charge of the protons will make the quark matter phase highly negatively charged at the onset of the mixed phase at point '1' and vice versa at point '2'.

9.2.4 Pasta in the Core

We note from Figure 9.5 that the pressure increases along the phase coexistence line. Hence, there exists a nonvanishing pressure gradient in the mixed phase for the Gibbs construction. In this case, the mixed phase will not shrink to zero size, as for the Maxwell construction. The generalized Gibbs construction for phase equilibrium will allow for an extended mixed phase inside compact stars, if a first-order phase transition is present.

The mixed phase consists of separate regions of the two phases, which can form geometric structures. The situation is in analogy to the one discussed for the transition of the inner crust to the outer core, where the phase transition can form the pasta in the crust, see Section 7.2.4. Aptly, one can call the structures for a first-order Gibbs phase transition in the core the pasta in the core. The extension of the pasta phase in the core could be much larger compared to the small one for the inner crust. The thickness of the inner crust is of the order of a few hundred meters and the pasta in the crust occupies just a fraction of the inner crust. The thickness of the core is in the range of a few kilometers and can be the major part of the entire neutron star. The mixed phase could occupy most of the core, possibly presenting a major component of a neutron star.

As we know from the discussion of the pasta in the crust, the geometric structures are stabilized by balancing the Coulomb energy and the surface energy. Let us estimate the size of the pasta structures by looking at the case of charged droplets with a radius R. The surface energy of a droplet is given by

$$E_S = 4\pi\sigma R^2, \tag{9.14}$$

where σ is the unknown surface tension between the two phases. The charged droplet is in a charged medium with the opposite charge, so the net charge depends on the difference in the charge density of the two phases

$$\Delta Z = \frac{4\pi}{3} R^3 \left(n_Q^{(\mathrm{I})} - n_Q^{(\mathrm{II})} \right) . \tag{9.15}$$

The Coulomb energy of the droplet is then

$$E_C = \frac{3}{5} \frac{(e\Delta Z)^2}{4\pi R} = \frac{16\pi^2 \alpha_{\mathrm{EM}}}{15} \left(n_Q^{(\mathrm{I})} - n_Q^{(\mathrm{II})} \right)^2 R^5 , \tag{9.16}$$

where $\alpha_{\mathrm{EM}} = e^2/(4\pi)$ is the electromagnetic fine structure constant. Minimizing the total energy results in the condition $E_S = 2E_C$, which we have derived before for the pasta in the crust, see Eq. (7.33). Now one can estimate a value of the radius of the droplets of

$$R = \left(\frac{15\sigma}{8\pi\alpha_{\mathrm{EM}} \left(n_Q^{(\mathrm{I})} - n_Q^{(\mathrm{II})} \right)^2} \right)^{1/3} = 5.7\,\mathrm{fm} \left(\frac{\sigma}{10\,\mathrm{MeV\,fm}^{-2}} \right)^{1/3} \left(\frac{n_Q^{(\mathrm{I})} - n_Q^{(\mathrm{II})}}{0.15\,\mathrm{fm}^{-3}} \right)^{-2/3} . \tag{9.17}$$

The precise value of the surface tension is presently unknown. Recall that if there is a phase transition at high density in QCD, it is dictated by chiral symmetry. So, one needs to know the surface tension between the chirally broken and the chirally restored phase at high density. A maximum value for the chiral phase transition of $\sigma = 10\,\mathrm{MeV\,fm}^{-2}$ has been obtained within a chiral model by Palhares and Fraga (2010) and Fraga et al. (2019). For comparison, the surface tension for droplets of nuclear matter in vacuum, that is, for nuclei, is about $\sigma_N = 1\,\mathrm{MeV\,fm}^{-2}$ an order of magnitude smaller, which will result in a droplet radius of $R = 2.6\,\mathrm{fm}$.

The surface and the Coulomb energy of the droplets will increase the total energy of the system, so the presence of structures will make the mixed phase energetically less favored. The extension of the mixed phase will shrink, therefore, if finite size effects are taken into account. For a critical value of the surface tension, the formation of a pasta phase with charged structures might turn even less favorable compared to two separate uncharged phases, and one is back to a Maxwell construction without a mixed phase at all. The critical surface tension has been estimated to be around $\sigma = 70\,\mathrm{MeV\,fm}^{-2}$, see for example, Heiselberg et al. (1993). For higher values of the surface tension, no mixed phase can form as it is not energetically favored anymore. The limit given for the surface tension of $\sigma = 10\,\mathrm{MeV\,fm}^{-2}$ would imply that a mixed phase could exist in the core of neutron stars.

9.3 A Third Family of Compact Stars

The possible presence of a first-order phase transition has another surprising feature to offer: the possible existence of a new class of compact stars. As this new solution would coexist with the other known compact star solutions of white dwarfs and neutron stars, it is also called the third family of compact stars.

9.3.1 Twin Stars

Figure 9.6 sketches the mass–radius diagram for compact stars with a third family. Hereby, we are making use of the stability analysis for the mass–radius diagram as summarized in Table 7.4. Part of the mass–radius diagram, for white dwarfs and neutron stars, has been discussed before in connection with Figure 7.11. In Figure 9.6, the stable modes numbered '0', '1', and '2' are depicted by hollow boxes, the unstable ones by filled boxes at various sequences of the mass–radius relation. Following the mass–radius relation with increasing central pressure, one has to start at large radii, that is, at the right end of the figure. The line from the right side of the figure to point A stands for the mass–radius relation of white dwarfs, which is stable, so all modes are shown by hollow boxes. The white dwarf sequence is stabilized by the Fermi pressure of nonrelativistic electrons. At point A, the sequence of white dwarfs reaches the maximum mass, the Chandrasekhar mass limit. The lowest mode '0' becomes unstable beyond point A, shown by the filled

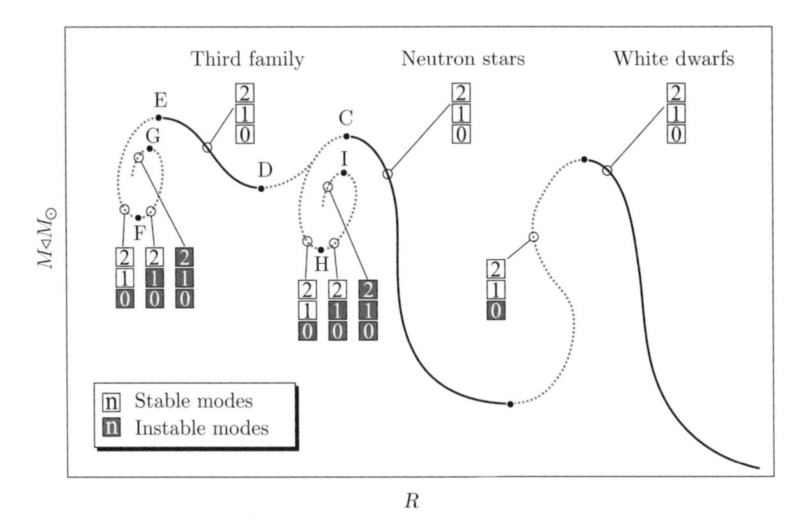

Figure 9.6 Stability regions for compact stars for white dwarfs, neutron stars, and a third family of compact stars. Reprinted from Schertler et al. (2000), copyright (2000), with permission from Elsevier

box as the maximum with increasing central pressure is passed counterclockwise. The mass–radius relation from point A to point B is then an unstable sequence.[1]

Point B stands for the lowest stable configuration of a neutron star, which is stabilized by the repulsive interaction between nucleons. The minimum in point B is passed clockwise, so the lowest mode turns stable again. Hence, the neutron star sequence from point B to point C, the Landau maximum mass, is stable. The maximum at point C is passed counterclockwise, so the lowest mode becomes unstable. For ordinary neutron star matter without a phase transition, the mass–radius relation continues from point C to point H, a minimum that is passed counterclockwise. So, another mode turns unstable, in this case the mode '1'. At point I, a maximum that is passed counterclockwise, mode '2' becomes unstable and so on.

Another possibility is depicted for the case that the sequence continues from point C to point D in a similar fashion as for point A to B. Point D is a minimum that is passed clockwise so that the unstable mode switches to a stable one. The sequence from point D to E stands for new stable configurations, a new class of compact stars besides the sequence of white dwarfs up to point A and the sequence of neutron stars between points B and C. The maximum at point E is passed counterclockwise, so the sequence of stable configurations ends here. Beyond point E, the same situation for points F and G is illustrated as for points H and I. No further stable sequence occurs beyond point E, where the spiral-like pattern at of the mass–radius relation results in more and more unstable modes.

Note that the third family is a new third solution to the TOV equations. The new stable sequence of the third family could be stabilized by a new phase appearing at high density, the chirally restored phase, or in simple terms the transition to quark matter, where now the repulsive interaction between quarks stabilizes the configurations. However, we stress that gravity is not sensitive to the microphysics involved. The only input for the TOV equation is the EOS in the form $P = P(\varepsilon)$.

The possible existence of a third family of compact stars was first proposed by Gerlach (1968) as part of his PhD work under the supervision of John Wheeler. Gerlach devised a concept to construct the EOS from the mass–radius relation. He showed that the feature needed in the EOS to arrive at a new stable sequence of compact stars has to be a transition region in the EOS followed by a sudden, drastic change in the speed of sound of the EOS at higher energy densities. The notion of a third family was rediscovered by Kämpfer (1981a), who succeeded in modeling a third family of solutions. His work was motivated by two different possibilities for a strong first-order phase transition at high densities for compact stars: pion condensation and the transition to quark matter. Finally, but many years

[1] For simplicity, we omit other extrema that are present between point A and B, see the discussion of Figure 7.11.

later, Glendenning and Kettner (2000), and independently Schertler et al. (2000), reopened the research on the third family of compact stars by finding stable new sequences using the Massachusetts Institute of Technology (MIT) bag model for the high-density quark phase. As the new stable branch had similar masses as the neutron star branch, they dubbed the third family solution twin stars. A unique signal for twin stars, or a third family, is to find two different compact stars with similar masses but different radii.

9.3.2 The Seidov Criterion

In order to have a third family of compact stars in the mass–radius diagram, one needs an unstable region first that then turns into a new stable one. The criterion for causing an unstable region by a first-order phase transition in neutron stars was considered first by Seidov (1971). He extended the Newtonian stability analysis of Lighthill (1950), see also Ramsey (1950), for small planetary cores by including effects from general relativity. The criterion was rediscovered by Kämpfer (1981b) when discussing phase transitions in neutron stars.

Consider an EOS with a phase transition at the pressure P_0 with a jump in energy density from ε_1 to ε_2. The details of the EOS below P_0 will be not essential for the following discussion. All that is needed is the jump in energy density at P_0.

Let us start with the compact star configuration at a central pressure just below the phase transition, that is, at $P_c = P_0$ and $\varepsilon_c = \varepsilon_1$. We take a look at the TOV equation and consider the behavior of the pressure, which we will denote as $P_-(r)$, for small radii. The TOV equation can be written as

$$\frac{\mathrm{d}P}{\mathrm{d}r} = -G\frac{P+\varepsilon}{r(r-2Gm_r)}(4\pi r^3 P + m_r), \tag{9.18}$$

with the equation for mass conservation

$$\frac{\mathrm{d}m_r}{\mathrm{d}r} = 4\pi r^2 \varepsilon. \tag{9.19}$$

For small radii we can consider ε to be constant. So, the mass for small radii simply reads

$$m_- = \frac{4\pi}{3} r^3 \varepsilon_1. \tag{9.20}$$

The correction from the Schwarzschild factor can be ignored for small radii as the effects from the curvature of space-time should be small. Putting the expression for the mass into the TOV equation, one finds for small radii

$$\frac{\mathrm{d}P_-}{\mathrm{d}r} = -G\frac{P_0+\varepsilon_1}{r^2}\left(4\pi r^3 P_0 + \frac{4\pi}{3}\varepsilon_1 r^3\right)$$
$$= -\frac{4\pi G}{3}(P_0+\varepsilon_1)(P_0+3\varepsilon_1)r. \tag{9.21}$$

Integration gives for the solution of the pressure for small radii

$$P_-(r) = P_0 - \frac{2\pi G}{3}(P_0+\varepsilon_1)(P_0+3\varepsilon_1)r^2. \tag{9.22}$$

Next, we take a compact star configuration with a central pressure P_c, which is slightly above P_0 by an amount δ:

$$P_c = P_0 + \delta. \tag{9.23}$$

As it is assumed now that with $P_c > P_0$ a small nucleus of the high-density phase will form with a radius r_n. At the boundary of the nucleus, the pressure is $P_+(r_n) = P_0$ and the mass is given by the mass of the nucleus:

$$m_+(r_n) = \frac{4\pi}{3}r_n^3\varepsilon_2, \tag{9.24}$$

where we use now the energy density of the high-density phase. The gradient in the pressure changes to

$$\frac{\mathrm{d}P_+}{\mathrm{d}r} = -G\frac{P_0+\varepsilon_1}{r^2}\left(4\pi r^3 P_0 + \frac{4\pi}{3}\varepsilon_2 r^3\right)$$
$$= -\frac{4\pi G}{3}(P_0+\varepsilon_1)(P_0+3\varepsilon_2)r. \tag{9.25}$$

Note that at the boundary of the nucleus, the energy density is ε_1 and that ε_2 only enters via inserting the mass of the nucleus M_+. Next, we need to relate the size of the nucleus to the increase in the pressure δ. The pressure difference is given by integrating the TOV equation from the center to the boundary of the nucleus. Again neglecting the small effects from the Schwarzschild factor, one finds that

$$\delta = -\frac{2\pi G}{3}(P_0+\varepsilon_2)(P_0+3\varepsilon_2)r_n^2. \tag{9.26}$$

The solution of the pressure for a central pressure slightly above the phase transition $P_+(r)$ should be a small deviation from the solution for a central pressure just below the phase transition $P_-(r)$. The difference of the two solutions is assumed to be a perturbation function

$$\Pi(r) = P_+(r) - P_-(r). \tag{9.27}$$

In analogy to Lighthill (1950), the solution of the differential equation for the perturbed function $\Pi(r)$ can be found by substituting the above ansatz into the

TOV equation. For a small linear perturbation, terms of higher power in $\Pi(r)$ and derivatives in $\Pi(r)$ can be neglected. Then, the solution has the form (Seidov, 1971)

$$\Pi(r) = A + \frac{B}{r}, \tag{9.28}$$

with two coefficients A and B to be determined from the boundary conditions at the radius of the nucleus r_n. The one for the pressure at r_n reads

$$\Pi(r_n) = A + \frac{B}{r_n} = P_+(r_n) - P_-(r_n) = \frac{2\pi G}{3}(P_0 + \varepsilon_1)(P_0 + 3\varepsilon_1)r_n^2, \tag{9.29}$$

where we used Eq. (9.22) for $r = r_n$ and that $P_+(r_n) = P_0$. The gradient of the pressure $\Pi(r)$ evaluated at the radius r_n can be read off from Eqs. (9.21) and (9.25) to be

$$\frac{\mathrm{d}\Pi}{\mathrm{d}r} = -\frac{B}{r_n^2} = \frac{\mathrm{d}P_+}{\mathrm{d}r} - \frac{\mathrm{d}P_-}{\mathrm{d}r} = -\frac{4\pi G}{3}(P_0 + \varepsilon_1)(2P_0 + 3\varepsilon_1 + 3\varepsilon_2)r_n. \tag{9.30}$$

The two equations of the boundary conditions can be solved for the coefficient A:

$$A = \frac{2\pi G}{3} r_n^2 (P_0 + \varepsilon_1)(3P_0 + 3\varepsilon_1 - 2\varepsilon_2) r_n. \tag{9.31}$$

Now we can use Eq. (9.26) to replace the radius of the nucleus r_n and find a relation to the shift of the central pressure δ

$$A = \frac{(P_0 + \varepsilon_1)(3P_0 + 3\varepsilon_1 - 2\varepsilon_2)}{(3P_0 + \varepsilon_2)(P_0 + \varepsilon_2)} \cdot \delta. \tag{9.32}$$

The total mass of the compact star is determined by the vanishing pressure at the radius of the compact star. For large radii, the term proportional to B can be ignored, so the solution for $P_+(r)$ will look like the solution for $P_-(r)$, with the central pressure shifted by the amount A instead of δ for large radii. So, if the central pressure is increased by δ, the resulting pressure profile will be shifted by A for large radii with a corresponding effect on the change of the total mass. The change of the total mass of the compact star with the change of the central pressure for the two configurations considered will then be given by the relation between A and δ, so

$$\frac{\mathrm{d}M_+}{\mathrm{d}P_c} = \frac{(P_0 + \varepsilon_1)(3P_0 + 3\varepsilon_1 - 2\varepsilon_2)}{(3P_0 + \varepsilon_2)(P_0 + \varepsilon_2)} \cdot \frac{\mathrm{d}M_-}{\mathrm{d}P_c}. \tag{9.33}$$

It is important to note that the change of the mass with the central pressure will change sign for

$$3P_0 + 3\varepsilon_1 - 2\varepsilon_2 < 0. \tag{9.34}$$

A change of sign means that a stable sequence where the mass increases with the central pressure will become an unstable one as the mass decreases with the central pressure. Hence, finally, we arrive at the Seidov criterion for an instability in the mass–radius relation for a first-order phase transition:

$$\frac{\varepsilon_2}{\varepsilon_1} > \frac{3}{2}\left(1 + \frac{P_0}{\varepsilon_1}\right). \tag{9.35}$$

The Newtonian limit is given by $\varepsilon_2/\varepsilon_1 > 3/2$. So, the mass–radius relation will show an unstable sequence if there is a jump in the energy density, which is larger than given by the Seidov criterion. For example, if there is a phase transition at, say, twice saturation energy density $2\varepsilon_{\mathrm{nm}}$ with a pressure of $P(2\varepsilon_{\mathrm{nm}})$, then the end of the phase transition has to be at an energy density of more than $3\varepsilon_{\mathrm{nm}} + 3P(2\varepsilon_{\mathrm{nm}})/2$ to produce an instability.

9.3.3 Classification of Hybrid Stars

We consider now the possible realizations of hybrid star configurations with a phase transition in the mass–radius diagram following the work of Alford et al. (2013). For this purpose, we have to specify the EOS for the low density and high density phases. The phase transition between the two phases is assumed to be characterized by the critical pressure P_{trans} at the transition energy density $\varepsilon_{\mathrm{trans}}$ and the jump in energy density $\Delta\varepsilon$ to the high density phase. The phase transition can then be defined by the ratios of $P_{\mathrm{trans}}/\varepsilon_{\mathrm{trans}}$ and $\Delta\varepsilon/\varepsilon_{\mathrm{trans}}$, which are the two free parameters to be varied. We can study the possible characteristic mass–radius curves in the plane of these two parameters. First, we consider the Seidov criterion for an unstable mass–radius sequence in terms of the two parameters, which reads

$$\frac{\Delta\varepsilon}{\varepsilon_{\mathrm{trans}}} > \frac{1}{2} + \frac{3}{2}\frac{P_{\mathrm{trans}}}{\varepsilon_{\mathrm{trans}}}. \tag{9.36}$$

So, the Seidov criterion is a straight line in the plane of the two parameters. Above the line, the mass–radius sequence gets unstable right at the onset of the phase transition.

To proceed further, we need to specify now the high-density EOS. For generating stable twin star configurations, the unstable sequence has to be turned to a stable one in the presence of the high-density phase. The generation of twin star sequence should, therefore, be most favorable for the stiffest possible EOS in the high density phase, which we know is of the form

$$P = s \cdot (\varepsilon - \varepsilon_0), \tag{9.37}$$

with the speed of sound squared $s = 1$, which we studied in detail in Section 4.3.5. This form of the high-density EOS for general values of s for hybrid stars has been also called the constant-sound-speed parameterization and has been used to study stable hybrid star configurations (Alford et al., 2013). The ansatz for the EOS for hybrid stars is then given by a low-density EOS up to the transition pressure followed by a constant-sound-speed EOS for pressures above the transition pressure:

$$P = \begin{cases} P_{\text{low}}(\varepsilon) & P \leq P_{\text{trans}} \\ P_{\text{trans}} + s \cdot (\varepsilon - (\varepsilon_{\text{trans}} + \Delta\varepsilon)) & P > P_{\text{trans}}. \end{cases} \tag{9.38}$$

Recall that the pressure is a continuous function of the radius for compact star configurations, while the energy density can jump discontinuously.

The solutions turn out to be quite generic and can be classified in four different types. In some cases, no stable twin star configuration is present.

If the central pressure reaches the transition pressure, the mass–radius diagram becomes unstable when the Seidov criterion is fulfilled. Then, there are two possibilities. It could be that a stable configuration cannot be established at higher central pressures at all. In this case, the high density phase is absent for all stable sequence, which constitutes class (A). If a high density phase manages to stabilize the mass–radius configuration at higher central pressures, the stable hybrid star branch will be disconnected from the original one, which characterizes class (D).

If the Seidov criterion is not fulfilled, then there are two more possibilities. In both cases, stable hybrid star configurations are present in the first branch. A maximum in the mass is reached then when the high density phase is present in the core of the hybrid star. Then, either there will be no further stable branch beyond that maximum mass, so the stable hybrid star configurations are all connected to the original branch (class C), or there appears another stable hybrid star branch beyond the maximum after an unstable sequence. Then there are two stable sequences with a high density phase, one connected to the original branch and one disconnected. So, both cases are realized in this last class (B). The latter class (B) is interesting insofar as it allows for twin star configurations, despite the fact that the Seidov criterion is not fulfilled.

Figure 9.7 illustrates the four different ABCD classes in the plane of the two parameters characterizing the phase transition. The straight line marks the Seidov criterion. Above the line, the Seidov criterion is fulfilled, below the line, it is violated. For the two classes A and D to be found above the Seidov line, one sees that a kink in the mass–radius relation is present that becomes unstable afterward. Class A is located at high transitional pressures, while class D is to be found at lower values of the transitional pressure. The absence of the hybrid branch in class

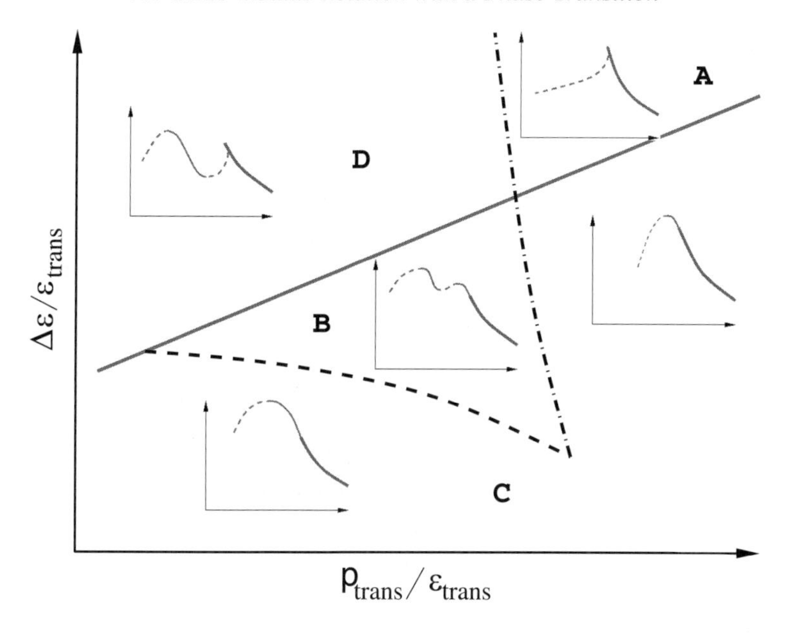

Figure 9.7 Classification of hybrid star mass–radius configurations in the plane of the jump in energy density and the critical pressure relative to the critical energy density. The four classes are: (A) absent hybrid branch, (B) both connected and disconnected hybrid branches, (C) connected hybrid branch, and (D) disconnected hybrid branch. Reprinted figure with permission from Alford et al. (2015). Copyright (2015) by the American Physical Society

A originates from the fact that the transition appears too close to the maximum mass configuration, so a new stable branch cannot be established anymore. The remarkable class B appears at intermediate values of the transitional pressure and the jump in energy density just below the Seidov line. Hence, the sequences of compact star configurations with central pressures slightly above the transition pressure will be just about stable. As the parameters of the phase transition are close to the Seidov line, the mass increases only slightly with the central pressure and the mass–radius relation becomes unstable with a small increase in the central pressure. If the jump in energy density is just about right, a high-density core develops with increasing central pressure, which then is able to stabilize the hybrid star configurations again.

9.4 Mass–Radius Relation with a Phase Transition

We have seen that the mass–radius relation for hybrid stars can show interesting patterns. Most notably, the existence of a new, third stable branch for compact stars appears for a pronounced phase transition in dense matter. As discussed in the previous section, the EOS producing such a new solution of the TOV equation has a jump in the energy density at some critical transition pressure. Such a jump occurs only for a Maxwell construction. For a Gibbs construction, a mixed phase will appear so that the jump is smoothened. Interestingly, the third family of compact stars has been seen also for a Gibbs construction, see Glendenning and Kettner (2000) and Schertler et al. (2000). Hence, the form of the EOS needed to produce a third family of compact stars must only have a sudden softening of the EOS at some transition pressure, accompanied by a sudden stiffening at higher density, see for example, Alvarez-Castillo and Blaschke (2015). We stress again that gravity is not sensitive to the particle composition, so the underlying microscopic origin of such a pattern of a sudden softening accompanied by a sudden stiffening cannot be read off from the mass–radius relation. However, the extraction of a nontrivial behavior in the EOS from the observation of an unusual pattern in the mass–radius diagram of neutron stars would be thought-provoking to say the least.

If one is lucky, the measurement of the radius and the mass from a few neutron stars with a sufficient degree of precision might be enough to find two separate branches in the mass–radius diagram. However, there is also the opportunity that the measurement of the mass and the radius of just one neutron star might be enough, which we outline now.

In Figure 9.2 we considered the region in the mass–radius relation for hybrid stars where the known low-density EOS is smoothly connected to the high-density EOS of pQCD. The two additional constraints were that the maximum mass has to be at least two solar masses and that the EOS is causal for all energy densities. Looking at the ABCD classifications of hybrid stars, the constraint on the region in the mass–radius relation applies to hybrid stars of class C only, that is, for configurations without a third family. We drop now the constraint that there is a smooth EOS and allow for sudden changes including a jump in the energy density following the procedure outlined by Tews et al. (2018, 2019).

The constraint from causality gives us a limit on the stiffness of the EOS and a limit on the maximum mass. To get the most massive configuration, we switch from the low-density EOS to the Zel'dovich EOS, with the speed of sound of $c_s^2 = 1$. This sudden switch will not be smooth, but will produce a kink in the relation of the pressure as a function of the energy density. The attentive reader might notice that such a type of EOS has been discussed before in Section 4.3.4. The resulting limit on the maximum mass is the Rhoades–Ruffini mass limit. The corresponding curve in the mass–radius diagram will set an upper bound on the radius for a given mass.

The lower limit in the radius for a given mass is given by the softest possible EOS, which is compatible with the two solar mass constraint on the maximum mass. The softest possible EOS is one with the lowest value for the speed of sound, that is, $c_s^2 = 0$. Matching such an EOS to the low-density EOS implies a

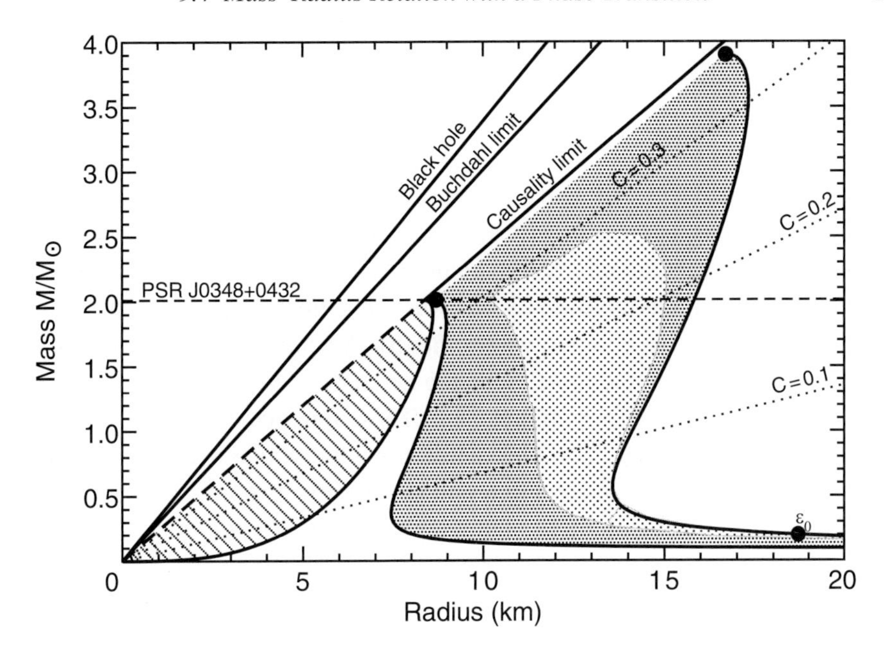

Figure 9.8 Mass-radius diagram for hybrid stars with a phase transition. The dark shaded region is allowed by causality, the pure neutron matter EOS, and for a maximum mass of at least $2M_{\odot}$. For comparison, the light shaded region shows the case for smooth EOSs without a phase transition, taken from Figure 9.2.

jump in the energy density at constant pressure. In order to stabilize the hybrid star configuration, the EOS has to stiffen at higher energy densities. Hence, we encounter the situation for producing a third family of compact stars. We know that the optimum choice for producing a third stable branch in the mass–radius diagram is to switch suddenly to an EOS with $c_s^2 = 1$. Hence, we use the constant-sound-speed parameterization for finding the most compact configuration. The transition pressure is given by the lowest bound of the pressure at the fiducial energy density of the pure neutron matter EOS. The jump in the energy density is chosen such that the maximum mass is equal to two solar masses. The resulting mass–radius relation constitutes the lower bound on the radius for a given mass for hybrid stars with a phase transition.

Figure 9.8 shows the mass–radius diagram for hybrid stars, allowing for a phase transition in the EOS, generalizing Figure 9.2 for hybrid stars with a smooth EOS. On the left side of the figure, several limiting lines are shown: the limit from general relativity for nonrotating black holes with a compactness of $C = 1/2$, the Buchdahl limit for incompressible fluids of $C = 4/9$, and the limit from the causal EOS of $C = 0.354$. All compact star configurations with a causal EOS have to be located to the right of this latter line. The dashed region excludes combinations of mass and radius for compact stars with causal EOSs, taking into account the two solar mass constraint from pulsar mass measurements. The light shaded region marks the combination of mass and radius for hybrid stars with an EOS smoothly connected to the low density EOS and that of pQCD copied from Figure 9.2.

The dark shaded region delineates the combinations of mass and radius bounded by the two limiting curves for EOSs with a phase transition. The left one is calculated for a constant-sound-speed parametrization with a jump in energy density tuned to arrive at a maximum mass of two solar masses. One sees that the mass–radius curve approaches at high masses that for the causal limit. The radius of the maximum mass configuration of $R = 8.6\,\mathrm{km}$ is very close to that for the causal limit at two solar masses, which is 8.3 km. The difference in the two mass–radius curves is just the outer layer of pure neutron matter, which becomes increasingly smaller for the more massive configurations. The effect is similar to the one seen for selfbound stars without a crust and with a crust shown in Figure 8.4, where the two mass–radius curves also approach each other with increasing masses.

The right limiting curve of the hybrid star configurations with a phase transition is that of Rhoades and Ruffini, now taking as fiducial energy density the one of pure neutron matter. The calculated maximum mass turns out to be $3.9M_\odot$ with a radius of 16.7 km and a central energy density of $3.6\varepsilon_0$. The value of the maximum mass is close to the Rhoades–Ruffini mass limit of $4.2M_\odot$ for vanishing fiducial pressure, see Eq. (4.74). The slightly lower maximum mass originates from the outer layer of pure neutron matter, which gives a small contribution to the pressure but a comparably larger one to the energy density.

One sees that the upper bound of the dark shaded region for the hybrid star configurations with a phase transition is close to the causality limit. One also realizes that the region in the mass–radius diagram for hybrid stars with a phase transition is considerably enlarged compared to that of hybrid stars without a phase transition. According to our analysis, the observation of a neutron star with a mass and radius outside the light shaded region but within the dark shaded region would imply that there exists a sudden change in the EOS of dense matter.

Exercises

(9.1) Derive an estimate for the surface tension of nuclei from the Bethe–Weizsäcker mass formula, Eq. (7.13), and the relation for the radius of nuclei, Eq. (7.14).

(9.2) Derive the TOV equation for a small perturbation of the pressure $\Pi(r)$, as outlined when deriving the Seidov criterion. Verify that the solution of the differential equation is given by $\Pi(r) = A + B/r$.

(9.3) Derive the criterion for an instability in the mass–radius diagram for a jump in the energy density for the Newtonian limit. Follow the steps outlined in the derivation of the Seidov criterion.

(9.4) Consider the case of the observation of a neutron star with a mass of $2.4M_\odot$. Calculate the implications for the maximum central energy density possible and the minimum radius of the neutron star using the scaling relations for the causal EOS with $c_s^2 = 1$.

(9.5) Assume that the low-density EOS is known up to $2\varepsilon_{\rm nm}$. What are the implications for the maximum mass possible for hybrid stars, its radius, and the central energy density? Use scaling relations and neglect effects from the transition pressure.

10

Gravitational Waves

Gravitational waves are ripples in space-time, a solution of the Einstein equations in vacuum in linearized theory. They share features of the electromagnetic waves but have properties that make them quite unique. The existence of gravitational waves was among the first predictions of general relativity made by Einstein himself in 1916 (Einstein, 1916b) and in a subsequent paper in 1918 (Einstein, 1918), where he corrected a 'regrettable' error of the previous work. It took nearly 100 years to confirm Einstein's famous last-to-be-confirmed prediction of general relativity by the direct detection of gravitational waves. The first empirical evidence for gravitational waves came from the measurement of a decrease of the orbital period of the Hulse–Taylor pulsar PSR B1913+16, which matched the prediction of general relativity (Taylor et al., 1979). The first direct gravitational wave event measured was GW150914 on September 14, 2015, observed by the two gravitational wave detectors of LIGO (Laser Interferometer Gravitational-Wave Observatory) at Hanford, WA, and at Livingston, LA (Abbott et al., 2016a). It constitutes also the first observation of the merger of two black holes. The analysis of the wave pattern revealed that the two black holes had masses of $36^{+5}_{-4}M_{\odot}$ and $29^{+4}_{-4}M_{\odot}$. The final black hole mass is found to be $62^{+4}_{-4}M_{\odot}$, so an energy equivalent to the missing mass of a whopping $3.0^{+0.5}_{-0.5}M_{\odot}$ has been released by the emission of gravitational waves from a distance of 410^{+160}_{-180} Mpc. For comparison, the energy released by the Sun in its entire lifetime will be of the order of $10^{-3}M_{\odot}$ and the total energy released by a supernova is of the order of $10^{-1}M_{\odot}$, which, however, is emitted mainly by neutrinos. So, GW150914 was at that time by far the most energetic astrophysical event detected. Still, the wiggles of space-time were extremely tiny when the gravitational wave passed through Earth only noticed by the highly sensitive gravitational wave detectors. Several more black hole mergers have been detected by the LIGO scientific collaboration, which includes also the gravitational wave detectors Virgo at Pisa in Italy and GEO600 at Hannover in Germany. In 2020, the gravitational wave detector KAGRA in the Kamioka mine in Japan will join the measurement of future gravitational wave events. Thus, gravitational wave astronomy is born, opening a new window to observing the universe. A more appropriate statement would be that we are able to listen to the universe now.

For compact star physics, a milestone was the observation of the gravitational wave event GW170817 (Abbott et al., 2017b) with the near-simultaneous detection of the gamma-ray burst GRB170817A by the satellite missions Fermi and INTEGRAL (International Gamma-Ray Astrophysics Laboratory) (Abbott et al., 2017c) followed by an astronomical transient called AT2017gfo (Abbott et al., 2017d). The measurement of an astronomical event over the entire range of electromagnetic spectrum from radio to gamma-rays including the measurement of gravitational waves constitutes a prime example of multi-messenger astronomy. The supernova SN1987A is another one where electromagnetic signals as well as neutrinos have been detected. The event seen as GW170817 could be identified as the merger of two neutron stars with a total mass of $M_{\text{total}} = 2.74^{+0.04}_{-0.01} M_\odot$ at a distance of 40^{+8}_{-14}pc. The increase of the amplitude and the frequency as well as the timescale matched the predictions of general relativity for the inspiral phase. The later phase in particular the merger phase of the two neutron stars could not be observed as the frequency was to high to be seen with the present sensitivity of LIGO. The gamma-ray burst GRB170817A was detected 1.7s after the gravitational wave passed Earth and has been associated to GW170817A. The gamma-ray burst is compatible with a jet formed from the collapse of the merging neutron stars to a black hole. The electromagnetic transient AT2017gfo also supports the interpretation of a neutron star merger. Most notably, the measurement of the typical pattern of a so-called kilo-nova in the afterglow corroborates the production of heavy elements by the rapid capture of neutrons, dubbed r-process nucleosynthesis, from the neutron-rich environment of the neutron star merger. The kilo-nova is about a thousand times more energetic than a typical nova, hence the name, and not as bright as a supernova. It is powered by the radioactive decay of unstable heavy elements synthesized in the ejecta of the neutron star merger. According to our present understanding of the generation of elements, this is the dominant process for generating heavy elements, as gold, in the universe, including the ones found on Earth.

In this chapter, we will discuss the gravitational wave pattern of the merger of two compact objects, as black holes or neutron stars, as predicted from general relativity. We will also devise two constraints on the properties of neutron stars that can be extracted from the non-observation of gravitational waves emitted directly by single pulsars and the limit of the tidal deformability of neutron stars from the neutron star merger event GW170817. For these purposes, we start with describing gravitational waves in terms of linearized gravity. For more details we refer to the many excellent textbooks on that topic, as for example, the two volumes of Maggiore (2007, 2018).

10.1 Linearized Gravity

In linearized theory, one expands Einstein's equation around the flat Minkowski metric $\eta_{\mu\nu}$, keeping only the first term in the expansion, so the equations become linear in the perturbation from flat space-time. So the ansatz for the metric is

$$g_{\mu\nu} = \eta_{\mu\nu} + h_{\mu\nu} \quad \text{with} \quad |h_{\mu\nu}| \ll 1, \tag{10.1}$$

where $h_{\mu\nu}$ stands for the small space-time dependent deviation from flat space-time. The requirement of smallness of the components of $h_{\mu\nu}$ is, of course, not independent of the chosen coordinate system. It states that one imposes a reference frame, where the components of $h_{\mu\nu}$ are small. Let us look at coordinate transformations and see what we mean by that last statement.

General relativity is invariant under all possible coordinate transformations $x^\mu \rightarrow x'^\mu(x)$. It is the expression of gauge symmetry of general relativity that the physical results do not change by choosing a specific reference frame of coordinates. The transformation law for the components of the full metric is that of a tensor of rank two:

$$g'_{\mu\nu}(x') = \frac{\partial x^\kappa}{\partial x'^\mu}\frac{\partial x^\lambda}{\partial x'^\nu} g_{\kappa\lambda}(x), \tag{10.2}$$

which holds for any coordinate transformation. In linearized theory, there is a special coordinate transformation, which does not change the relation of equation (10.1):

$$x'^\mu = x^\mu + \xi^\mu(x) \quad \text{with} \quad |\partial_\mu \xi_\nu| \ll 1. \tag{10.3}$$

The change of the full metric under this transformation results in the change for the perturbation $h_{\mu\nu}$:

$$h'_{\mu\nu}(x') = h_{\mu\nu}(x) - \left(\partial_\mu \xi_\nu + \partial_\nu \xi_\mu\right), \tag{10.4}$$

when keeping only the first-order terms, which you are asked to verify in the exercises. We see now that the requirement for the transformation that $|\partial_\mu \xi_\nu| \ll 1$ ensures that the transformed perturbation stays small, that is, $|h'_{\mu\nu}| \ll 1$. So the form of the defining relation for $h_{\mu\nu}$, Eq. (10.1), remains intact for the transformation, Eq. (10.4). This property is the residual gauge symmetry of the linearized theory. Note that the analog of Eq. (10.4) in electromagnetism is $A'^\mu = A^\mu - \partial^\mu \theta$.

For the metric in linearized gravity, one can compute now the Einstein equations in vacuum. To simplify the expressions, one introduces the quantity

$$\bar{h}_{\mu\nu} = h_{\mu\nu} - \frac{1}{2} \cdot \eta_{\mu\nu} \quad \text{with} \quad h = \eta^{\kappa\lambda} h_{\kappa\lambda}. \tag{10.5}$$

Note that in linearized theory, indices are lowered and raised with the metric of flat space-time $\eta_{\mu\nu}$ as the use of the full metric would involve higher order terms beyond linear order. Hence, h being the trace of $h_{\mu\nu}$ is defined by using $\eta_{\mu\nu}$. The Einstein equation in linearized theory then reads

$$\Box\bar{h}_{\mu\nu} + \eta_{\mu\nu}\partial^\kappa\partial^\lambda\bar{h}_{\kappa\lambda} - \partial^\kappa\partial_\nu\bar{h}_{\mu\kappa} - \partial^\kappa\partial_\mu\bar{h}_{\nu\kappa} = -16\pi G T_{\mu\nu}. \tag{10.6}$$

Recalling the analog situation in electromagnetism, one can choose a specific gauge for the electromagnetic field A^μ to simplify the equation of motion, the Lorenz gauge $\partial_\mu A^\mu = 0$. The corresponding choice for the gravitational field is then the Lorenz gauge in the form

$$\partial^\nu\bar{h}_{\mu\nu} = 0. \tag{10.7}$$

Plugging the Lorenz gauge into the equation of motion, we arrive at the equation for gravitational waves in linearized theory in the simple form

$$\Box\bar{h}_{\mu\nu} = -16\pi G T_{\mu\nu}. \tag{10.8}$$

Here, the d'Alembert operator is that of flat space-time $\Box = \partial_\mu\partial^\mu$. As $g_{\mu\nu}$ is symmetric in the indices, $\bar{h}_{\mu\nu}$ is also symmetric in the indices and has ten relevant components. The Lorenz gauge gives four extra conditions, so $\bar{h}_{\mu\nu}$ has six independent components in general.

In vacuum, the wave equation reduces to

$$\Box\bar{h}_{\mu\nu} = 0 \tag{10.9}$$

in analogy to electromagnetism where the wave equation reads $\Box A_\mu = 0$. A plane wave solution with the ansatz

$$h_{\mu\nu} \propto \epsilon_{\mu\nu}\exp(ik_\lambda x^\lambda) \quad \text{with} \quad k_\lambda = (\omega, \boldsymbol{k}), \tag{10.10}$$

where $\epsilon_{\mu\nu}$ is the polarization tensor, results in the energy–momentum relation for gravitational waves in the form

$$-\omega^2 + \boldsymbol{k}^2 = 0 \quad \text{or} \quad \omega = |\boldsymbol{k}|. \tag{10.11}$$

We note that the energy–momentum relation is that of massless particles propagating at the speed of light. The corresponding particles of gravity in analogy to photons for electromagnetism are called gravitons. So gravitons are massless and propagate at the speed of light, similar to photons.

Let us count now the degrees of freedom of gravitons. In vacuum, the residual gauge symmetry, Eq. (10.4), gives four additional constraints. In fact, a residual gauge transformation does not change the Lorenz gauge condition if

$$\begin{aligned}\partial^\nu \bar{h}'_{\mu\nu} &= \partial^\nu \bar{h}_{\mu\nu} - (\partial^\nu\partial_\mu\xi_\nu + \partial^\nu\partial_\nu\xi_\mu - \partial_\mu\partial_\kappa\xi^\kappa) \\ &= \partial^\nu \bar{h}_{\mu\nu} - \Box\xi_\mu \\ &= \partial^\nu \bar{h}_{\mu\nu},\end{aligned} \tag{10.12}$$

that is, for residual gauge transformations fulfilling the condition $\Box\xi_\mu = 0$. So, one can choose ξ_μ to give four additional conditions on $\bar{h}_{\mu\nu}$ so that one is left with only two independent components for $\bar{h}_{\mu\nu}$. The standard choice is to set the trace $\bar{h} = 0$ by fixing ξ_0, so $\bar{h}_{\mu\nu} = h_{\mu\nu}$, and to set $h_{0i} = 0$ by fixing ξ_i. The Lorenz condition for the time-like component $\mu = 0$ results in

$$\partial^0 h_{00} + \partial^i h_{0i} = \partial^0 h_{00} = 0, \tag{10.13}$$

where we used that $h_{0i} = 0$. One sets then $h_{00} = 0$. The Lorenz condition for the space-like components $\mu = i$ gives another three conditions that read

$$\partial^\nu h_{i\nu} = \partial^j h_{ij} = 0, \tag{10.14}$$

where we used again that $h_{0i} = 0$. In summary, one has the following eight conditions for gravitational waves in vacuum

$$h_{0\mu} = 0 \qquad h = h^i_i = 0 \qquad \partial^j h_{ij} = 0, \tag{10.15}$$

which is called the transverse-traceless gauge, or the TT gauge for short, with the notation h^{TT}_{ij} for the gravitational wave in the TT gauge. Note that the plane wave in the TT gauge h^{TT}_{ij} is transverse to the direction of propagation as

$$\partial^j h_{ij} = 0 \quad \to \quad n^j h_{ij} = 0 \quad \text{with} \quad \hat{n} = \boldsymbol{k}/|\boldsymbol{k}|. \tag{10.16}$$

The solution of the wave equation, Eq. (10.9), in the TT gauge can be written in the form of a plane wave

$$h^{\mathrm{TT}}_{ij}(t,z) = \begin{pmatrix} h_+ & h_\times & 0 \\ h_\times & -h_+ & 0 \\ 0 & 0 & 0 \end{pmatrix}_{ij} \cos[\omega(t-z)] \tag{10.17}$$

for choosing the direction of propagation $\hat{n}$ along the z-axis. The two independent components h_+ and $h_\times$ denote the two polarizations of the gravitational wave. Note that the gravitational wave is described by a tensor of rank 2. Correspondingly, the graviton has a spin equal to two $s = 2$. For comparison, the photon is described by a vector and its spin is equal to one $s = 1$. Also, the photon has two independent components corresponding to two degrees of freedom for the polarization. In fact, this is generic. Massless particles with a nonzero spin have only two independent degrees of freedom for the polarizations. These two states are the projection of the spin along or in the opposite direction of propagation and are called helicity states. Photons have the two helicity states ± 1 and gravitons have the two helicity states ± 2.

Let us look at the behavior of test masses under the influence of a gravitational wave passing by. We consider a test mass at $\tau = 0$ initially at rest in the TT gauge. The geodesic equation reads

$$\frac{\mathrm{d}^2 x^i}{\mathrm{d}\tau^2} = -\left[\Gamma^i_{\nu\rho}\frac{\mathrm{d}x^\nu}{\mathrm{d}\tau}\frac{\mathrm{d}x^\rho}{\mathrm{d}\tau}\right]_{\tau=0} = -\left[\Gamma^i_{00}\frac{\mathrm{d}x^0}{\mathrm{d}\tau}\frac{\mathrm{d}x^0}{\mathrm{d}\tau}\right]_{\tau=0}, \tag{10.18}$$

as initially $\mathrm{d}x^i/\mathrm{d}\tau = 0$. The Christoffel symbols to linear order in the perturbation are given by

$$\Gamma^\mu_{\nu\rho} = \frac{1}{2}\eta^{\mu\sigma}\left(\partial_\nu h_{\rho\sigma} + \partial_\rho h_{\nu\sigma} - \partial_\sigma h_{\nu\rho}\right), \tag{10.19}$$

so in the TT gauge

$$\Gamma^i_{00} = \frac{1}{2}\eta^{ij}\left(2\partial_0 h_{0j} - \partial_j h_{00}\right) = 0. \tag{10.20}$$

The vanishing Christoffel symbol Γ^i_{00} implies that the acceleration of the test particle vanishes. Hence, the test particle remains at rest, the spatial position in the chosen space-time coordinates do not change. The same will be true for the distance between two test masses, so a gravitational wave will not change the coordinate distance between two test masses. However, the proper distance does change between two test masses. The proper distance in x-direction at equal times is given by the metric by using Eq. (10.17) as

$$\mathrm{d}s^2 = \left[1 + h_+^{TT}\cos(\omega t)\right]\mathrm{d}x^2. \tag{10.21}$$

For two test masses being a length L apart, the integrated proper distance is

$$s = L\cdot\left[1 + h_+^{TT}\cos(\omega t)\right]^{1/2} \approx L\cdot\left[1 + \frac{1}{2}h_+^{TT}\cos(\omega t)\right]. \tag{10.22}$$

Hence, the proper distance changes periodically in time when a gravitational wave passes through two test masses. The proper distance can be measured by the travel time of a laser beam running between the two test masses, which is the basic principle of gravitational wave detectors such as LIGO, VIRGO, GEO600, and KAGRA.

10.2 Production of Gravitational Waves

10.2.1 Energy of Gravitational Waves

In order to measure the energy of gravitational waves, one needs to separate gravitational waves from a background so that a gauge transformation, that is, a change of coordinates, does not change the result. As such, the curvature or change of the background can be distinguished by a separation of scales. Gravitational waves with a wavelength λ much shorter than the characteristic curvature scale of the background L_B or a frequency f much larger than the characteristic timescale of the change of the background f_B can be measured so that it cannot be gauged away by a coordinate transformation. This separation can be achieved by averaging over a spatial volume with length scale l larger than the wavelength but smaller than the characteristic curvature scale. Similarly, one can also average for a timescale that is between the two timescales of the gravitational wave and that of the background. In the following, we implicitly assume that such an averaging is performed for measuring gravitational waves.

As the electromagnetic waves carry energy, so do gravitational waves. In electromagnetism, the energy–momentum tensor is proportional to the derivative of the electromagnetic field $\partial_\mu A_\nu$ squared. We expect a similar pattern for gravitational waves. From Einstein's equation, we know that the energy–momentum tensor is proportional to the Einstein tensor. This expectation bodes well with the fact that the Einstein tensor as well as the Riemann tensor for gravitational waves depends on partial derivatives of the perturbation $h_{\mu\nu}$ to second order. All we have to do now is to sort out the right order of the indices of the partial derivatives and the perturbation tensor. Within the Lorenz gauge, repeated indices of the partial derivative with one of the indices of the perturbation tensor will vanish. Let us also choose the coordinate system so that the trace of $h_{\mu\nu}$ vanishes. Hence, the only combination that remains is where the indices of the two perturbation tensors are repeated. In fact, the energy–momentum tensor of gravitational waves turns out to be

$$t_{\mu\nu} = \frac{1}{32\pi G} \left\langle \partial_\mu h_{\alpha\beta} \partial_\nu h^{\alpha\beta} \right\rangle, \tag{10.23}$$

where the brackets stand for a proper averaging, either over a properly chosen length scale or over a properly chosen timescale. In the exercises, you are asked to show that this expression is indeed gauge invariant. We are then free to choose the TT gauge. The energy density of gravitational waves reads then

$$\frac{\mathrm{d}E}{\mathrm{d}t\,\mathrm{d}A} = t^{00} = \frac{1}{32\pi G} \left\langle \dot{h}_{ij}^{TT} \dot{h}_{ij}^{TT} \right\rangle. \tag{10.24}$$

The energy passing through a unit area per unit time $\mathrm{d}A = r^2 \mathrm{d}\Omega$, where $\mathrm{d}\Omega$ is the unit steradian, is proportional to the energy density of the gravitational wave.

The power is defined by the energy emitted per unit time, which is given by integrating over the area A as

$$P_{\rm gw} = \frac{{\rm d}E}{{\rm d}t} = \int {\rm d}A\, t^{00} = \frac{r^2}{32\pi G} \int {\rm d}\Omega \left\langle \dot{h}_{ij}^{TT} \dot{h}_{ij}^{TT} \right\rangle. \tag{10.25}$$

Note that a gravitational wave amplitude traveling radially outward has the form

$$h_{ij}^{TT} = \frac{1}{r} f_{ij}(t-r) \tag{10.26}$$

as the solution to the wave equation in spherical coordinates. This result is in close analogy to an electromagnetic wave, with $t_{\rm ret} = t - r$ being the retarded time. The power emitted will then be independent of the radial distance r, as the factors of r cancel out when inserting Eq. (10.26) in to Eq. (10.25).

10.2.2 Einstein's Quadrupole Formula

Weak gravitational fields involve typically also small velocities. For a gravitationally bound system, the virial theorem results in the equality $E_{\rm kin} = -U/2$ where U is the Newtonian potential. Hence, for two masses

$$\frac{1}{2}\mu v^2 = \frac{1}{2}\frac{G\mu m}{d} \quad \text{or} \quad v^2 = \frac{R_s}{2d}, \tag{10.27}$$

where μ is the reduced mass, m is the total mass of the system and d is the distance between the two masses. For weak fields with $R_S/d \ll 1$, one sees that $v \ll 1$. For the angular frequency of the source ω_s, it follows that $\omega_s = v/d$, so a typical wavelength of the emitted wave with $\lambda \sim 1/\omega_s = d/v \gg d$ is much larger than the source size. Hence, one can expand the emission in multipole moments and the lowest multipole moments will be the dominant one, similarly for electromagnetic waves.

In the following, we consider a source of gravitational waves to the lowest order in the multipole expansion. Also, we take the distance to the source to be much larger than the size of the source $r \gg d$. The general solution of the wave equation, Eq. (10.8), with a source term can be written down as

$$\bar{h}_{\mu\nu}(t, \boldsymbol{x}) = 4G \int {\rm d}^3x' \frac{1}{|\boldsymbol{x} - \boldsymbol{x}'|} T_{\mu\nu}(t - |\boldsymbol{x} - \boldsymbol{x}'|, \boldsymbol{x}') \tag{10.28}$$

in analogy to the wave equation for electromagnetism where the energy-momentum tensor is replaced by the charge current as the source term. Note that the integration runs only over the source, so the integration over x' is bounded by the source size. For $r \gg d$, we can approximate $|\boldsymbol{x} - \boldsymbol{x}'| \approx |\boldsymbol{x}| = r$ and the integral simplifies to

$$\bar{h}_{\mu\nu}(t,\boldsymbol{x}) = \frac{4G}{r} \int \mathrm{d}^3x'\, T_{\mu\nu}(t - r, \boldsymbol{x}'). \tag{10.29}$$

Let us consider now the first three mass moments of the source, which we define as

$$M(t) = \int \mathrm{d}^3x'\, T^{00}(t,\boldsymbol{x}') \tag{10.30}$$

$$M^i(t) = \int \mathrm{d}^3x'\, T^{00}(t,\boldsymbol{x}') \cdot x'^i \tag{10.31}$$

$$M^{ij}(t) = \int \mathrm{d}^3x'\, T^{00}(t,\boldsymbol{x}') \cdot x'^i x'^j. \tag{10.32}$$

Using energy–momentum conservation and partial integration with vanishing boundary terms, one can show that

$$\dot{M} = 0 \qquad \ddot{M}^i = 0, \tag{10.33}$$

which can be interpreted as mass and momentum conservation. For the second mass moment, one finds that

$$\ddot{M}^{ij} = 2 \int \mathrm{d}^3x'\, T^{ij}(t,\boldsymbol{x}') \tag{10.34}$$

by consecutive use of energy–momentum conservation and partial integration. So, we can rewrite the spatial parts of the equation for the produced gravitational wave amplitude as

$$\bar{h}_{ij}(t,\boldsymbol{x}) = \frac{2G}{r} \ddot{M}_{ij}(t - r). \tag{10.35}$$

For a gravitational wave far away from the source, we can take the TT gauge, so only the spatial components of the energy–momentum tensor have to be considered under the integral. Also, only the traceless part of the mass moment will enter the expression for the gravitational wave amplitude. We introduce the corresponding traceless quantity as the quadrupole moment

$$Q_{ij} = M_{ij} - \frac{1}{3} M_{kk} \delta_{ij}. \tag{10.36}$$

We transform the expression for the gravitational wave amplitude to the TT gauge by multiplying both sides with a projection operator so that

$$h_{ij}^{TT}(t,\boldsymbol{x}) = \Lambda_{ij,kl} h_{kl}(t,\boldsymbol{x}). \tag{10.37}$$

Here, $\Lambda_{ij,kl}$ is the projection operator that projects a tensor to its transverse-traceless form. The explicit expression for the projection operator is

$$\Lambda_{ij,kl} = P_{ik}P_{jl} - \frac{1}{2}P_{ij}P_{kl} \quad \text{with} \quad P_{ij} = \delta_{ij} - n_i n_j, \tag{10.38}$$

where n is a unit vector pointing along $\boldsymbol{x}$. So, one gets for the gravitational wave amplitude in the TT gauge the formula

$$h_{ij}^{TT}(t, \boldsymbol{x}) = \frac{2G}{r}\Lambda_{ij,kl}\ddot{Q}_{kl}(t - r). \tag{10.39}$$

Finally, we are in the position to calculate the power emitted in gravitational waves by setting the result for the gravitational wave amplitude into Eq. (10.25)

$$P_{\text{gw}} = \frac{G}{8\pi}\int \mathrm{d}\Omega\, \Lambda_{ij,kl}\left\langle \dddot{Q}_{ij}\dddot{Q}_{kl}\right\rangle. \tag{10.40}$$

Note that the dependence on the distance r drops out in the expression. Integrating over the angles, we arrive at Einstein's quadrupole formula for the power emitted in gravitational waves

$$P_{\text{gw}} = \frac{G}{5}\left\langle \dddot{Q}_{ij}\dddot{Q}_{ij}\right\rangle. \tag{10.41}$$

Let us recapitulate the steps leading to Einstein's quadrupole formula. Assuming wavelengths larger than the source size, the gravitational wave amplitude is proportional to the integral of the energy–momentum tensor over the source. In the TT gauge, the integral of the spatial components of the energy momentum tensor is rewritten as the integral over the second derivative of the traceless quadrupole moment of the mass density. The monopole and the dipole moment of the mass distribution vanish due to mass and momentum conservation. The power is proportional to the time derivative of the gravitational wave amplitude squared. Hence, the power emitted in gravitational waves is proportional to the triple time derivative of the quadrupole moment squared.

10.2.3 Detecting Gravitational Waves

LIGO has a design sensitivity of about $h \sim 10^{-23}$ for frequencies around 100 Hz (Abbott et al., 2016b). From Eq. (10.35), we can make simple order-of-magnitude estimates on the expected gravitational wave amplitude for gravitational wave detectors and see whether or not the gravitational wave can be measured with LIGO.

First, note that the gravitational wave amplitude decreases linearly with the distance of the source to the observer. Hence, an increase in the sensitivity of the gravitational wave detector by a factor two will increase the volume, which can be observed by a factor eight. This feature is different compared to electromagnetic observations, where the flux is measured instead, which decreases with the distance squared. So, an increase of the sensitivity of a telescope by factor two will increase the volume that can be observed only by a factor $\sqrt{8} \approx 2.8$.

In simple terms, the quadrupole moment can be approximated to be $Q \sim MR^2$, where M is the typical mass scale and R the typical length scale of the system. The gravitational wave amplitude according to Eq. (10.35) then can be estimated to be

$$h \sim \frac{2G}{r} MR^2\omega_s^2 \sim \frac{R_s}{r} v^2, \tag{10.42}$$

where $R_s = 2GM$ is the Schwarzschild radius for the typical mass scale and $v = \omega_s R$ the typical velocity of the system.

Let us look at two examples that will be discussed in the following sections in more detail. A rotating neutron star, a pulsar, with a period of about $f = 100\,\text{Hz}$ and a radius of about $R = 10\,\text{km}$ has a typical velocity at the surface of about $v = 2\pi f R \approx 0.02c$. Most of the pulsars known are located within our galaxy. For distances within our galaxy, one can adopt a canonical distance of 10 kpc. The estimate of the gravitational wave amplitude for a wobbling pulsar with a typical mass scale of $1M_\odot$ would be then

$$h \sim 4 \times 10^{-21}\epsilon \left(\frac{R_S}{3\text{ km}}\right)\left(\frac{10\text{ kpc}}{r}\right)\left(\frac{v}{0.02}\right)^2, \tag{10.43}$$

where ϵ measures the deviation from a perfect sphere, which is responsible for the gravitational wave emission, that is, the ellipticity of the rotating neutron star. The typical frequency would be the rotation frequency of pulsars, which is between a few Hertz to a few hundred Hertz. So, LIGO is able to probe gravitational wave emissions from pulsars and set limits on the ellipticity ϵ.

The second example is two neutron stars orbiting each other. The Hulse–Taylor pulsar system has an orbital period of about eight hours. Using Kepler's third law, the typical velocity is estimated to be

$$v = \omega_s R = \omega_s \left(\frac{GM}{\omega_s^2}\right)^{1/3} = (GM\omega_s)^{1/3} = (GM2\pi f)^{1/3} \approx 10^{-3} \tag{10.44}$$

for a mass scale of $M = 1M_\odot$. The gravitational wave amplitude from binary pulsars in our galaxy is then estimated to be about

$$h \sim 10^{-23}\left(\frac{R_S}{3\text{ km}}\right)\left(\frac{10\text{ kpc}}{r}\right)\left(\frac{v}{10^{-3}}\right)^2, \tag{10.45}$$

with a characteristic frequency of $f \sim 10^{-4}$ Hz for an orbital period of about a few hours. The amplitude is close to the limit of the sensitivity of LIGO. However, the frequency is too low to be seen by LIGO. The future space mission LISA

(Large Interferometer Space Antenna) can measure gravitational wave frequencies of 10^{-4} to 10^{-1} Hz and will be able to detect directly gravitational waves from binary neutron stars (Audley et al., 2017).

Merging neutron stars would emit gravitational waves with frequencies corresponding to the scale of the size of neutron stars, that is, of about a few hundred Hertz, as seen for the gravitational wave event GW170817. However, the estimated merger rate for our galaxy is only about 1 event per 5×10^4 years. So, one needs to be able to observe at the order of 10^5 galaxies to see one neutron star merger event per year. The closest galaxy cluster to our galaxy is the Virgo cluster, with 1,000 galaxies about 20 Mpc away. At a distance of about 100 Mpc lies the large Coma cluster of galaxies, with thousands of galaxies. So, our estimate for the gravitational wave amplitude for a neutron star merger with say $v \sim 0.1c$ (for $f = 100\,\text{Hz}$) at the distance of the Coma cluster comes out to be

$$h \sim 10^{-23} \left(\frac{R_S}{3\text{ km}}\right) \left(\frac{100\text{ Mpc}}{r}\right) \left(\frac{v}{0.1}\right)^2, \tag{10.46}$$

which is just with in reach to be measured with LIGO. Note that the source of the neutron star merger event GW170817 was located at a distance of about 40 Mpc, so it could be well measured with LIGO.

10.3 Ellipticity of Neutron Stars

We note a fundamental difference to electromagnetic waves, where the lowest moment for the power emitted is the second time derivative of the dipole squared, see Eq. (6.17). As charge is conserved, there is no monopole emission in electromagnetic waves. However, something like a conserved charged momentum does not exist, so the lowest moment for electromagnetic radiation is the dipole not the quadrupole. In Fourier space, the power emitted in electromagnetic dipole radiation is therefore proportional to the angular frequency to the fourth power, $P_{\text{em}} \propto \omega^4$. The power emitted in gravitational quadrupole radiation is proportional to the angular frequency to the sixth power, $P_{\text{gw}} \propto \omega^6$. We used these dependencies for the energy loss of pulsars, see Section 6.5.5, which result in a different braking index for pulsars for gravitational wave emission compared to electromagnetic emission.

In this subsection, we derive the energy loss from gravitational wave emission from pulsars using the quadrupole formula. We assume that the pulsar has a non-vanishing quadrupole moment by a small deviation from a perfect sphere as a 'mountain' on the surface of the pulsar. For simplicity, we will describe the mass distribution by an ellipsoid. With present gravitational wave detectors, the emission of gravitational waves from pulsars due to quadrupole radiation can be detected directly. In fact, strong limits on the ellipticity of pulsars have been derived from the non-observation of gravitational waves from known pulsars from LIGO (Abbott et al., 2019a, 2019b).

We start by recalling the definition of the moment of inertia tensor from classical mechanics

$$I_{\mathrm{ij}} = \int \mathrm{d}^3x \, \rho(x) \left(r^2\delta_{ij} - x_i x_j\right), \tag{10.47}$$

where $\rho(x)$ is the distribution of the mass density. We realize that the moment of inertia tensor is just the negative of the quadrupole moment defined in Eq. (10.36), that is, $I_{\mathrm{ij}} = -Q_{\mathrm{ij}}$, when inserting the definition of the second mass moment, Eq. (10.32). The moment of inertia can be diagonalized with three eigenvalues I_1, I_2, and I_3. Assuming an ellipsoid with a semiaxis of a along x, b along y, and c along z and constant mass density ρ, that is, for an incompressible fluid, the moment of inertia I_3 is given by

$$I_3 = \int \mathrm{d}^3x \, \rho(x) \left(x^2 + y^2\right) = abc\rho \int \mathrm{d}x'\mathrm{d}y'\mathrm{d}z' \, \left(a^2x'^2 + b^2y'^2\right), \tag{10.48}$$

where we set $x' = x/a$, $y' = y/b$, and $z' = z/c$. Using spherical coordinates, one gets

$$\begin{aligned} I_3 &= abc\rho \int_0^1 \mathrm{d}r' \, r'^2 \int_0^{2\pi} \mathrm{d}\phi \int_0^{\pi} \mathrm{d}\theta \sin\theta \, \left(a^2r'^2\cos^2\phi + b^2r'^2\sin^2\phi\right)\sin^2\theta \\ &= abc\rho \int_0^1 \mathrm{d}r' r'^4 \int_0^{2\pi} \mathrm{d}\phi \int_0^{\pi} \mathrm{d}\theta \sin^3\theta \, \left(a^2\cos^2\phi + b^2\sin^2\phi\right) \\ &= \frac{4\pi}{15}\rho abc(a^2 + b^2) = \frac{1}{5}M(a^2 + b^2), \end{aligned} \tag{10.49}$$

where $M = 4\pi abc\rho/3$ is the total mass of the ellipsoid.

Consider a neutron star rotating uniformly around the z-axis with an angular frequency ω_{rot}. The moment of inertia in the frame of the observer I' are related to that in the body frame I by the rotation matrix

$$R = \begin{pmatrix} \cos(\omega_{\mathrm{rot}}t) & \sin(\omega_{\mathrm{rot}}t) & 0 \\ -\sin(\omega_{\mathrm{rot}}t) & \cos(\omega_{\mathrm{rot}}t) & 0 \\ 0 & 0 & 1 \end{pmatrix}. \tag{10.50}$$

As the moment of inertia is a tensor of rank two, the transformation from the body frame to the observer's frame reads $I' = RIR^T$. In explicit form, the components of the transformed moment of inertia tensor are given by

$$I_{11} = I_1 \cos^2(\omega_{\rm rot} t) + I_2 \sin^2(\omega_{\rm rot} t)$$
$$= \frac{I_1 + I_2}{2} + \frac{I_1 - I_2}{2} \cos(2\omega_{\rm rot} t) \tag{10.51}$$
$$I_{12} = (I_2 - I_1) \sin(\omega_{\rm rot} t) \cos(\omega_{\rm rot} t)$$
$$= -\frac{I_1 - I_2}{2} \sin(2\omega_{\rm rot} t) \tag{10.52}$$
$$I_{22} = I_1 \sin^2(\omega_{\rm rot} t) + I_2 \cos^2(\omega_{\rm rot} t)$$
$$= \frac{I_1 + I_2}{2} - \frac{I_1 - I_2}{2} \cos(2\omega_{\rm rot} t), \tag{10.53}$$

where we used standard trigonometric identities. One realizes that the time-dependent arguments of the trigonometric functions depend on the characteristic angular frequency $\omega_{\rm gw} = 2\omega_{\rm rot}$. The gravitational wave amplitude depends on the second time derivative of the quadrupole moment or the negative of the moment of inertia. To project on the TT gauge for the gravitational wave amplitude, the transverse-traceless form of the quadrupole moment has to be taken by using

$$\Lambda_{ij,kl} \ddot{Q}_{kl} = \begin{pmatrix} (\ddot{Q}_{11} - \ddot{Q}_{22})/2 & \ddot{Q}_{12} & 0 \\ \ddot{Q}_{21} & -(\ddot{Q}_{11} - \ddot{Q}_{22})/2 & 0 \\ 0 & 0 & 0 \end{pmatrix}. \tag{10.54}$$

One can assign the h_+-amplitude to the diagonal components and the $h_\times$-amplitude to be the off-diagonal ones, as for plane waves, see Eq. (10.17). The two polarizations of the gravitational wave amplitude are then given by

$$h_+ = \frac{G}{r}(\ddot{Q}_{11} - \ddot{Q}_{22}) = \frac{G\omega_{\rm gw}^2}{r}(I_1 - I_2)\cos(\omega_{\rm gw} t) \tag{10.55}$$

$$h_\times = \frac{2G}{r}\ddot{Q}_{12} = \frac{G\omega_{\rm gw}^2}{r}(I_1 - I_2)\sin(\omega_{\rm gw} t). \tag{10.56}$$

One realizes that the gravitational wave amplitudes depend only on the difference of the moment of inertia $I_1 - I_2$. We introduce the ellipticity by defining $\epsilon = (I_1 - I_2)/I_3$, which measures the deviation of the ellipsoid from a perfect sphere. Note that $\epsilon \approx (b - a)/a$ for $b \approx a$. The order of magnitude of the typical moment of inertia of neutron stars can be estimated by taking typical values of the mass $M = 1.4 M_\odot$ and the radius $R = 10\,\rm km$. Assuming a sphere of constant density, the moment of inertia for a neutron star is about

$$I_3 \approx \frac{2}{5} M R^2 = 1 \times 10^{38}\ \rm kg\ m^2 \left(\frac{M}{1.4\, M_\odot}\right)\left(\frac{R}{10\,\rm km}\right)^2. \tag{10.57}$$

The gravitational wave amplitude can be rewritten in terms of the frequency of the gravitational wave $f_{\rm gw} = \omega_{\rm gw}/(2\pi)$ as

$$h_+ = h_0 \cos(2\pi f_{\rm gw} t) \qquad h_\times = h_0 \sin(2\pi f_{\rm gw} t), \tag{10.58}$$

with a typical overall magnitude for pulsars of

$$h_0 = \frac{4\pi^2 G}{r} I_3 f_{\rm gw}^2 \epsilon = 10^{-21} \epsilon \left(\frac{10\ {\rm kpc}}{r}\right) \left(\frac{I_3}{10^{38}\ {\rm kg\ m^2}}\right) \left(\frac{f_{\rm gw}}{100\ {\rm Hz}}\right)^2 . \tag{10.59}$$

Here we used most of the known pulsars are located within our galaxy, with a typical distance scale of $r = 10\,$kpc. For comparison, the distance of our Sun to the center of our galaxy is about 8 kpc. Also, we set as the typical rotation frequency of pulsar $f = 100\,$Hz. We find that the gravitational wave frequency emitted from wobbling pulsars should be twice the rotation frequency $f_{\rm gw} = 2 f_{\rm rot}$, corresponding to typical frequencies of a few hundred Hertz. Gravitational wave amplitudes with frequencies of a few hundred Hertz and of a magnitude of $h \sim 10^{-21}$ can be measured with the present sensitivity of LIGO. Observation of more than 200 known pulsars by LIGO shows no sign of gravitational wave emission (Abbott et al., 2019b). For each pulsar, the non-detection of gravitational waves can be cast into a limit of the ellipticity. The lowest limit found is reported to be $\epsilon \leq 2 \times 10^{-9}$ (Abbott et al., 2019a). For a radius of $R = 10\,$km, this limit corresponds to a maximum deviation from a perfect sphere on the level of only 20μm!

The power emitted in gravitational waves from wobbling pulsars is given by the quadrupole formula and reads

$$\frac{{\rm d}E_{\rm gw}}{{\rm d}t} = \frac{G}{5}\left\langle \dddot{Q}_{11}^2 + \dddot{Q}_{12}^2 + \dddot{Q}_{22}^2 + \dddot{Q}_{21}^2 \right\rangle = \frac{32G}{5}\epsilon^2 I_3^2 \omega_{\rm rot}^6 . \tag{10.60}$$

Using the characteristic value for I_3 from Eq. (10.57), one can set limits on the energy loss of pulsars via emission of gravitational waves. In several cases, LIGO is able to set limits below the observed energy loss of pulsars from the change of angular frequency, that is, below the spin-down limit of pulsars. The most stringent limit comes from the observation of the Crab pulsar, where the spin-down limit was beaten by

$$\frac{\dot{E}_{\rm gw}}{\dot{E}_{\rm spin\text{-}down}} \leq 0.017\%. \tag{10.61}$$

Hence, less than 0.017% of the energy lost by the Crab pulsar can come from the emission of gravitational waves due to a nonvanishing quadrupole moment. The second best limit is for the Vela pulsar, with a spin-down limit of 0.18% (Abbott et al., 2019b)

10.4 Neutron Star Mergers

10.4.1 The Chirp Mass

Consider two masses m_1 and m_2 in a circular orbit with a radius R and an angular frequency ω_s in the Newtonian limit. From the balance of the centrifugal force to the gravitational force, one gets Kepler's law

$$\frac{v^2}{R} = \frac{Gm}{R^2} \longrightarrow \omega_s^2 = \frac{Gm}{R^3}, \tag{10.62}$$

with the total mass $m = m_1 + m_2$ and the velocity v of the two masses. The reduced mass of the system is defined as $\mu = m_1 m_2/m$. From the virial theorem, the total orbital energy of the bound system is given by half the potential energy, that is,

$$E_{\text{orbit}} = -\frac{Gm_1m_2}{2R} = -\frac{G\mu m}{2R}. \tag{10.63}$$

Using Kepler's law to replace the radius R, one finds that

$$E_{\text{orbit}} = -\frac{G^{2/3}\mu m^{2/3}\omega_s^{2/3}}{2} = -\left(\frac{G^2 M_c^5 \omega_s^2}{8}\right)^{1/3}, \tag{10.64}$$

where we introduced the mass scale characterizing the system, the so-called chirp mass defined as

$$M_c = \mu^{3/5} m^{2/5} = \frac{(m_1 m_2)^{3/5}}{(m_1 + m_2)^{1/5}}. \tag{10.65}$$

We note that the chirp mass is the only mass scale determining the energy of a gravitationally bound binary system.

10.4.2 Gravitational Wave Emission from Binary Systems

We calculate now the gravitational wave amplitude from a binary system with masses m_1 and m_2 using Eq. (10.39). The recipe is as follows: set up the energy–momentum tensor, calculate the second mass moment, rewrite it in terms of the quadrupole moment, and take the second time derivative. We look at the system well before the two masses merge, so the masses move at nonrelativistic velocities.

The trajectory of the masses is described in the center-of-mass system as

$$x_1(t) = R \sin \omega_s t \qquad x_2(t) = R \cos \omega_s t \qquad x_3(t) = 0, \tag{10.66}$$

where we put the orbit in the $x_1 - x_2$ plane. We can treat the system so that there is a point mass with the reduced mass μ of the system orbiting a point mass with the total mass of the system m at the origin, see for example, Eq. (10.63). Using, for example, Eq. (2.96), the energy-momentum tensor can then be written as

$$T^{\mu\nu}(t,\boldsymbol{r}) = \frac{p^\mu p^\nu}{\mu}\delta^3(\boldsymbol{r} - \boldsymbol{r_0}(t)), \tag{10.67}$$

where $\boldsymbol{r_0} = (x_1(t), x_2(t), x_3(t))$. The 00-component is given by

$$T^{00}(t,\boldsymbol{r}) = \mu\delta^3(\boldsymbol{r} - \boldsymbol{r_0}(t)). \tag{10.68}$$

The second mass moment reads then

$$M_{ij}(t) = \int \mathrm{d}^3x\, x_i\, x_j\, T^{00}(t,\boldsymbol{r}) = \mu \int \mathrm{d}^3x\, x_i\, x_j\, \delta(\boldsymbol{r} - \boldsymbol{r_0}(t)) = \mu\, x_i(t)\, x_j(t). \tag{10.69}$$

The quadrupole moment is the traceless expression of the second mass moment

$$Q_{ij}(t) = \mu\left(x_i(t)x_j(t) - \frac{1}{3}R^2\delta_{ij}\right), \tag{10.70}$$

where $R = |\boldsymbol{r_0}|$ is the distance of the two masses, which is constant for a circular orbit. Hence, the quadrupole moment and the second mass moment differ only by a constant and we can also take the time derivative of the second mass moment instead of the time derivative of the quadrupole moment. The nonvanishing second mass moments are given by

$$M_{11} = \mu R^2 \sin^2(\omega_s t) = \frac{1}{2}\mu R^2(1 - \cos(2\omega_s t)) \tag{10.71}$$

$$M_{12} = M_{21} = \mu R^2 \sin(\omega_s t)\cos(\omega_s t) = \frac{1}{2}\mu R^2 \sin(2\omega_s t) \tag{10.72}$$

$$M_{22} = \mu R^2 \cos^2(\omega_s t) = \frac{1}{2}\mu R^2(1 + \cos(2\omega_s t)). \tag{10.73}$$

We notice that the characteristic angular frequency for gravitational wave emission will be twice the angular frequency of the orbit $\omega_{\text{gw}} = 2\omega_s$. The second time derivative of the quadrupole moment is

$$\ddot{Q}_{11} = -\ddot{Q}_{22} = \frac{1}{2}\mu R^2\omega_{\text{gw}}^2 \cos(\omega_{\text{gw}} t) \tag{10.74}$$

$$\ddot{Q}_{12} = \ddot{Q}_{21} = -\frac{1}{2}\mu R^2\omega_{\text{gw}}^2 \sin(\omega_{\text{gw}} t). \tag{10.75}$$

According to the quadrupole formula, the total power emitted in gravitational waves is given by

$$\frac{dE_{gw}}{dt} = \frac{G}{5}\left\langle \dddot{Q}_{11}^2 + \dddot{Q}_{12}^2 + \dddot{Q}_{22}^2 + \dddot{Q}_{21}^2 \right\rangle = \frac{32G}{5}\mu^2 R^4 \omega_s^6 = \frac{32}{5G}\left(\frac{GM_c\omega_{GW}}{2}\right)^{10/3}. \tag{10.76}$$

We see that the next-to-last expression could have been derived also by using the result for a wobbling neutron star, Eq. (10.60), by replacing the moment of inertia with the one for a binary system in the center-of-mass frame. In the last step, we rewrote the result in terms of the chirp mass.

The emission of gravitational waves takes away energy from the binary system. We equate the power emitted in gravitational waves to the loss of the binding energy per unit time of the binary system

$$\frac{dE_{gw}}{dt} = -\frac{dE_{orbit}}{dt}. \tag{10.77}$$

Using Eq. (10.64), one arrives at a differential equation for the change of the angular orbital frequency

$$\dot{\omega}_{GW} = \frac{12 \cdot 2^{1/3}}{5} (GM_c)^{5/3}\, \omega_{GW}^{11/3}. \tag{10.78}$$

The orbital angular frequency will increase with time, so the orbit will shrink with time. At some time t_{merge}, the two neutron stars will merge. One can readily solve the differential equation with the result

$$\omega_{GW}(\tau) = 2\left(\frac{5}{256\tau}\right)^{3/8} (GM_c)^{-5/8}, \tag{10.79}$$

where τ is the time to merge $\tau = t_{merge} - t$. Hence, as expected, the frequency of the emitted gravitational waves will increase with time. Note that the frequency will formally diverge at the time of the merger, that is, for $\tau \to 0$. In reality, our weak field approximation will break down before the two neutron stars touch each other. As the frequency increases drastically close to the merger, one can imagine that it sounds like the chirping of a bird. This feature is the underlying reason why the mass M_c appearing in the expression of the gravitational wave frequency is called the chirp mass.

The time to the merger can be calculated by inverting the result for the gravitational wave angular frequency, Eq. (10.79), in terms of the time to merge τ:

$$\tau = \frac{5}{256}\left(\frac{\omega_{GW}}{2}\right)^{-8/3} (GM_c)^{-5/3} = 9.83\ \text{Myr} \cdot \left(\frac{P_{orbit}}{1\ \text{h}}\right)^{8/3} \left(\frac{M_c}{M_\odot}\right)^{-5/3}. \tag{10.80}$$

Table 10.1 *Properties of some known double neutron star systems. Masses are given in units of* $M_\odot$. *Note that J0737-3039 is the double pulsar system where both neutron stars are seen as pulsars. (data taken from Tauris et al., 2017. © AAS. Reproduced with permission)*

Pulsar	P_{orbit} (days)	e	M_{pulsar}	M_{comp}	M_{total}	M_c	τ_{gw} (Myr)
J0737−3039	0.102	0.088	1.338	1.249	2.587	1.125	86
B1534+12	0.421	0.274	1.333	1.346	2.678	1.166	2,730
J1756−2251	0.320	0.181	1.341	1.230	2.570	1.118	1,660
J1906+0746	0.166	0.085	1.291	1.322	2.613	1.137	309
B1913+16	0.323	0.617	1.440	1.389	2.828	1.231	301
B2127+11C	0.335	0.681	1.358	1.354	2.713	1.180	217

This estimate gives timescales of several tens of millions of years as the typical time to merge for double neutron star systems, which have orbital periods in the range of a few hours, see Table 10.1.

There is a correction term for elliptic orbits, which has been calculated for a given eccentricity e by Peters and Mathews (1963). The emitted power will be enhanced by a factor

$$f(e) = \left(1 - e^2\right)^{-7/2} \left(1 + \frac{73}{24}e^2 + \frac{37}{96}e^4\right), \tag{10.81}$$

which will shorten the time to merge. As the binary system loses angular momentum during inspiral, the eccentricity will eventually be negligibly small at the time of the merger. So, one needs to integrate the equation of motion over time with a time-dependent eccentricity, which results in a non-analytic expression. The time to merge for elliptic orbits can be approximated by the formula

$$\tau_{\text{gw}} \approx \frac{5}{256}\left(\frac{P_{\text{orbit}}}{2\pi}\right)^{8/3} (GM_c)^{-5/3} \cdot (1 - e^2)^{7/2}, \tag{10.82}$$

where e is the initial eccentricity. For $e = 0$, the approximate formula gives the result for circular orbits, Eq. (10.80). Table 10.1 lists known double neutron star systems in our galaxy that will merge in less than 50 Gyr, see Tauris et al. (2017). The orbital period is in the range of a few hours. One notices that the masses of the double neutron star systems are quite similar. The range of the chirp masses is just $M_c = 1.1\text{–}1.2 M_\odot$. The time to merge is in the range of about a hundred million years to a few gigayears. The reader is asked to check the approximation for the time to merge with the one given in Table 10.1 in the exercises.

Let us use the power of logarithmic derivatives to calculate how the orbital radius R shrinks with time. From Kepler's law, we know that the orbital radius R and the orbital angular frequency ω_s are related to each other, as $\omega_s^2 \propto R^{-3}$, see Eq. (10.62). The solution for the gravitational wave frequency as a function of time can be written as $\mathrm{d}\omega_{\mathrm{GW}}/\omega_{\mathrm{GW}} = -3/8(\mathrm{d}\tau/\tau)$, see Eq. (10.78). Then, one gets for the logarithmic derivative

$$\frac{\mathrm{d}R}{R} = -\frac{2}{3}\frac{\mathrm{d}\omega_{\mathrm{GW}}}{\omega_{\mathrm{GW}}} = \frac{1}{4}\frac{\mathrm{d}\tau}{\tau}. \tag{10.83}$$

Integration of the differential equation for $R(\tau)$ gives $R(\tau) \propto \tau^{1/4}$. The radius will decrease slowly first and then decrease rapidly close to the time of the merger. The phase of the drastic drop of the radius is called the plunge phase.

Finally, we consider the time dependence of the gravitational wave amplitude as it appears in gravitational wave detectors. Using Eqs. (10.55) and (10.56) and the expressions for the second time derivative of the quadrupole moments, Eqs. (10.74) and (10.75), one finds that

$$h_+ = \frac{G}{r}(\ddot{Q}_{11} - \ddot{Q}_{22}) = \frac{G\omega_{\mathrm{gw}}^2}{r}\mu R^2 \cos(\omega_{\mathrm{gw}} t) \tag{10.84}$$

$$h_\times = \frac{2G}{r}\ddot{Q}_{12} = -\frac{G\omega_{\mathrm{gw}}^2}{r}\mu R^2 \sin(\omega_{\mathrm{gw}} t). \tag{10.85}$$

We see that the two polarizations have the same overall magnitude, so we can write them in the form

$$h_+ = h_0 \cos(\omega_{\mathrm{gw}} t) \qquad h_\times = -h_0 \sin(\omega_{\mathrm{gw}} t) \tag{10.86}$$

with

$$h_0 = \frac{G\omega_{\mathrm{gw}}^2}{r}\mu R^2 = \frac{2^{4/3}}{r}(GM_c)^{5/3}\omega_{\mathrm{GW}}^{2/3}. \tag{10.87}$$

In the second step, we used Kepler's law again. However, so far we ignored that ω_{GW} is time-dependent. So, actually, the argument $\omega_{\mathrm{GW}} t$ needs to be replaced by the integral of $\omega_{\mathrm{GW}}(t)$ over the time. The time derivative of ω_{GW} will introduce additional terms. In our approximation, we assume quasi-circular motion, so the change of the angular frequency is small and can be ignored in a first approximation. So, we use the solution for ω_{GW} from Eq. (10.79) and get for the overall amplitude factor of the emitted gravitational wave from binary systems

$$h_0(\tau) = \frac{5^{1/4}}{r}\left(\frac{1}{\tau}\right)^{1/4}(GM_c)^{5/4}. \tag{10.88}$$

Hence, the amplitude of the gravitational wave grows with time as $\tau^{-1/4}$. We note that the increase of the gravitational wave amplitude is inversely proportional to the decrease of the orbital radius R, as $R \propto \tau^{1/4}$. The expression diverges for the time of the merger $\tau \to 0$, which is an artefact of our approximations used. The only quantity needed to fix the amplitude is again the chirp mass M_c, not the individual masses or any other combination thereof.

In summary, the gravitational wave emitted from binary systems will have an increase in the frequency, Eq. (10.79), as well as an increase in the amplitude with time, Eq. (10.88). The gravitational wave signal will sound like a chirping bird, starting with a low amplitude at low frequency and increasing rapidly to a high amplitude with a high frequency. Accordingly, the only dimensionful parameter that enters for both, the frequency and the amplitude, deserves to be called the chirp mass M_c. This dependency is the reason that the chirp mass of the neutron star merger GW170817 could be well determined to be $M_c = 1.188^{+0.004}_{-0.002} M_\odot$, while the total mass of $M_{\text{total}} = 2.74^{+0.04}_{-0.01} M_\odot$ is less well known (Abbott et al., 2017b).

10.4.3 Tidal Deformability

The gravitational wave pattern discussed so far for the inspiral phase assumes that the neutron stars are point-like objects. In reality, of course, neutron stars have a finite size that will leave an imprint on the form of the gravitational wave even before the two neutron stars touch each other. In the following, we consider the change of the gravitational wave pattern during the inspiral due to effects from the deformation of neutron stars. As we will see, one can extract a parameter called the tidal deformability from the gravitational wave signal of the inspiral phase, which is strongly tied to the equation of state (EOS) of neutron stars. This connection was first pointed out by Flanagan and Hinderer (2008), see also Hinderer (2008).

Let us start with Newtonian gravity. Also, we separate the two neutron stars of the neutron star merger by looking at one neutron star being tidally deformed by the gravitational field of the other neutron star. Later on, we add the contribution from the second neutron star. So, consider a neutron star that is exposed to a time-dependent external tidal field that we expand in the radial coordinate $\boldsymbol{x}$:

$$\Phi_{\text{ext}}(\boldsymbol{x}) = \Phi_{\text{ext}}(0) + \frac{\partial \Phi_{\text{ext}}}{\partial x^i} x^i + \frac{1}{2} \frac{\partial^2 \Phi_{\text{ext}}}{\partial x^i \partial x^j} x^i x^j + \cdots \tag{10.89}$$

The first term is a constant, which we choose to be zero. The second term induces just a translation of the center of mass and can be set to zero in the baryocenter frame. So, we can write the external potential as

$$\Phi_{\text{ext}}(\boldsymbol{x}) = \frac{1}{2} \frac{\partial^2 \Phi_{\text{ext}}}{\partial x^i \partial x^j} x^i x^j + \cdots = \frac{1}{2} \varepsilon_{ij} x^i x^j + \cdots , \tag{10.90}$$

where ε_{ij} stands for the external quadrupolar tidal field. The external tidal field will generate a time-dependent perturbation in the mass density distribution of the neutron star. The perturbed gravitational potential of the neutron star can be calculated by using

$$\Phi_{\rm star}(\boldsymbol{x}) = -G \int \mathrm{d}^3x' \frac{\rho}{|\boldsymbol{x} - \boldsymbol{x}'|}. \tag{10.91}$$

We expand the denominator of the integrand in multipole moments

$$\frac{1}{|\boldsymbol{x} - \boldsymbol{x}'|} = \frac{1}{r} + \frac{1}{r^3} x_i x_i' + \frac{3}{2r^5}\left(x_i x_j - \frac{1}{3} r^2 \delta_{ij}\right) x_i' x_j' + \cdots, \tag{10.92}$$

where $r = |\boldsymbol{x}|$. Inserting the expansion in Eq. (10.91) and integrating gives the standard Newtonian potential from the first term in the expansion. The second term is the dipole term that vanishes in the integration. The third term is the quadrupole term, which gives the first correction to the Newtonian potential due to the second mass moment. As the quadrupole term projects on the traceless form of the mass density, the second mass moment can be replaced by the quadrupole moment of the mass density. So, one finds that the perturbed gravitational potential of the neutron star looks like

$$\begin{aligned}\Phi_{\rm star}(\boldsymbol{x}) &= -\frac{GM}{r} - \frac{3G}{2r^5}\left(x_i x_j - \frac{1}{3} r^2 \delta_{ij}\right) \int \mathrm{d}^3x' \rho\, x_i' x_j' \\ &= -\frac{GM}{r} - \frac{3G}{2r^5}\left(x_i x_j - \frac{1}{3} r^2 \delta_{ij}\right) Q_{ij},\end{aligned} \tag{10.93}$$

where M is the total mass of the neutron star and Q_{ij} the induced quadrupole moment of the neutron star. We define the tidal deformability λ as the proportionality coefficient between the external quadrupolar tidal field and the induced quadrupole moment of the neutron star:

$$Q_{ij} = -\lambda \varepsilon_{ij}. \tag{10.94}$$

Using this relation to replace the induced quadrupole moment by the external quadrupolar field, the total potential can be expressed as

$$\Phi = \Phi_{\rm ext} + \Phi_{\rm star} = -\frac{GM}{r} + \frac{1}{2}\varepsilon_{ij} x_i x_j \left(1 + \frac{3G}{r^5}\lambda\right) + \cdots \tag{10.95}$$

The tidal deformation in Newtonian theory was introduced by Love (1909) when studying the deformability of Earth and the Moon. He introduced coefficients of the tidal deformation for arbitrary values of the moment l as k_l. The coefficient for the tidal quadrupole moment is then the Love number k_2 connected to λ by the relation

$$k_2 = \frac{3G}{2R^5}\lambda, \tag{10.96}$$

where R is the radius of the neutron star. Note that k_2 is dimensionless. The prefactor $3/2$ cancels the corresponding prefactor for the quadrupole term in the multipole expansion of the neutron star potential. The perturbed metric coefficient g_{00} in the Newtonian approximation can then be written in terms of the Love number k_2 as

$$g_{00} = -(1 + 2\Phi) = -1 + \frac{2GM}{r} - \varepsilon_{ij}x_i x_j \left(1 + 2k_2 \left(\frac{R}{r}\right)^5\right) + \cdots \tag{10.97}$$

Let us switch now to the treatment in general relativity. Far outside the neutron star, the perturbed metric given is the asymptotic solution to which the full solution will be matched on. Thereby, one gets a connection between the Love number k_2 seen at the far outside with the property of the neutron star, that is, the EOS of neutron stars. We start with perturbing the full metric of space-time around the equilibrium metric from the solution of the Tolman–Oppenheimer–Volkoff equation

$$g_{\mu\nu} = g_{\mu\nu}^{(0)} + h_{\mu\nu}, \tag{10.98}$$

where $h_{\mu\nu}$ is the metric perturbation that we treat as a linearized theory of gravity to linear order. The ansatz for the metric perturbation adopts spherical harmonics appropriate for $l = 2$ perturbations, that is, spherical harmonics Y_{lm} with $l = 2$. As the energy–momentum tensor contains only diagonal terms, the perturbed metric will also be diagonal. The symmetry in the angle ϕ in spherical Schwarzschild coordinates will reduce the number of radial functions in the perturbations from four to three, so we can write the perturbed metric as

$$h_{\mu\nu} = \mathrm{diag}[-e^{\alpha(r)}H_0(r), e^{\beta(r)}H_2(r), r^2K(r), r^2\sin^2\theta K(r)]Y_{20}(\theta,\phi), \tag{10.99}$$

where $\alpha(r)$ and $\beta(r)$ are the metric functions from the equilibrium solution, that is, from the solution of the TOV equation. The perturbations of the energy–momentum tensor are given by

$$\delta T_0^0 = -\delta\varepsilon Y_{20}(\theta,\phi) \qquad \delta T_i^i = \delta p Y_{20}(\theta,\phi), \tag{10.100}$$

where $\delta\varepsilon$ and δp are the perturbations in the energy density and the pressure, respectively. These perturbations are related to each other by the relations

$$\delta p = \frac{\mathrm{d}p}{\mathrm{d}\varepsilon}\delta\varepsilon = c_s^2\delta\varepsilon, \tag{10.101}$$

where c_s is the energy density dependent speed of sound of the neutron star matter. Inserting the ansatz for the perturbed metric and the perturbed energy–momentum tensor in the perturbed Einstein equations

$$\delta G_\nu^\mu = 8\pi G\delta T_\nu^\mu \tag{10.102}$$

results in the following relations for the metric functions of the perturbed metric, see Hinderer (2008) and Hinderer et al. (2010):

$$-H_2(r) = H_0(r) = H(r) \qquad K'(r) = H'(r) + 2H(r)\alpha'(r), \tag{10.103}$$

where the prime denotes a derivative with respect to the radial Schwarzschild coordinate r. So, one is left with one radial function $H(r)$, which describes the tidal perturbation of the neutron star. The perturbed Einstein equations result in the following differential equation for $H(r)$:

$$H''(r) + H'(r)\left\{\frac{2}{r} + e^{\beta(r)}\left[\frac{2Gm(r)}{r^2} + 4\pi r G(p-\varepsilon)\right]\right\}$$
$$+ H(r)\left[-\frac{6e^{\beta(r)}}{r^2} + 4\pi G e^{\beta(r)}\left(5\varepsilon + 9p + \frac{\varepsilon + p}{c_s^2}\right) - \alpha'(r)^2\right] = 0. \tag{10.104}$$

For small values of r, the regular solution of $H(r)$ has the form $H(r) = a_0 r^2 + \cdots$, where a_0 is a constant. For large values of $H(r)$, the solution has to be matched to the asymptotic solution given in Eq. (10.97). It turns out that the matching requires knowledge of the combination

$$y(r) = \frac{r\,H'(r)}{H(r)}, \tag{10.105}$$

where $y(r)$ is a solution of the differential equation (Postnikov et al., 2010):

$$ry'(r) + y(r)^2 + y(r)e^{\beta(r)}\left[1 + 4\pi r^2 G(p-\varepsilon)\right]$$
$$+ r^2\left[-\frac{6e^{\beta(r)}}{r^2} + 4\pi G e^{\beta(r)}\left(5\varepsilon + 9p + \frac{\varepsilon + p}{c_s^2}\right) - \alpha'(r)^2\right] = 0. \tag{10.106}$$

The boundary condition for $y(r)$ is $y(0) = 2$ as $H(r) = a_0 r^2 + \cdots$ for $r \to 0$. The matching to the asymptotic form results in the analytic expression of the Love number k_2 in terms of $y_R = y(R)$, where R is the radius of the neutron star. If there is a discontinuity at the surface $r = R$, the matching of the solution to the asymptotic one leads to an extra term for y_R (Damour and Nagar, 2009; Hinderer et al., 2010):

$$y_R = \frac{RH'(R)}{H(R)} - \frac{4\pi R^3 \varepsilon(R)}{M}. \tag{10.107}$$

Finally, the Love number can be expressed in terms of y_R and the compactness of the neutron star $C = GM/R$, where M is the mass of the neutron star:

$$k_2 = \frac{8C^5}{5}(1-2C)^2[2+2C(y_R-1)-y_R] \times \Big\{2C(6-3y_R + 3C(5y_R-8)) + 4C^3[13-11y_R+C(3y_R-2)+2C^2(1+y_R)] + 3(1-2C)^2[2-y_R+2C(y_R-1)]\log(1-2C)\Big\}^{-1}. \quad (10.108)$$

We note that the Love number vanishes for a compactness of $C = 1/2$, which corresponds to the that a black hole. Hence, for a Schwarzschild black hole, $k_2 = 0$.

Let us now consider the Newtonian limit that corresponds to $C \to 0$, $p \ll \varepsilon$, $c_s^2 \ll 1$, and $G\varepsilon r^2 \ll 1$. The differential equation for $y(r)$ reads in the Newtonian limit:

$$ry'(r) + y(r)^2 + y(r) - 6 + 4\pi r^2 G\frac{\varepsilon}{c_s^2} = 0. \quad (10.109)$$

Expanding the expression for k_2 in terms of the compactness C results in the Newtonian relation for k_2:

$$k_2^{\text{Newton}} = \frac{1}{2}\left(\frac{2-y_R}{y_R+3}\right). \quad (10.110)$$

There are analytically known solutions for polytropes of the form $p \propto n^{1+1/n}$ for a polytropic indices of $n = 0$, which corresponds to an incompressible fluid, and of $n = 1$, which corresponds to an interaction-dominated EOS, see Section 4.4. In the case $n = 0$, the solution for $y(r)$ is simply a constant $y(r) = 2$ fixed by the boundary condition at $r = 0$. However, at the surface of the star, there is a discontinuity as the energy density jumps from $\varepsilon(R)$ to zero. According to Eq. (10.107), the value of y_R needs to be shifted by $y_R = 2 - 3 = -1$, so that

$$y_R = -1 \qquad k_2^{\text{Newton}} = 3/4 \qquad (n = 0) \quad (10.111)$$

for a polytrope with $n = 0$. For the case of a polytrope with $n = 1$, one finds that (Hinderer, 2008)

$$y_R = \frac{\pi^2 - 9}{3} \approx 0.290 \qquad k_2^{\text{Newton}} = \frac{15-\pi^2}{2\pi^2} \approx 0.260 \qquad (n = 1). \quad (10.112)$$

Note that the Love number for the case of an incompressible fluid corresponds to the upper limit for the value of k_2 not only in the Newtonian limit. An incompressible fluid is the stiffest possible fluid reacting to an external quadrupole potential. It cannot be compressed so it is squished and flows in reaction to the external quadrupole potential. On the contrary, if a fluid can be highly compressed, it forms a high-density core and a low-density mantle around the core. The fluid can react to an

Table 10.2 *A sample of the Love numbers k_2 for solar system bodies. Reprinted by permission from Springer Nature (Lainey, 2016), © (2016).*

	Mercury	Venus	Moon
k_2	0.451 ± 0.014	0.295 ± 0.066	0.02405 ± 0.000176
	Mars	Saturn	Titan
k_2	0.173 ± 0.009	0.390 ± 0.024	0.589 ± 0.150

external potential by increasing the energy density in the core so that only a small quadrupole moment is induced. Hence, by determining the Love number, one learns something about the compressibility of the fluid, that is, the softness or stiffness of the EOS.

The determination of the Love number gives also details about the inner structure of celestial spheres of fluid. Values of the Love number for Newtonian systems have been determined for objects in our solar system that are listed in Table 10.2. The Love numbers for solar system bodies can be determined by the observation of moons or by flybys of satellites. However, modeling planets or moons is nontrivial. One knows in principle the material, iron, silicate rock, or water is usually taken, but one does not know the layer structure nor which material is present at which depth. Extracting the Love number for Earth is further complicated by the presence of an ocean and convection in the mantle of Earth. The nominal value of the Love number for an elastic Earth is about $k_2 \approx 0.3$. The Love number of our Moon is quite small, which can be modeled by a mixture of a fluid and a solid core. Titan has a particularly large value for the Love number, requiring that Titan is highly deformable, which is consistent with a global ocean under the ice crust of Titan (Iess et al., 2012).

In general relativity, the Love number will be smaller compared to the Newtonian case due to the correction terms from the compactness of the star C, see Eq. (10.108). Figure 10.1 shows a collection of numerical results for the Love number k_2 for different EOSs as a function of the compactness $C = GM/R$. The gray lines show the results for polytropes of the form $p \propto n^{1+1/n}$ for different polytropic indices n and for power laws in terms of the energy density $p \propto \varepsilon^{1+1/n}$. The Love number decreases roughly inversely proportional with the compactness. The Love numbers for neutron star EOSs are shown by solid lines (consisting of nucleons and leptons) and by dash-dotted lines (with hyperons or quark matter in the core). For the range of compactnesses relevant for neutron stars, $C = 0.1 - 0.3$, the Love number decreases also roughly inversely proportional to the compactness. The values for the Love number are between $k_2 = 0.14$ and 0.01 for a compactness between $C = 0.1$ and 0.3. For small values of C, the Love number tends to zero

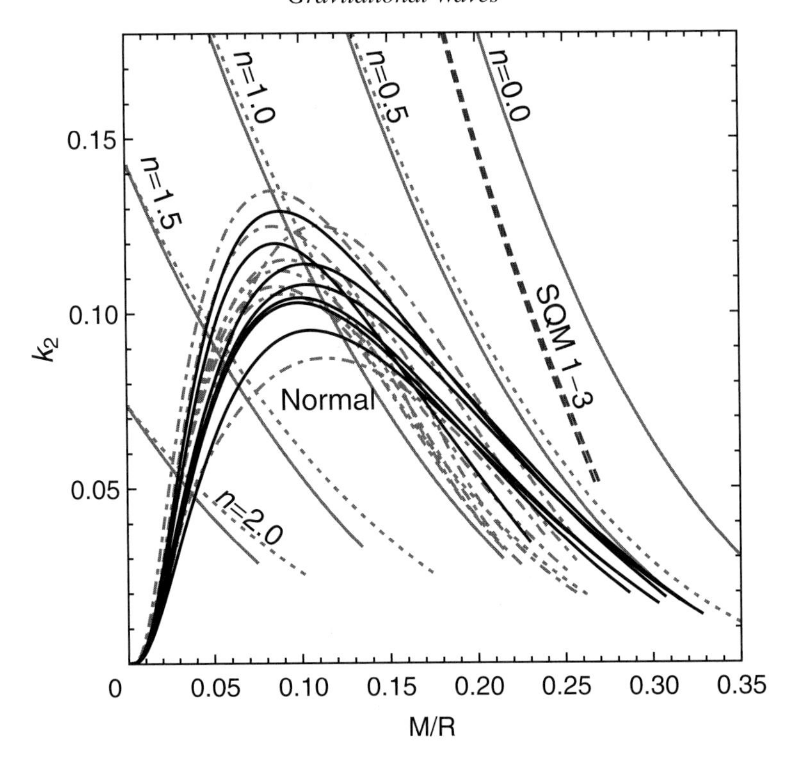

Figure 10.1 The Love number k_2 versus the compactness $C = GM/R$ (gravitational units are used for the label so that $G = 1$) for different EOSs: polytropes with the polytropic index n (solid lines), energy density power law $p \propto \varepsilon^{1+1/n}$ (small dashed lines), selfbound quark stars with different vacuum terms labeled SQM, and a sample of normal neutron star equations of state consisting of nucleons and leptons only (solid thick lines) or with exotic components as hyperons and quarks (dot-dashed lines). Reprinted figure with permission from Hinderer et al. (2010). Copyright (2010) by the American Physical Society

for all neutron star EOSs. In addition, the result is shown for a selfbound star using the Massachusetts Institute of Technology (MIT) bag model with different values for the vacuum term. The results are lying on one line due to the scaling behavior of the solution for the MIT bag model, see Eqs. (4.66) and (4.67). Note that both k_s and C are dimensionless, so the scaling relation results in a single curve. This will change if the Love number is plotted versus a dimensionful quantity, as just the mass or the radius. The values of k_2 for the selfbound star are close to the results for a polytrope with $n = 0$, that is, that of an incompressible fluid. Actually, in the limit of $C \to 0$, the results for k_2 will coincide for the selfbound star and an incompressible fluid. In fact, we know that selfbound matter has a vanishing pressure at a nonvanishing value of the energy density. So, without corrections from general relativity, a selfbound star has a constant energy density and behaves as a sphere of an incompressible fluid.

The tidal deformability λ and thereby the Love number can be determined from the gravitational wave signal of neutron star merger by a change of the phase. The effect from the tidal deformation of the two neutron stars, λ_1 and λ_2, respectively, add up linearly to the total tidal phase change (Flanagan and Hinderer, 2008)

$$\delta\psi = -\frac{9v^5}{16\mu(GM)^4}\left(\frac{m_1+12m_2}{m_1}\lambda_1 + \frac{m_2+12m_1}{m_2}\lambda_2\right) = -\frac{117v^5}{8\mu(GM)^4}\tilde{\lambda}, \tag{10.113}$$

with the orbital velocity $v = (GM\omega)^{1/3}$, the reduced mass μ, and the total mass $M = m_1 + m_2$. Here we introduced the weighted average of the tidal deformability $\tilde{\lambda}$:

$$\tilde{\lambda} = \frac{1}{26}\left(\frac{m_1+12m_2}{m_1}\lambda_1 + \frac{m_2+12m_1}{m_2}\lambda_2\right). \tag{10.114}$$

The prefactor is chosen so that $\tilde{\lambda} = \lambda_1 = \lambda_2$ for equal masses $m_1 = m_2$. One can define now dimensionless tidal deformability parameters by

$$\Lambda_1 = \frac{G\lambda_1}{(Gm_1)^5} \qquad \Lambda_2 = \frac{G\lambda_2}{(Gm_2)^5}, \tag{10.115}$$

with the weighted average

$$\tilde{\Lambda} = \frac{16}{13}\frac{(m_1+12m_2)m_1^4\Lambda_1 + (m_2+12m_1)m_2^4\Lambda_2}{(m_1+m_2)^5}. \tag{10.116}$$

Again, the prefactor is chosen so that $\tilde{\Lambda} = \Lambda_1 = \Lambda_2$ for equal masses $m_1 = m_2$. Note that the weighted dimensionless tidal deformability parameter can be also written as

$$\tilde{\Lambda} = \frac{G\tilde{\lambda}}{(G\tilde{m})^5}, \tag{10.117}$$

where $\tilde{m} = (m_1 + m_2)/2$ is the average mass.

For equal mass binaries with $m_1 = m_2 = m$, $R_1 = R_2 = R$, and $\lambda_1 = \lambda_2$, the Love number can be recast in the form

$$k_2(C) = \frac{3G\lambda}{2R^5} = \frac{3}{2}\tilde{\Lambda}\left(\frac{Gm}{R}\right)^5 = \frac{3}{2}\tilde{\Lambda}C^5. \tag{10.118}$$

One sees that the tidal deformability parameter

$$\tilde{\Lambda} = \frac{2k_2(C)}{3C^5} \tag{10.119}$$

has a strong dependence on the compactness C. As the Love number depends roughly inversely proportionally on the compactness for the range $C = 0.1-0.3$, the dependence on the compactness is about $\tilde{\Lambda} \propto C^{-6}$. For a known neutron star mass m, this dependence relates to a strong dependence on the radius of the neutron star R. Hence, an upper limit on the tidal deformability relates to a strong upper limit on the radius of the neutron star. The limit on the radius can be cast in the form

$$R = \left(\frac{3\tilde{\Lambda}}{2k_2(C)}\right)^{1/5} Gm. \tag{10.120}$$

For the neutron star merger event GW170817, the LIGO collaboration has constrained the tidal deformability parameter to

$$\tilde{\Lambda} = 300^{+420}_{-230} \qquad \text{(GW170817)}, \tag{10.121}$$

assuming a low spin for the merging neutron stars (Abbott et al., 2019c). The total mass extracted from GW170817 was $M = 2.73^{+0.04}_{-0.01}$. In order to get an upper limit on the radius, we need to get a lower limit on the Love number $k_2(C)$. Figure 10.1 seems to indicate that the Love number for a polytrope with $n = 1$ can serve as a lower limit of the Love number for a large sample of neutron star EOSs. Using $k_2(0.15) = 0.0776$ from Hinderer (2008, see the table in the erratum), we find the following expression for the upper limit on the radius of a neutron star

$$R \leq 13.6\ \text{km} \left(\frac{\tilde{\Lambda}}{720}\right)^{1/5} \left(\frac{k_2}{0.0776}\right)^{-1/5} \left(\frac{m}{1.37M_\odot}\right), \tag{10.122}$$

assuming equal mass binary neutron stars (as a cross check $C = Gm/R = 0.149$, so we used the appropriate limiting value for k_2). The limit on the radius of a neutron star implied by the gravitational wave signal GW170817 has been scrutinized in several publications by more sophisticated calculations using samples of up to a million EOSs of piecewise polytropes (see Abbott et al., 2017b, 2018b; Annala et al., 2018; Most et al., 2018; De et al., 2018; Tews et al., 2018). The overall finding is that GW170817 constrains the radius of a $m = 1.4M_\odot$ neutron star to $R \leq 13.2-13.7$ km, which coincides with our rough estimate.

Hence, just this one measurement of the gravitational wave of the inspiral of merging neutron stars results in a strong constraint on the radius of a neutron star with a mass of around $1.4M_\odot$. Looking at the mass–radius diagram for neutron stars, Figure 9.2, we see that the radius limit from GW170817 rules out the neutron star configurations to the right of the mass–radius curve labeled as II for $m = 1.4M_\odot$. Stiffer EOSs resulting in large radii, such as the curve labeled III, are ruled out. Softer EOSs, such as the curve labeled I, are compatible with the radius constraint. We note that hybrid star configurations with a phase transition, shown in Figure 9.8 by the dark gray shaded area, can have lower radii compared to neutron star configurations. This opens this exciting possibility to have a possible smoking gun for the existence of a phase transition inside compact stars by finding a tidal deformability too small to be explainable by ordinary neutron star configurations without a phase transition. As of this writing, a new observing run for gravitational waves just started with the prospect of detecting several neutron star merger events per year (Abbott et al., 2018c). Maybe, while reading these lines, a gravitational wave from a compact star merger might have just passed through you ....

Exercises

(10.1) Derive the infinitesimal invariance of the perturbation $h_{\mu\nu}$, Eq. (10.4), from the coordinate transformation, Eq. (10.3), and the transformation law of the metric $g_{\mu\nu}$.

(10.2) Show that the wave equation $\Box h_{\mu\nu} = 0$ has plane wave solutions proportional to $\exp(ik_\mu x^\mu)$. Show that the four-vector is light-like ($k_\mu k^\mu = 0$). Derive the dispersion relation for gravitational waves $\omega = k_0 c = |\boldsymbol{k}|c$ and demonstrate that gravitational waves travel with the speed of light.

(10.3) Show that the expression for the energy–momentum tensor of gravitational waves, Eq. (10.23), does not change under a gauge transformation, Eq. (10.4). Hint: under the averaging brackets, use integration by parts and neglect the boundary terms.

(10.4) The Λ-tensor is defined as

$$\Lambda_{ij,kl} = P_{ik}P_{jl} - \frac{1}{2}P_{ij}P_{kl},$$

with the projection operator defined as

$$P_{ij} = \delta_{ij} - n_i n_j,$$

where n is a unit vector in the direction of propagation of the gravitational wave. Show that $\Lambda_{ij,kl}$ is a projection operator (acting the operator twice gives the original operator):

$$\Lambda_{ij,kl}\Lambda_{kl,mn} = \Lambda_{ij,kl},$$

that it is transverse in all indices $n^i \Lambda_{ij,kl} = 0\ldots$, and that it is traceless $\Lambda_{ii,kl} = \Lambda_{ij,kk} = 0$ so that it can be used to project a tensor into its transverse-traceless form (as e.g., the tensor $h_{\mu\nu}$).

(10.5) For the gravitational wave event GW170817, one has also seen a gamma-ray burst at 1.7 s after the gravitational wave arrived on Earth. By triangulation, the minimum distance has been determined to be $D = 26\,\text{Mpc}$. Assume that the gravitational wave has a velocity of v_{GW} different from the velocity of light v_{EM}. Show that the relative difference in the velocity can be written as

$$\frac{\Delta v}{v_{\text{EM}}} \approx v_{\text{EM}} \frac{\Delta t}{D},$$

where $\Delta v = v_{\text{GW}} - v_{\text{EM}}$. Assume that the photons were emitted at the same time or 10 seconds after the gravitational wave, respectively. What are the limits on the relative velocity difference of gravitational waves and photons?

(10.6) Show that from the definition of the mass moments, Eqs. (10.30), (10.31), and (10.32), the relations $\dot{M} = 0$ and $\ddot{M}^i = 0$ can be interpreted as mass and momentum conservation and derive the relation given in Eq. (10.34).

(10.7) Project the second mass moment into the TT gauge by using the Λ projector. Show first that for an arbitrary matrix A_{kl}

$$\Lambda_{ij,kl} A_{kl} = (PAP)_{ij} - \frac{1}{2} P_{ij} \text{Tr}(PA).$$

Then, project along the z-axis with $P = \text{diag}(1, 1, 0)$ and show that

$$\Lambda_{ij,kl} \ddot{M}_{kl} = \begin{pmatrix} (\ddot{M}_{11} - \ddot{M}_{22})/2 & \ddot{M}_{12} & 0 \\ \ddot{M}_{21} & -(\ddot{M}_{11} - \ddot{M}_{22})/2 & 0 \\ 0 & 0 & 0 \end{pmatrix}.$$

(10.8) Compare the time to merge given in Table 10.1 with the one given by the approximation equation (10.82). Compare the time to merge for the given ellipticity and for circular orbits, that is, for $e = 0$. Calculate the enhancement factor for the emitted power for the Hulse–Taylor pulsar, see Eq. (10.81).

References

Abbott, B. P., Abbott, R., Abbott, T. D., et al. 2016a. Observation of Gravitational Waves from a Binary Black Hole Merger. *Phys. Rev. Lett.*, **116**(6), 061102.

Abbott, B. P., Abbott, R., Abbott, T. D., et al. 2016b. Sensitivity of the Advanced LIGO Detectors at the Beginning of Gravitational Wave Astronomy. *Phys. Rev. D*, **93**(11), 112004. [Addendum: ibid. *D* **97** (2018) 059901].

Abbott, B. P., Abbott, R., Abbott, T. D., et al. 2017a. First Search for Gravitational Waves from Known Pulsars with Advanced LIGO. *Astrophys. J.*, **839**(1), 12. [Erratum: ibid. **851** (2017) 71].

Abbott, B. P., Abbott, R., Abbott, T. D., et al. 2017b. GW170817: Observation of Gravitational Waves from a Binary Neutron Star Inspiral. *Phys. Rev. Lett.*, **119**(16), 161101.

Abbott, B. P., Abbott, R., Abbott, T. D., et al. 2017c. Gravitational Waves and Gamma-Rays from a Binary Neutron Star Merger: GW170817 and GRB 170817A. *Astrophys. J.*, **848**(2), L13.

Abbott, B. P., Abbott, R., Abbott, T. D., et al. 2017d. Multi-messenger Observations of a Binary Neutron Star Merger. *Astrophys. J.*, **848**(2), L12.

Abbott, B. P., Abbott, R., Abbott, T. D., et al. 2018a. GWTC-1: A Gravitational-Wave Transient Catalog of Compact Binary Mergers Observed by LIGO and Virgo during the First and Second Observing Runs. *Phys. Rev. X* **9**, 031040.

Abbott, B. P., Abbott, R., Abbott, T. D., et al. 2018b. GW170817: Measurements of Neutron Star Radii and Equation of State. *Phys. Rev. Lett.*, **121**(16), 161101.

Abbott, B. P., Abbott, R., Abbott, T. D., et al. 2018c. Prospects for Observing and Localizing Gravitational-Wave Transients with Advanced LIGO, Advanced Virgo and KAGRA. *Living Rev. Rel.*, **21**(1), 3.

Abbott, B. P., Abbott, R., Abbott, T. D., et al. 2019a. Searches for Continuous Gravitational Waves from Fifteen Supernova Remnants and Fomalhaut b with Advanced LIGO. *Astrophys. J.*, **875**(2), 122.

Abbott, B. P., Abbott, R., Abbott, T. D., et al. 2019b. Searches for Gravitational Waves from Known Pulsars at Two Harmonics in 2015–2017 LIGO Data. *Astrophys. J.*, **879**(1), 10.

Abbott, B. P., Abbott, R., Abbott, T. D., et al. 2019c. Properties of the Binary Neutron Star Merger GW170817. *Phys. Rev. X*, **9**(1), 011001.

Abelev, B. I., Aggarwal, M. M., Ahammed, Z., et al. 2007. Strangelet Search at RHIC. *Phys. Rev., C*, **76**, 011901.

Acharya, S. *et al.* [ALICE Collaboration] 2019. Study of the Λ-Λ interaction with femtoscopy correlations in pp and p-Pb collisions at the LHC. *Phys. Lett. B* **797**, 134822.

Ade, P. A. R., Aghanim, N., Arnaud, M., et al. 2016. Planck 2015 Results. XIII. Cosmological Parameters. *Astron. Astrophys.*, **594**, A13.

Agnihotri, Pratik, Schaffner-Bielich, Jürgen, and Mishustin, Igor N. 2009. Boson Stars with Repulsive Selfinteractions. *Phys. Rev. D*, **79**, 084033.

Akiyama, Kazunori, Alberdi, Antxon, Alef, Walter, et al. 2019. First M87 Event Horizon Telescope Results. I. The Shadow of the Supermassive Black Hole. *Astrophys. J.*, **875**(1), L1.

Alcock, Charles, Farhi, Edward, and Olinto, Angela. 1986. Strange Stars. *Astrophys. J.*, **310**, 261.

Alford, Mark G., Burgio, G. F., Han, Sophia, Taranto, Gabriele, and Zappalà, Dario. 2015. Constraining and Applying a Generic High-Density Equation of State. *Phys. Rev. D*, **92**(8), 083002.

Alford, Mark G., Han, Sophia, and Prakash, Madappa. 2013. Generic Conditions for Stable Hybrid Stars. *Phys. Rev. D*, **88**(8), 083013.

Alvarez-Castillo, D. E., and Blaschke, D. 2015. Mixed Phase Effects on High-Mass Twin Stars. *Phys. Part. Nucl.*, **46**(5), 846–848.

Ambartsumyan, V. A., and Saakyan, G. S. 1960. The Degenerate Superdense Gas of Elementary Particles. *Sov. Astron.*, **4**, 187.

Ambartsumyan, V. A., and Saakyan, G. S. 1962. On Equilibrium Configurations of Superdense Degenerate Gas Masses. *Soviet Astronomy*, **5**, 601.

Annala, Eemeli, Gorda, Tyler, Kurkela, Aleksi, and Vuorinen, Aleksi. 2018. Gravitational-Wave Constraints on the Neutron-Star-Matter Equation of State. *Phys. Rev. Lett.*, **120**(17), 172703.

Antoniadis, John, Freire, Paulo C. C., Wex, Norbert, et al. 2013. A Massive Pulsar in a Compact Relativistic Binary. *Science*, **340**, 6131.

Archibald, Anne M., Gusinskaia, Nina V., Hessels, Jason W. T., et al. 2018. Universality of Free Fall from the Orbital Motion of a Pulsar in a Stellar Triple System. *Nature*, **559**(7712), 73–76.

Arzoumanian, Zaven, Brazier, Adam, Burke-Spolaor, Sarah, et al. 2018. The NANOGrav 11-Year Data Set: High-Precision Timing of 45 Millisecond Pulsars. *Astrophys. J. Suppl.*, **235**(2), 37.

Audley, Heather, Babak, Stanislav, Baker, John, et al. 2017. *Laser Interferometer Space Antenna*. www.elisascience.org/files/publications/LISA_L3_20170120.pdf.

Baade, W. 1942. The Crab Nebula. *Astrophys. J.*, **96**, 188.

Baade, Walter, and Zwicky, Fritz. 1934a. Supernovae and Cosmic Rays. *Phys. Rev.*, **45**, 138.

Baade, Walter, and Zwicky, Fritz. 1934b. Cosmic Rays from Super-Novae. *Proc. Natl. Acad. Sci. USA*, **20**, 259.

Baluni, Varouzhan. 1978. Nonabelian Gauge Theories of Fermi Systems: Chromotheory of Highly Condensed Matter. *Phys. Rev. D*, **17**, 2092.

Bardeen, J. M., Thorne, K. S., and Meltzer, D. W. 1966. A Catalogue of Methods for Studying the Normal Modes of Radial Pulsation of General-Relativistic Stellar Models. *Astrophys. J.*, **145**, 505.

Barstow, Martin Adrian, Bond, Howard E., Holberg, J. B., Burleigh, M. R., Hubeny, I., and Koester, D. 2005. Hubble Space Telescope Spectroscopy of the Balmer lines in Sirius B. *Mon. Not. Roy. Astron. Soc.*, **362**, 1134–1142.

Baym, G., and Chin, S. A. 1976. Can a Neutron Star Be a Giant MIT Bag? *Phys. Lett. B*, **62**, 241–244.

Baym, Gordon, Pethick, Christopher, and Sutherland, Peter. 1971. The Ground State of Matter at High Densities: Equation of State and Stellar Models. *Astrophys. J.*, **170**, 299–317.

Bazavov, A., Bhattacharya, T., DeTar, C., et al. 2014. Equation of State in (2+1)-Flavor QCD. *Phys. Rev. D*, **90**, 094503.

Bell-Burnell, S. J. 1977. Petit Four. *Annals of the New York Academy of Sciences*, **302**, 685.

Bell Burnell, J. 2009. Reflections on the Discovery of Pulsars. Page 14 of: Proceedings of the Special Session "Accelerating the Rate of Astronomical Discovery" of the 27th IAU General Assembly. August 11–14, 2009. Rio de Janeiro, Brazil.

Berry, D. K., Caplan, M. E., Horowitz, C. J., Huber, Greg, and Schneider, A. S. 2016. "Parking-Garage" Structures in Nuclear Astrophysics and Cellular Bio-physics. *Phys. Rev. C*, **94**(5), 055801.

Bertotti, B., Iess, L., and Tortora, P. 2003. A Test of General Relativity Using Radio Links with the Cassini Spacecraft. *Nature*, **425**, 374–376.

Bethe, H. A., and Bacher, R. F. 1936. Nuclear Physics A. Stationary States of Nuclei. *Rev. Mod. Phys.*, **8**, 82–229.

Bodmer, A. R. 1971. Collapsed Nuclei. *Phys. Rev. D*, **4**, 1601.

Boguta, J., and Stöcker, Horst. 1983. Systematics of Nuclear Matter Properties in a Nonlinear Relativistic Field Theory. *Phys. Lett. B*, **120**, 289–293.

Bond, H. E., Bergeron, P., and Bedard, A. 2017b. Astrophysical Implications of a New Dynamical Mass for the Nearby White Dwarf 40 Eridani B. *Astrophys. J.*, **848**, 16.

Bond, H. E., Gilliland, R. L., Schaefer, G. H., et al. 2015. Hubble Space Telescope Astrometry of the Procyon System. *Astrophys. J.*, **813**, 106.

Bond, H. E., Schaefer, G. H., Gilliland, R. L., et al. 2017a. The Sirius System and Its Astrophysical Puzzles: Hubble Space Telescope and Ground-Based Astrometry. *Astrophys. J.*, **840**, 70.

Buchdahl, Hans A. 1959. General Relativistic Fluid Spheres. *Phys. Rev.*, **116**, 1027.

Burgay, Marta, D'Amico, N., Possenti, A., et al. 2003. An Increased Estimate of the Merger Rate of Double Neutron Stars from Observations of a Highly Relativistic System. *Nature*, **426**, 531–533.

Burwitz, Vadim, Haberl, F., Neuhaeuser, R., et al. 2003. The Thermal Radiation of the Isolated Neutron Star RX J1856.5-3754 Observed with Chandra and XMM-Newton. *Astron. Astrophys.*, **399**, 1109–1114.

Busza, W., Jaffe, R. L., Sandweiss, J., and Wilczek, Frank. 2000. Review of Speculative "Disaster Scenarios" at RHIC. *Rev. Mod. Phys.*, **72**, 1125–1140.

Cameron, A. G. W. 1959. Neutron Star Models. *Astrophys. J.*, **130**, 884.

Chabanat, E., Meyer, J., Bonche, P., Schaeffer, R., and Haensel, P. 1997. A Skyrme Parametrization from Subnuclear to Neutron Star Densities. *Nucl. Phys. A*, **627**, 710–746.

Chandrasekhar, S. 1931a. The Highly Collapsed Configurations of a Stellar Mass. *Mon. Not. Roy. Astron. Soc.*, **91**(5), 456–466.

Chandrasekhar, S. 1931b. The Maximum Mass of Ideal White Dwarfs. *Astrophys. J.*, **74**, 81–82.

Chandrasekhar, Subrahmanyan. 1939. *An Introduction to the Study of Stellar Structure*. Chicago: The University of Chicago Press.

Chatterjee, Debarati, and Vidaña, Isaac. 2016. Do Hyperons Exist in the Interior of Neutron Stars? *Eur. Phys. J. A*, **52**(2), 29.

Chodos, A., Jaffe, R. L., Johnson, K., Thorn, Charles B., and Weisskopf, V. F. 1974. A New Extended Model of Hadrons. *Phys. Rev. D*, **9**, 3471–3495.

Chowdhury, P. Roy., and Basu, D. N. 2006. Nuclear Matter Properties with the Re-evaluated Coefficients of Liquid Drop Model. *Acta Phys. Polon. B*, **37**, 1833–1846.

Cocke, W. J., Disney, M. J., and Taylor, D. J. 1969. Discovery of Optical Signals from Pulsar NP 0532. *Nature*, **221**, 525.

Colpi, M., Shapiro, S. L., and Wasserman, I. 1986. Boson Stars: Gravitational Equilibria of Selfinteracting Scalar Fields. *Phys. Rev. Lett.*, **57**, 2485–2488.

Cromartie, H. Thankful, Fonseca, E., Ransom, S. M., et al. 2019. A Very Massive Neutron Star: Relativistic Shapiro Delay Measurements of PSR J0740+6620, *Nat. Astron.*, **4**(1), 72.

Damour, Thibault, and Nagar, Alessandro. 2009. Relativistic Tidal Properties of Neutron Stars. *Phys. Rev. D*, **80**, 084035.

Danysz, M., and Pniewski, J. 1953. Delayed Disintegration of a Heavy Nuclear Fragment. *Philos. Mag.*, **44**, 348.

Davidson, K., and Fesen, R. A. 1985. Recent Developments Concerning the Crab Nebula. *Ann. Rev. Astron. Astrophys.*, **23**, 119–146.

De, Soumi, Finstad, Daniel, Lattimer, James M., et al. 2018. Tidal Deformabilities and Radii of Neutron Stars from the Observation of GW170817. *Phys. Rev. Lett.*, **121**(9), 091102.

DeGrand, Thomas A., Jaffe, R. L., Johnson, K., and Kiskis, J. E. 1975. Masses and Other Parameters of the Light Hadrons. *Phys. Rev. D*, **12**, 2060.

Demorest, Paul, Pennucci, Tim, Ransom, Scott, Roberts, Mallory, and Hessels, Jason. 2010. Shapiro Delay Measurement of a Two Solar Mass Neutron Star. *Nature*, **467**, 1081–1083.

Derrick, G. H. 1964. Comments on Nonlinear Wave Equations as Models for Elementary Particles. *J. Math. Phys.*, **5**, 1252–1254.

Drischler, C., Carbone, A., Hebeler, K., and Schwenk, A. 2016. Neutron Matter from Chiral Two- and Three-Nucleon Calculations Up to N^3LO. *Phys. Rev. C*, **94**(5), 054307.

Duerr, Hans-Peter. 1956. Relativistic Effects in Nuclear Forces. *Phys. Rev.*, **103**, 469–480.

Duflo, J., and Zuker, A. P. 1995. Microscopic Mass Formulae. *Phys. Rev. C*, **52**, R23.

Dufour, P., Blouin, S., et al. 2017. The Montreal White Dwarf Database: A Tool for the Community. Page 3 of: Tremblay, P.-E., Gaensicke, B., and Marsh, T. (eds.), *20th European White Dwarf Workshop*. Astronomical Society of the Pacific Conference Series, vol. 509. San Francisco: The Astronomical Society of the Pacific.

Dutra, M., Lourenco, O., Avancini, S. S., et al. 2014. Relativistic Mean-Field Hadronic Models under Nuclear Matter Constraints. *Phys. Rev. C*, **90**(5), 055203.

Dutra, M., Lourenco, O., Sa Martins, J. S., et al. 2012. Skyrme Interaction and Nuclear Matter Constraints. *Phys. Rev. C*, **85**, 035201.

Duyvendak, J. J. L. 1942. Further Data Bearing in the Identification of the Crab Nebula with the Supernova of 1054 A. D. Part I. The Ancient Oriental Chronicles. *Proc. Astr. Soc. Pac.*, **54**, 91.

Einstein, Albert. 1916a. Die Grundlage der allgemeinen Relativitiätstheorie. *Annalen Phys.*, **49**(7), 769–822.

Einstein, Albert. 1916b. Näherungsweise Integration der Feldgleichungen der Gravitation. *Sitzungsber. Preuss. Akad. Wiss. Berlin (Math. Phys.)*, **1916**, 688–696.

Einstein, Albert. 1918. Über Gravitationswellen. *Sitzungsber. Preuss. Akad. Wiss. Berlin (Math. Phys.)*, **1918**, 154–167.

Elliott, J. B., Lake, P. T., Moretto, L. G., and Phair, L. 2013. Determination of the Coexistence Curve, Critical Temperature, Density, and Pressure of Bulk Nuclear Matter from Fragment Emission Data. *Phys. Rev. C*, **87**(5), 054622.

Ellis, John R., Giudice, Gian, Mangano, Michelangelo L., Tkachev, Igor, and Wiedemann, Urs. 2008. Review of the Safety of LHC Collisions. *J. Phys. G*, **35**, 115004.

Espinoza, C. M., Lyne, A. G., and Stappers, B. W. 2017. New Long-Term Braking Index Measurements for Glitching Pulsars Using a Glitch-Template Method. *Mon. Not. Roy. Astron. Soc.*, **466**(1), 147–162.

Espinoza, C. M., Lyne, A. G., Stappers, B. W., and Kramer, M. 2011. A Study of 315 Glitches in the Rotation of 102 Pulsars. *Mon. Not. Roy. Astron. Soc.*, **414**, 1679–1704.

Farhi, E., and Jaffe, R. L. 1984. Strange Quark Matter. *Phys. Rev. D*, **30**, 2379.

Flanagan, Eanna E., and Hinderer, Tanja. 2008. Constraining Neutron Star Tidal Love Numbers with Gravitational Wave Detectors. *Phys. Rev. D*, **77**, 021502.

Foldy, L. L. 1978. Electrostatic Stability of Wigner and Wigner–Dyson Lattices. *Phys. Rev. B*, **17**, 4889–4894.

Fonseca, Emmanuel, Stairs, Ingrid H., and Thorsett, Stephen E. 2014. A Comprehensive Study of Relativistic Gravity using PSR B1534+12. *Astrophys. J.*, **787**, 82.

Fowler, R. H. 1926. On Dense Matter. *Mon. Not. Roy. Astron. Soc.*, **87**, 114–122.

Fraga, Eduardo S., Hippert, Maurício, and Schmitt, Andreas. 2019. Surface Tension of Dense Matter at the Chiral Phase Transition. *Phys. Rev. D*, **99**(1), 014046.

Fraga, Eduardo S., Kurkela, Aleksi, and Vuorinen, Aleksi. 2014. Interacting Quark Matter Equation of State for Compact Stars. *Astrophys. J. Lett.*, **781**, L25.

Fraga, Eduardo S., Pisarski, Robert D., and Schaffner-Bielich, Jürgen. 2001. Small, Dense Quark Stars from Perturbative QCD. *Phys. Rev. D*, **63**, 121702(R).

Freedman, Barry A., and McLerran, Larry D. 1977. Fermions and Gauge Vector Mesons at Finite Temperature and Density. 3. The Ground State Energy of a Relativistic Quark Gas. *Phys. Rev. D*, **16**, 1169.

Friedberg, R., Lee, T. D., and Pang, Y. 1987. Mini-Soliton Stars. *Phys. Rev. D*, **35**, 3640.

Friedman, B., and Pandharipande, V. R. 1981. Hot and Cold, Nuclear and Neutron Matter. *Nucl. Phys. A*, **361**, 502–520.

Furnstahl, R. J., and Serot, Brian D. 2000. Large Lorentz Scalar and Vector Potentials in Nuclei. *Nucl. Phys. A*, **673**, 298–310.

Gal, A., Hungerford, E. V., and Millener, D. J. 2016. Strangeness in Nuclear Physics. *Rev. Mod. Phys.*, **88**(3), 035004.

Gerlach, Ulrich H. 1968. Equation of State at Supranuclear Densities and the Existence of a Third Family of Superdense Stars. *Phys. Rev.*, **172**, 1325–1330.

Gillessen, S., Eisenhauer, F., Fritz, T. K., et al. 2009. The Orbit of the Star S2 around SgrA* from VLT and Keck Data. *Astrophys. J.*, **707**, L114–L117.

Glendenning, Norman K. 1992. First Order Phase Transitions with More Than One Conserved Charge: Consequences for Neutron Stars. *Phys. Rev. D*, **46**, 1274–1287.

Glendenning, Norman K. 2000. *Compact Stars – Nuclear Physics, Particle Physics, and General Relativity*. Second edn. New York: Springer.

Glendenning, Norman K., and Kettner, Christiane. 2000. Non-identical Neutron Star Twins. *Astron. Astrophys.*, **353**, L9.

Gold, T. 1968. Rotating Neutron Stars as the Origin of the Pulsating Radio Sources. *Nature*, **218**, 731.

Goldreich, Peter, and Julian, William H. 1969. Pulsar Electrodynamics. *Astrophys. J.*, **157**, 869.

Gorda, Tyler, Kurkela, Aleksi, Romatschke, Paul, Säppi, Matias, and Vuorinen, Aleksi. 2018. Next-to-Next-to-Next-to-Leading Order Pressure of Cold Quark Matter: Leading Logarithm. *Phys. Rev. Lett.*, **121**(20), 202701.

Greiner, Carsten, Koch, Peter, and Stöcker, Horst. 1987. Separation of Strangeness from Antistrangeness in the Phase Transition from Quark to Hadron Matter: Possible Formation of Strange Quark Matter in Heavy-Ion Collisions. *Phys. Rev. Lett.*, **58**, 1825.

Greiner, Carsten, Rischke, Dirk H., Stöcker, Horst, and Koch, Peter. 1988. The Creation of Strange Quark Matter Droplets as a Unique Signature for Quark-Gluon Plasma Formation in Relativistic Heavy-Ion Collisions. *Phys. Rev. D*, **38**, 2797.

Grill, Fabrizio, Pais, Helena, Providencia, Constanca, Vidaña, Isaac, and Avancini, Sidney S. 2014. Equation of State and Thickness of the Inner Crust of Neutron Stars. *Phys. Rev. C*, **90**, 045803.

Haensel, P., Zdunik, J. L., and Schaeffer, R. 1986. Strange Quark Stars. *Astron. Astrophys.*, **160**, 121.

Hansen, Brad M. S., Brewer, James, Fahlman, Greg G., et al. 2002. The White Dwarf Cooling Sequence of the Globular Cluster Messier 4. *Astrophys. J.*, **574**, L155–L158.

Harrison, B. K., Thorne, K. S., Wakano, M., and Wheeler, J. A. 1965. *Gravitation Theory and Gravitational Collapse*. Chicago: The Unversity of Chicago Press.

Hasenfratz, P., Horgan, R. R., Kuti, J., and Richard, J. M. 1980. The Effects of Colored Glue in the QCD Motivated Bag of Heavy Quark – Anti-quark Systems. *Phys. Lett. B*, **95**, 299–305.

Heiselberg, H., Pethick, C. J., and Staubo, E. F. 1993. Quark Matter Droplets in Neutron Stars. *Phys. Rev. Lett.*, **70**, 1355–1359.

Hempel, Matthias, and Schaffner-Bielich, Jürgen. 2008. Mass, Radius, and Composition of the Outer Crust of Nonaccreting Cold Neutron Stars. *J. Phys. G*, **35**, 014043.

Hessels, Jason W. T., Ransom, Scott M., Stairs, Ingrid H., et al. 2006. A Radio Pulsar Spinning at 716-hz. *Science*, **311**, 1901–1904.

Hester, J. J., Mori, K., Burrows, D., et al. 2002. Hubble Space Telescope and Chandra Monitoring of the Crab Synchrotron Nebula. *Astrophys. J. Lett.*, **577**, L49–L52.

Hewish, A., Bell, S. J., Pilkington, J. D. H., Scott, P. F., and Collins, R. A. 1968. Observation of a Rapidly Pulsating Radio Source. *Nature*, **217**, 709.

Hinderer, Tanja. 2008. Tidal Love Numbers of Neutron Stars. *Astrophys. J.*, **677**, 1216–1220. [Erratum: ibid. **697** (2009) 964].

Hinderer, Tanja, Lackey, Benjamin D., Lang, Ryan N., and Read, Jocelyn S. 2010. Tidal Deformability of Neutron Stars with Realistic Equations of State and Their Gravitational Wave Signatures in Binary Inspiral. *Phys. Rev. D*, **81**, 123016.

Hobbs, G., Coles, W., Manchester, R. N., et al. 2012. Development of a Pulsar-Based Time-Scale. *Mon. Not. Roy. Astron. Soc.*, **427**, 2780–2787.

Hornick, Nadine, Tolos, Laura, Zacchi, Andreas, Christian, Jan-Erik, and Schaffner-Bielich, Jürgen. 2018. Relativistic Parameterizations of Neutron Matter and Implications for Neutron Stars. *Phys. Rev. C*, **98**(6), 065804.

Horowitz, C. J., Schneider, A. S., and Berry, D. K. 2010. Crystallization of Carbon Oxygen Mixtures in White Dwarf Stars. *Phys. Rev. Lett.*, **104**, 231101.

Huang, W. J., Audi, G., Wang, M., et al. 2017. The AME2016 Atomic Mass Evaluation (I). Evaluation of Input Data; and Adjustment Procedures. *Chinese Physics C*, **41**, 030002.

Hugenholtz, N. M., and van Hove, L. 1958. A Theorem on the Single Particle Energy in a Fermi Gas with Interaction. *Physica*, **24**, 363–376.

Iess, L., Jacobson, R. A., Ducci, M., et al. 2012. The Tides of Titan. *Science*, **337**, 457.

Itoh, N. 1970. Hydrostatic Equilibrium of Hypothetical Quark Stars. *Prog. Theor. Phys.*, **44**, 291.

Ivanenko, D. D., and Kurdgelaidze, D. F. 1965. Hypothesis Concerning Quark Stars. *Astrophys.*, **1**, 251.

Kämpfer, Burkhard. 1981a. On the Possibility of Stable Quark and Pion Condensed Stars. *J.Phys. A*, **14**, L471–L475.

Kämpfer, Burkhard. 1981b. On Stabilizing Effects of Relativity in Cold Spheric Stars with a Phase Transition in the Interior. *Phys. Lett. B*, **101**, 366–368.

Kaup, David J. 1968. Klein–Gordon Geon. *Phys. Rev.*, **172**, 1331–1342.

Kramer, M., Stairs, I. H., Manchester, R. N., et al. 2006. Tests of General Relativity from Timing the Double Pulsar. *Science*, **314**, 97–102.

Kramer, M., and Wex, N. 2009. The Double Pulsar System: A Unique Laboratory for Gravity. *Class. Quant. Grav.*, **26**, 073001.

Krüger, T., Tews, I., Hebeler, K., and Schwenk, A. 2013. Neutron Matter from Chiral Effective Field Theory Interactions. *Phys. Rev. C*, **88**, 025802.

Kurkela, Aleksi, Fraga, Eduardo S., Schaffner-Bielich, Jürgen, and Vuorinen, Aleksi. 2014. Constraining Neutron Star Matter with Quantum Chromodynamics. *Astrophys. J.*, **789**, 127.

Lainey, V. 2016. Quantification of Tidal Parameters from Solar System Data. *Celest. Mech. Dyn. Astron.*, **126**, 145–156.

Landau, Lev D. 1932. On the Theory of Stars. *Physik. Zeits. Sowjetunion*, **1**, 285.

Large, M. I., Vaughan, A. E., and Mills, B. Y. 1968. A Pulsar Supernova Association? *Nature*, **220**, 340–341.

Lattimer, James M., and Prakash, Madappa. 2005. The Ultimate Energy Density of Observable Cold Matter. *Phys. Rev. Lett.*, **94**, 111101.

Lattimer, James M., and Prakash, Madappa. 2011. What a Two Solar Mass Neutron Star Really Means. Pages 275–304 of: Lee, Sabine (ed.), *From Nuclei to Stars: Festschrift in Honor of Gerald E Brown*. Singapore: World Scientific.

Lighthill, M. J. 1950. On the Instability of Small Planetary Cores (II). *Mon. Not. Roy. Astron. Soc.*, **110**, 339.

Lindblom, L. 1984. Limits on the Gravitational Redshift from Neutron Stars. *Astrophys. J.*, **278**, 364–368.

Lindblom, Lee. 2018. Causal Representations of Neutron-Star Equations of State. *Phys. Rev. D*, **97**(12), 123019.

Lorimer, D. R. 2011. Blind Surveys for Radio Pulsars and Transients. *AIP Conf. Proc.*, **1357**, 11–18.

Lorimer, D. R., Bailes, M., McLaughlin, M. A., Narkevic, D. J., and Crawford, F. 2007. A Bright Millisecond Radio Burst of Extragalactic Origin. *Science*, **318**, 777.

Love, A. E. H. 1909. The Yielding of the Earth to Disturbing Forces. *Proc. R. Soc. Lond. A*, **82**, 73–88.

Lyne, A. G., Burgay, M., Kramer, M., et al. 2004. A Double-Pulsar System – A Rare Laboratory for Relativistic Gravity and Plasma Physics. *Science*, **303**, 1153–1157.

Lyne, Andrew, Jordan, Christine, Graham-Smith, Francis, Espinoza, Cristobal, Stappers, Ben, and Weltrvrede, Patrick. 2015. 45 Years of Rotation of the Crab Pulsar. *Mon. Not. Roy. Astron. Soc.*, **446**, 857–864.

Maggiore, Michele. 2007. *Gravitational Waves. Vol. 1: Theory and Experiments*. Oxford Master Series in Physics. Oxford: Oxford University Press.

Maggiore, Michele. 2018. *Gravitational Waves. Vol. 2: Astrophysics and Cosmology*. Oxford: Oxford University Press.

Manchester, R. N., Hobbs, G. B., Teoh, A., and Hobbs, M. 2005. The Australia Telescope National Facility Pulsar Catalogue. *Astron. J.*, **129**, 1993.

Martinez, J. G., Stovall, K., Freire, P. C. C., et al. 2015. Pulsar J0453+1559: A Double Neutron Star System with a Large Mass Asymmetry. *Astrophys. J.*, **812**(2), 143.

Mason, B. D., Hartkopf, W. I., and Miles, K. N. 2017. Binary Star Orbits. V. The Nearby White Dwarf/Red Dwarf Pair 40 Eri BC. *Astron. J.*, **154**, 200.

Mayall, N. U., and Oort, J. H. 1942. Further Data Bearing in the Identification of the Crab Nebula with the Supernova of 1054 A. D. Part II. The Astronomical Aspects. *Proc. Astr. Soc. Pac.*, **54**, 95.

Millener, D. J., Dover, C. B., and Gal, A. 1988. Lambda Nucleus Single Particle Potentials. *Phys. Rev. C*, **38**, 2700.

Miller, M. C. et al. 2019. PSR J0030+0451 Mass and Radius from NICER Data and Implications for the Properties of Neutron Star Matter. Astrophys. J. **887**, L24

Möller, P., Sierk, A. J., Ichikawa, T., and Sagawa, H. 2016. Nuclear Ground-State Masses and Deformations: FRDM(2012). *Atom. Data Nucl. Data Tabl.*, **109–110**, 1–204.

Möller, Peter, Myers, William D., Sagawa, Hiroyuki, and Yoshida, Satoshi. 2012. New Finite-Range Droplet Mass Model and Equation-of-State Parameters. *Phys. Rev. Lett.*, **108**(5), 052501.

Most, Elias R., Weih, Lukas R., Rezzolla, Luciano, and Schaffner-Bielich, Jürgen. 2018. New Constraints on Radii and Tidal Deformabilities of Neutron Stars from GW170817. *Phys. Rev. Lett.*, **120**(26), 261103.

Narain, Gaurav, Schaffner-Bielich, Jürgen, and Mishustin, Igor N. 2006. Compact Stars Made of Fermionic Dark Matter. *Phys. Rev. D*, **74**, 063003.

Negele, J. W., and Vautherin, D. 1973. Neutron Star Matter at Subnuclear Densities. *Nucl. Phys. A*, **207**, 298–320.

Okamoto, Minoru, Maruyama, Toshiki, Yabana, Kazuhiro, and Tatsumi, Toshitaka. 2012. Three Dimensional Structure of Low-Density Nuclear Matter. *Phys. Lett. B*, **713**, 284–288.

Oppenheimer, J. R., and Volkoff, G. M. 1939. On Massive Neutron Cores. *Phys. Rev.*, **55**, 374–381.

Özel, F., and Freire, P. 2016. Masses, Radii, and the Equation of State of Neutron Stars. *Annu. Rev. Astron. Astrophys.*, **54**, 401–440.

Özel, Feryal, Baym, Gordon, and Guver, Tolga. 2010. Astrophysical Measurement of the Equation of State of Neutron Star Matter. *Phys. Rev. D*, **82**, 101301.

Özel, Feryal, and Psaltis, Dimitrios. 2009. Reconstructing the Neutron-Star Equation of State from Astrophysical Measurements. *Phys. Rev. D*, **80**, 103003.

Pacini, F. 1967. Energy Emission from a Neutron Star. *Nature*, **216**, 567.

Pais, Helena, and Stone, Jirina R. 2012. Exploring the Nuclear Pasta Phase in Core-Collapse Supernova Matter. *Phys. Rev. Lett.*, **109**, 151101.

Palhares, Leticia F., and Fraga, Eduardo S. 2010. Droplets in the Cold and Dense Linear Sigma Model with Quarks. *Phys. Rev. D*, **82**, 125018.

Peters, P. C., and Mathews, J. 1963. Gravitational Radiation from Point Masses in a Keplerian Orbit. *Phys. Rev.*, **131**, 435–439.

Pethick, C. J., Ravenhall, D. G., and Lorenz, C. P. 1995. The Inner Boundary of a Neutron Star Crust. *Nucl. Phys. A*, **584**, 675–703.

Pietrzynski, G., Thompson, I. B., Gieren, W., e al. 2010. Accurate Dynamical Mass Determination of a Classical Cepheid in an Eclipsing Binary System. *Nature*, **468**(542).

Postnikov, Sergey, Prakash, Madappa, and Lattimer, James M. 2010. Tidal Love Numbers of Neutron and Self-Bound Quark Stars. *Phys. Rev. D*, **82**, 024016.

Potekhin, Alexander Y., and Chabrier, Gilles. 2000. Equation of State of Fully Ionized Electron – Ion plasmas. 2. Extension to Relativistic Densities and to the Solid Phase. *Phys. Rev. E*, **62**, 8554–8563.

Provencal, J. L., Shipman, H. L., Koester, D., Wesemael, F., and Bergeron, P. 2002. Procyon B: Outside the Iron Box. *Astrophys. J.*, **568**, 324–334.

Ramsey, W. H. 1950. On the Instability of Small Planetary Cores (I). *Mon. Not. Roy. Astron. Soc.*, **110**, 325.

Ransom, S. M., Stairs, I. H., Archibald, A. M., et al. 2014. A Millisecond Pulsar in a Stellar Triple System. *Nature*, **505**, 520.

Ravenhall, D. G., Pethick, C. J., and Wilson, J. R. 1983. Structure of Matter Below Nuclear Saturation Density. *Phys. Rev. Lett.*, **50**, 2066.

Read, Jocelyn S., Lackey, Benjamin D., Owen, Benjamin J., and Friedman, John L. 2009. Constraints on a Phenomenologically Parameterized Neutron-Star Equation of State. *Phys. Rev. D*, **79**, 124032.

Rhoades, C. E., and Ruffini, R. 1974. Maximum Mass of a Neutron Star. *Phys. Rev. Lett.*, **32**, 324–327.

Riley, T. E. et al. 2019. A NICER View of PSR J0030+451: Millisecond Pulsar Parameter Estimation. Astrophys. J. **887**, L21

Rosenfeld, Léon. 1974. Discussion of the Report of D. Pines. Page 174: *Astrophysics and Gravitation: Proceedings of the Sixteenth Solvay Conference on Physics*. Brusells: Editions de l'Université Bruxelles.

Ruffini, Remo, and Bonazzola, Silvano. 1969. Systems of Selfgravitating Particles in General Relativity and the Concept of an Equation of State. *Phys. Rev.*, **187**, 1767–1783.

Rüster, Stefan B., Hempel, Matthias, and Schaffner-Bielich, Jürgen. 2006. The Outer Crust of Non-accreting Cold Neutron Stars. *Phys. Rev. C*, **73**, 035804.

Sagawa, Hiroyuki, and Möller, Peter. 2017. New Mass Model FRDM 2012 and Symmetry Energy. *JPS Conf. Proc.*, **14**, 010804.

Sahu, Kailash C., Anderson, J., Casertano, S., et al. 2017. Relativistic Deflection of Background Starlight Measures the Mass of a Nearby White Dwarf Star. *Science*, **356**(6342), 1046–1050.

Salpeter, E. E. 1960. Matter at High Densities. *Ann. Phys. (N.Y.)*, **11**, 393.

Schaffner-Bielich, J. 2008. Hypernuclear Physics for Neutron Stars. *Nucl. Phys. A*, **804**, 309–321.

Schertler, K., Greiner, C., Schaffner-Bielich, J., and Thoma, M. H. 2000. Quark Phases in Neutron Stars and a "Third Family" of Compact Stars as a Signature for Phase Transitions. *Nucl. Phys. A*, **677**, 463.

Schneider, A. S., Berry, D. K., Briggs, C. M., Caplan, M. E., and Horowitz, C. J. 2014. Nuclear "Waffles." *Phys. Rev. C*, **90**(5), 055805.

Schuetrumpf, B., Klatt, M. A., Iida, K., et al. 2015. Appearance of the Single Gyroid Network Phase in "Nuclear Pasta" Matter. *Phys. Rev. C*, **91**(2), 025801.

Schutz, Bernhard. 2009. *A First Course in General Relativity*. Cambridge: Cambridge University Press.

Schwarzschild, Karl. 1916a. Über das Gravitationsfeld einer Kugel aus inkompressibler Flüssigkeit nach der Einsteinschen Theorie. Pages 424–434 of: *Sitzungsber. K. Preuss. Akad. Wiss.*

Schwarzschild, Karl. 1916b. Über das Gravitationsfeld eines Massenpunktes nach der Einsteinschen Theorie. Pages 189–196 of: *Sitzungsber. K. Preuss. Akad. Wiss.*

Schwenk, A., and Pethick, C. J. 2005. Resonant Fermi Gases with a Large Effective Range. *Phys. Rev. Lett.*, **95**, 160401.

Sedrakian, Armen, and Clark, John W. 2018. Superfluidity in Nuclear Systems and Neutron Stars *Eur. Phys. J.*, **155**(9), 167.

Seidov, Z. F. 1971. The Stability of a Star with a Phase Change in General Relativity Theory. *Soviet Astronomy*, **15**, 347.

Shapiro, Stuart L., and Teukolsky, Saul A. 1983. *Black Holes, White Dwarfs, and Neutron Stars: The Physics of Compact Objects*. New York: John Wiley & Sons.

Shlomo, S., Kolomietz, V. M., and Colo, G. 2006. Deducing the Nuclear-Matter Incompressibility Coefficient from Data on Isoscalar Modes. *European Physical Journal A*, **30**, 23–30.

Siemens, P. J., and Pandharipande, V. R. 1971. Neutron Matter Computations in Bruckner and Variational Theories. *Nucl. Phys. A*, **173**, 561–570.

Skyrme, T. 1959. The Effective Nuclear Potential. *Nucl. Phys.*, **9**, 615–634.

Staelin, D. H., and Reifenstein, III, E. C. 1968. Pulsating Radio Sources Near the Crab Nebula. *Science*, **162**, 1481–1483.

Steiner, Andrew W., Lattimer, James M., and Brown, Edward F. 2010. The Equation of State from Observed Masses and Radii of Neutron Stars. *Astrophys. J.*, **722**, 33–54.

Stone, J. R., and Reinhard, P. G. 2007. The Skyrme Interaction in Finite Nuclei and Nuclear Matter. *Prog. Part. Nucl. Phys.*, **58**, 587–657.

Straumann, Norbert. 2013. *General Relativity*. Second edn. Dordrecht: Springer.

Tanabashi, M., Nagoya, U., Nagoya, K. M. I., et al. 2018. Review of Particle Physics. *Phys. Rev. D*, **98**(3), 030001.

Tauris, T. M., Kramer, M., Freire, P. C. C., et al. 2017. Formation of Double Neutron Star Systems. *Astrophys. J.*, **846**(2), 170.

Taylor, J. H., Fowler, L. A., and McCulloch, P. M. 1979. Measurements of General Relativistic Effects in the Binary Pulsar PSR 1913+16. *Nature*, **277**, 437–440.

Tews, I., Margueron, J., and Reddy, S. 2018. Critical Examination of Constraints on the Equation of State of Dense Matter Obtained from GW170817. *Phys. Rev. C*, **98**(4), 045804.

Tews, I., Margueron, J., and Reddy, S. 2019. Confronting Gravitational-Wave Observations with Modern Nuclear Physics Constraints. *Eur. Phys. J. A.*, **55**(6), 97.

Tews, Ingo, Lattimer, James M., Ohnishi, Akira, and Kolomeitsev, Evgeni E. 2017. Symmetry Parameter Constraints from a Lower Bound on Neutron-Matter Energy. *Astrophys. J.*, **848**(2), 105.

Thorne, Kip S. 1994. *Black Holes & Time Warps: Einstein's Outrageous Legacy*. New York: W. W. Norton & Company.

Thorsett, S. E., Arzoumanian, Z., Camilo, F., and Lyne, A. G. 1999. The Triple Pulsar System PSR B1620-26 in M4. *Astrophys. J.*, **523**, 763.

Tiengo, A., and Mereghetti, S. 2007. XMM-Newton Discovery of 7 s Pulsations in the Isolated Neutron Star RX J1856.5-3754. *Astrophys. J. Lett.*, **657**, L101–L104.

Todd-Rutel, B. G., and Piekarewicz, J. 2005. Neutron-Rich Nuclei and Neutron Stars: A New Accurately Calibrated Interaction for the Study of Neutron-Rich Matter. *Phys. Rev. Lett.*, **95**, 122501.

Toimela, T. 1985. Perturbative QED and QCD at Finite Temperatures and Densities. *Int. J. Theor. Phys.*, **24**, 901. [Erratum: *Int. J. Theor. Phys.* 26, 1021 (1987)].

Tolman, Richard C. 1934. *Relativity, Thermodynamics and Cosmology*. Oxford: Oxford University Press.

Tolman, Richard C. 1939. Static Solutions of Einstein's Field Equations for Spheres of Fluid. *Phys. Rev.*, **55**, 364–373.

Touboul, Pierre, Metris, Gilles, Rodrigues, Manuel, et al. 2017. MICROSCOPE Mission: First Results of a Space Test of the Equivalence Principle. *Phys. Rev. Lett.*, **119**(23), 231101.

Typel, S., Röpke, G., Klähn, T., Blaschke, D., and Wolter, H. H. 2010. Composition and Thermodynamics of Nuclear Matter with Light Clusters. *Phys. Rev. C*, **81**, 015803.

Wagner, T. A., Schlamminger, S., Gundlach, J. H., and Adelberger, E. G. 2012. Torsion-Balance Tests of the Weak Equivalence Principle. *Class. Quant. Grav.*, **29**, 184002.

Walecka, J. D. 1974. A Theory of Highly Condensed Matter. *Ann. Phys. (N.Y.)*, **83**, 491.

Walter, F. M., Eisenbeiß, T., Lattimer, J. M., et al. 2010. Revisiting the Parallax of the Isolated Neutron Star RX J185635-3754 Using HST/ACS Imaging. *Astrophys. J.*, **724**, 669–677.

Wang, M., Audi, G., Kondev, F. G., Huang, W. J., Naimi, S., and Xu, X. 2017. The AME2016 Atomic Mass Evaluation (II). Tables, Graphs and References. *Chinese Physics C*, **41**, 030003.

Weber, Fridolin. 2005. Strange Quark Matter and Compact stars. *Prog. Part. Nucl. Phys.*, **54**, 193–288.

Weisberg, Joel M., and Huang, Yuping. 2016. Relativistic Measurements from Timing the Binary Pulsar PSR B1913+16. *Astrophys. J.*, **829**(1), 55.

Weizsäcker, C. F. von. 1935. Zur Theorie der Kernmassen. *Zeitschrift f. Physik*, **96**, 431–458.

Wheeler, J. A. 1955. Geons. *Phys. Rev.*, **97**, 511–536.

Wheeler, J. A., and Ford, K. 1998. *Geons, Black Holes, and Quantum Foam: A Life in Physics*. New York: W. W. Norton & Company.

Williams, J. G. 1994. Contributions to the Earth's Obliquity Rate, Precession, and Mutation. *Astron. J.*, **108**, 711–724.

Witten, Edward. 1984. Cosmic Separation of Phases. *Phys. Rev. D*, **30**, 272.

Wolf, R. N., Beck, D., Blaum, K., et al. 2013. Plumbing Neutron Stars to New Depths with the Binding Energy of the Exotic Nuclide ^{82}Zn. *Phys. Rev. Lett.*, **110**(4), 041101.

Wolszczan, A. 1994. Confirmation of Earth-Mass Planets Orbiting the Millisecond Pulsar PSR B1257+12. *Science*, **264**, 538–542.

Wolszczan, A., and Frail, D. A. 1992. A Planetary System around the Millisecond Pulsar PSR1257 + 12. *Nature*, **355**, 145–147.

Yakovlev, Dima G., and Pethick, C. J. 2004. Neutron Star Cooling. *Ann. Rev. Astron. Astrophys.*, **42**, 169–210.

Yakovlev, Dmitry G., Haensel, Pawel, Baym, Gordon, and Pethick, Christopher J. 2013. Lev Landau and the Concept of Neutron Stars. *Phys. Usp.*, **56**, 289–295. [*Usp. Fiz. Nauk*, **183**, 307 (2013)].

Zel'dovich, Ya. B. 1961. The Equation of State at Ultrahigh Densities and Its Relativistic Limitations. *Zh. Eksp. Teoret. Fiz.*, **41**, 1609.

Zürn, G., Wenz, A. N., Murmann, S., et al. 2013. Pairing in Few-Fermion Systems with Attractive Interactions. *Physical Review Letters*, **111**(17), 175302.

Index